Marshall Brain's

MORE
How STUFF Works

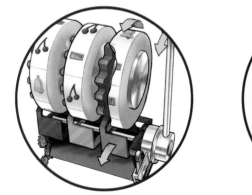 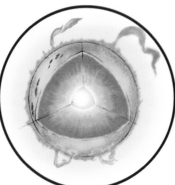 

Marshall Brain's
MORE
How STUFF Works

Marshall Brain
and the Staff at HowStuffWorks.com

Wiley Publishing, Inc.

For general information on our other products and services or to obtain technical support please contact our Customer Care Department within the U.S. at 800-762-2974, outside the U.S. at 317-572-3993 or fax 317-572-4002.

Wiley also publishes its books in a variety of electronic formats. Some content that appears in print may not be available in electronic books.

ISBN: 0-7645-6711-X
EAN: 9780764567117
UPC: 785555110816

Cataloging-in-Publication Data is available from the Library of Congress.

HowStuffWorks, Inc.
EDITOR-IN-CHIEF: Marshall Brain
EDITORIAL DIRECTOR: Katherine Fordham Neer
ILLUSTRATOR: Lee Dempsey
CONTRIBUTING WRITERS: Marshall Brain, Tom Harris, Kevin Bonsor, Craig Freudenrich, Ph.D. , Katherine Fordham Neer, Karim Nice, Jeff Tyson, Ann Meeker O'Connell, Shel Brannan, Debra Beller, Scott Aldous, Carl Bianco, M.D., Jim Bowen, Bob Broten, Todd Gould, Michael Morrisey, Matthew Weschler, and John Zavisa.
PHOTOGRAPHY: Roxanne Reid. We appreciate the kind cooperation of the American Red Cross, Bay Leaf Volunteer Fire Department and Queenscape Landscape Company for letting us take pictures of their stuff.

Manufactured in the United States of America
10 9 8 7 6 5 4 3 2 1

Book design by Michele Laseau and Holly Wittenberg
Cover design by José Almaguer and Michele Laseau

Preface

Hello,

Wow. Here we are again. It seems like yesterday that we were finishing the final edits for the first HowStuffWorks book and now, as you read this, you're holding a copy of its sequel. A lot has happened in the last 12 months. On one level, I've had some amazing changes in both my personal and professional life. And on another there have been certain events that have had an impact on people around the globe.

Personally, my wife and I were thrilled to welcome two more children to our happy family—our twin boys were born in April. Professionally, HowStuffWorks, Inc. has increased its presence across several media platforms. The Web site garnered a spot on TIME magazine's list "The 50 Best Web Sites." The completion of this book marks our fourth published work with Wiley Publishing, Inc. and the TBS Superstation plans to host the weekly program HowStuffWorks at the Movies. One of the ways we come up with topics is from the emails we receive every day. Directly after the September 11th incident, we had an overwhelming number of emails from people around the world. Almost everyone wanted to know how biological and chemical warfare, gas masks, and body armor work. As airports stepped up security, thousands of people were reading our article about metal detectors every day. Almost daily, we received hundreds of emails thanking us for the incisive, practical descriptions we have regarding how certain military devices work. Because we received such an incredible response to these articles, we created a special chapter for this book titled Police, Military, and Defense.

The events surrounding September 11th have had a profound effect on a huge number of people. A common sentiment expressed by many of our readers was that they had a sense of confusion and helplessness. As I was responding to our readers by writing some of these articles, I found that having an understanding of this technology helped. Although some of the information I found could be alarming, it was actually very helpful to understand it all. Knowledge is empowering.

Other chapters of this book are filled with all kinds of technology that I find fascinating. From fire extinguishers to fax machines, water guns to weightlessness, and snow makers to skyscrapers—I hope you enjoy learning about how this stuff works as much as I do.

Sincerely,

Contents

Marshall, Katherine, Tom, and Lee gaze at the gallery of Lee's handiwork.

Creating a book like this is an incredible experience. This book contains more than 130 different articles and over a 100 hand-drawn illustrations. At any given time, there were dozens of people working on its content or look and feel. The entire team worked through a series of processes to produce what you now hold in your hands. To create a book like this, you need to:

- Come up with the words.
- Come up with the art.
- Create the cover.
- Design the layout.
- Copyedit and proof everything.

In creating the first HowStuffWorks book last year, we came up with a basic layout design and cover that worked amazingly well with the content. So, this time we were able to work with and rely on those original concepts.

The process for creating the words was pretty similar to what we did last time, too. Of course we do have some really different content this time, including new categories like *Nature*, *Heavy Equipment*, and *Police, Military, and Defense*. A dozen or so people wrote the articles—including Karim, Kevin, Craig, Jeff, Tom, and myself. Tom, Katherine, and I then edited the articles for voice, reading level, and size.

The copyediting and proofing was a multiphase process handled mostly by Suzanne, Ben, and Katherine.

This explains the words, cover, and layout, and the copyediting and proofing processes, but what about the art?

Marshall and Tom.

As it turns out, the process of creating art for a book like this is fascinating! The actual process this time was technically the same, but instead of simply seeing pieces of art at various stages like with the last book, this time I was able to witness the entire art process—from start to finish. Every day I couldn't wait to come to work to see what new illustration Lee had created. We even have our own gallery of Lee's illustrations throughout the HowStuffWorks office—just outside the conference room are drawings of a fire engine, bulletproof vest, and a lie-detector test set-up. On the way to the company breakroom, you see a yo-yo, a bug zapper, and a cut-away of an aerosol can. As I'm writing this, there are sketches of a washing machine, a paintball gun, and a chain stitch for the sewing machine article plastered around Lee's desk.

An incredible amount of work goes into producing these illustrations and drawings. That's actually one of the coolest things about the first big HowStuffWorks book and it's the same for this book. The words and the art really must fit together to explain the inner workings of a subject. To make sure this happens, the first thing Lee does is read the article. After he's finished reading, Lee talks about the article with the author or editor. Together, they figure out what it is that Lee should illustrate.

Once the subject is decided upon, Lee has to do some more research—especially for drawings that show a 3-D view or a cut-away. He uses our Web site and other online resources, books, photos, and actual models to create his initial sketches. After he locates all of his reference material, he sketches out a diagram to establish a viewing angle and component locations.

Next, he uses that preliminary sketch to create what he calls a *working drawing*. At this stage, all of the components of an item are in their specific positions and the dimensions are finalized. Lee uses the photocopier (see "How Photocopiers Work" in *Around the Office*) to enlarge the working drawing so that he can then add in the *line weights* (the thickness of the line around the actual parts). He uses line weights to focus attention on certain areas or to make aspects of a drawing more clear.

This version of the illustration is then scanned (see "How Scanners Work" in *Around the Office*) to create a computer file. Using a graphics tablet, Lee colors everything in and voilà—the illustration is complete. When you look at the detail of something like the fire-engine illustration—just the sheer number of lines and curves Lee has put down on the sheet of paper—you realize how remarkable this process is and how rare his talent is. It is truly amazing to watch Lee work.

Katherine and Lee.

Scavenger Hunt

We write about thousands of topics for both the HowStuffWorks Web site and the HowStuffWorks books. In order to explain these things, we have to physically take them apart—to actually see inside them—so that we can figure out just how they work. Some objects, like a humidifier, are pretty easy to locate. For that, we simply went to a store around the corner. Other objects, however, aren't that easy to find—in fact, some objects are totally cost prohibitive or impossible to purchase at all! You can't just walk into someplace called "Firehouse Supplies R Us" for a fire engine!

So, how do we find all the stuff we need? It turns out it's a lot like being part of a scavenger hunt—the game where each player gets a huge list of all kinds of different objects and the players have to either locate the actual object, or at least a photo of it, for everything on the list. That's what we do. If it's too big or too expensive to buy, we send a lot of emails and make a lot of phone calls until we can locate something. Then a writer and a photographer go and check it out. The writer uses the information to create the article and then we either use the actual photos to illustrate the article or an artist will create an original drawing from the information he or she finds in the photos.

chapter one

LET'S PLAY

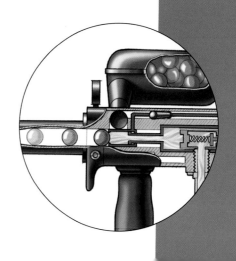

How **WATER GUNS** Work

Soon after its introduction, the water gun took its place among the most popular summer toys of all time, and it's easy to see why: When you're a kid, or a kid at heart, what better way to cool off on a hot afternoon than waging an epic water battle against your friends and family?

HSW Web Links

www.howstuffworks.com

How Paintball Works
How Snow Makers Work
How Water Heaters Work
How Force, Power Torque and Energy Works
How Water Towers Work
How SCUBA Works
How Toilets Work
How Washing Machines Work

Over the years, water guns have evolved considerably. Thirty years ago, a typical water warrior was armed only with a small squirt pistol that had a fairly short range and an even more limited ammunition reservoir. These days, you'll find an entire arsenal of water weapons at most toy stores, complete with water machine guns, water bazookas, and even water grenade launchers.

The Classic Water Gun

A classic squirt gun has just a few basic parts:

- A trigger lever, which activates a small pump.
- A pump that is attached to a plastic tube that draws water from the bottom of the reservoir. The pump forces this water down a narrow barrel and out a small hole at the gun's muzzle.
- A hole, or nozzle, which focuses the flowing water into a concentrated stream.

The only complex part in this design is the water pump, and it's about as simple as they come. The main moving element is a piston, housed inside a cylinder. Inside the cylinder is a small spring. To operate the pump:

1) You pull the trigger back, pushing the piston into the cylinder.
2) The piston compresses the spring, causing the spring to push the piston back out of the cylinder when you release the trigger.

These two strokes of the piston, into the cylinder and out again, constitute the entire pump cycle.

The downstroke—the piston pushing in—shrinks the volume of the cylinder, forcing water or air out of the pump. The upstroke, the spring pushing the piston back out, expands the cylinder volume, sucking water or air into the pump. In a water gun, you need to suck water in from the reservoir below and

Valve Closed

Valve Open

force it out through the barrel above. In order to get all the water moving through the barrel, the pump must only force water up—it cannot force water back into the reservoir. In other words, the water must move through the pump in only one direction.

The device that makes this possible is called a *one-way valve*. The one-way valve in a basic squirt pistol consists of a tiny rubber ball that rests neatly inside a small seal. There are actually two of these simple one-way valves: one between the reservoir and the pump, and another between the pump and the nozzle.

This pump design is beautiful in its simplicity, but it has two big limitations:

- The amount of water in each blast is limited.
- The duration of the blast is also limited.

Throughout the history of water guns, designers have been wrestling with these two problems to create a better pumping system.

Under Pressure

In 1982, a nuclear scientist named Lonnie Johnson came up with an ingenious solution to these problems. In his spare time, he was working on a new heat-pump system that would use moving water to regulate temperature. Late one night, he attached a model of the pumping mechanism to the bathroom sink and was startled by the powerful water blast that shot across the room. In that instant, he was struck by the idea for a water gun that would use compressed air to provide pressure for a water blast.

To make his idea a reality, Johnson enlisted the help of an accomplished inventor named Bruce D'Andrade. Together, D'Andrade and Johnson came up with the basic design that would become the Super Soaker.

Super Soakers are built around a pump mechanism, but moving the pump doesn't actually drive water out of the gun; it serves to build up water pressure before the blast. In the

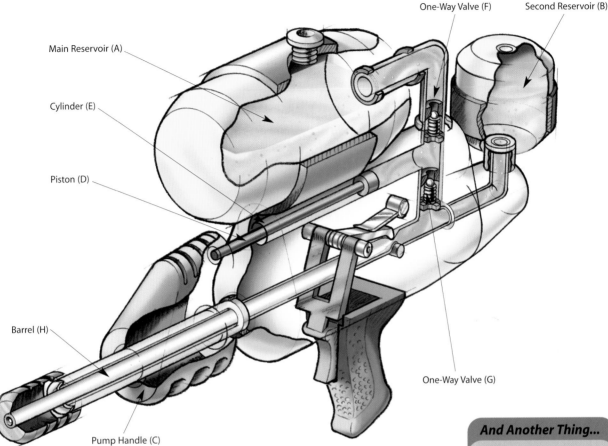

Main Reservoir (A)

Cylinder (E)

Piston (D)

Barrel (H)

Pump Handle (C)

One-Way Valve (F)

Second Reservoir (B)

One-Way Valve (G)

first wave of Super Soakers, you build up this pressure by pumping air directly into a single water reservoir. As you pump in more air, it becomes more and more compressed and applies greater pressure to the water inside.

In later models, you build pressure by pumping water instead of air. This sort of gun has two water reservoirs (labeled A and B), which are connected together via a network of tubes. To load the gun, you fill the larger reservoir (A) with water. To prime the gun for a blast, you pull the pump handle (C) in and out several times. The pump handle is connected to a long, narrow piston (D), which moves back and forth inside a cylinder (E). This pump is similar to the one in a squirt-gun pistol, and it relies on the same one way–valve system to control the direction of water flow. The first valve (F) is positioned between the large water reservoir and the pump mechanism, and the second valve (G) is positioned between the pump and the smaller water reservoir, which feeds into the barrel of the gun (H).

On the upstroke of the pump cycle, when you pull the pump handle out, the

receding piston pulls in water from the large reservoir above. The second one-way valve (G) keeps water from flowing up from the smaller reservoir (B). On the downstroke of the pump cycle, when you push the pump handle in, the plunging piston drives the water out of the cylinder, through the second one-way valve (G) and into the small reservoir (B). The first one-way valve (F) keeps the pressurized water from flowing back up into the large reservoir (A).

All Wet

The only thing keeping the water inside the gun is the trigger mechanism. The trigger is simply a lever secured to the gun housing. A stiff length of metal attached to the housing holds the top part of this lever against the flexible plastic tube leading to the gun's barrel, pinching it so no water can get through. When you pull the trigger back, the metal piece bends, and the lever releases the plastic tube. With this passageway open, the pressurized air can push all of the water out of the gun. If you build up enough pressure, the water is expelled at a very high velocity.

How **PAINTBALL** Works

Since its introduction in the 1980s, paintball has become a worldwide phenomenon. Enthusiasts have formed teams, set up leagues, and organized tournaments. Paintball is still a long way from the popularity of older sports like basketball or football, but new paintball playing fields and organizations are popping up all the time. The main thing that sets paintball apart from other sports is the equipment involved. To play paintball, you need a gun (called a marker*), a supply of paintballs, and a safety mask.*

HSW Web Links

www.howstuffworks.com

How Water Blasters Work
How Flintlock Guns Work
How Body Armor Works
How Military Camouflage
  Works
How Landmines Work
How Force, Power, Torque
  and Energy Work
How Air Bags Work

Cool Facts

Originally, paintball markers weren't intended for sport. The first markers were developed in the 1970s for use in forestry and agriculture. Foresters used the markers to tag certain trees (for research or for planning trails, for example). Farmers also used markers to tag cattle.

At some point, it occurred to a few foresters or farmers to shoot the markers at each other, and the game of paintball was born. But things didn't really get going until 1981, when a group of 12 weekend warriors used some forester markers to play a grown-up version of capture the flag.

Paintballs have the same basic design as gel-cap pills or bath-oil beads. They consist of a glob of colored liquid encased in a gelatin capsule and measure about half an inch in diameter. The "paint," which comes in many colors, is non-toxic, biodegradable, and water soluble (so it washes off skin and clothing). The capsule is durable enough to hold up to normal handling, but if you throw it against something, it will burst.

The Way of the Marker

A paintball marker shoots a paintball out of the barrel at about 300 feet per second (about 90 meters per second). Most paintball markers do this using a compressed gas source, such as a tank of liquid carbon dioxide or compressed air.

At the beginning of the firing cycle, a sear piece holds the bolt back in the cocked position. With the bolt in this position, a single paintball can fall out of the ammunition hopper and land in the *breech* (the area behind the barrel). A small catch holds the paintball in place.

When you pull the trigger, it pushes the sear. As the sear moves back, it unlocks the bolt. The bolt spring throws the bolt forward, toward the gas valve.

When the marker is cocked, a spring-loaded plunger blocks the passageway between the valve and the rest of the marker. When the bolt slides forward, it hits a pin connected to this plunger. This impact pushes the plunger against the spring, opening up the gas passageway to the rest of the marker. At the same time, the front section of the bolt pushes the paintball forward and blocks the hopper port.

As soon as the bolt hits the plunger pin, the compressed gas rushes out and flows around the valve box to the area behind the paintball. The substantial air pressure forces the paintball down the barrel, out of the marker.

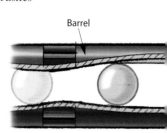

Barrel

A raised portion on the plunger pin initially blocks the gas passageway to the rear of the marker. But as the pin keeps moving forward, the raised portion clears the passageway, and the gas flows backward. This pressure pushes the bolt all the way back, and the sear locks it into position again. As the front of the bolt slides backward, it opens up the hopper port, and another paintball falls into position. The marker is ready for another shot.

Changing Firing Speeds

Many markers have a simple tool for varying firing speed: A screw in the breech. Tightening the screw constricts the gas passageway, reducing gas flow. This means less gas pressure, and therefore less force acting on the paintball. Loosening the screw allows greater gas flow, increasing the force on the paintball. Some markers use more sophisticated pressure regulators.

A number of modern marker designs have built-in electric motors, solenoids, and electronic controls. In an electric semiautomatic marker, the electronics primarily control the gas valves and the pressure regulators. In automatic models, the electronics might also control motors to continually cock and fire the marker. As long as the player keeps the trigger depressed, the marker will keep launching paintballs.

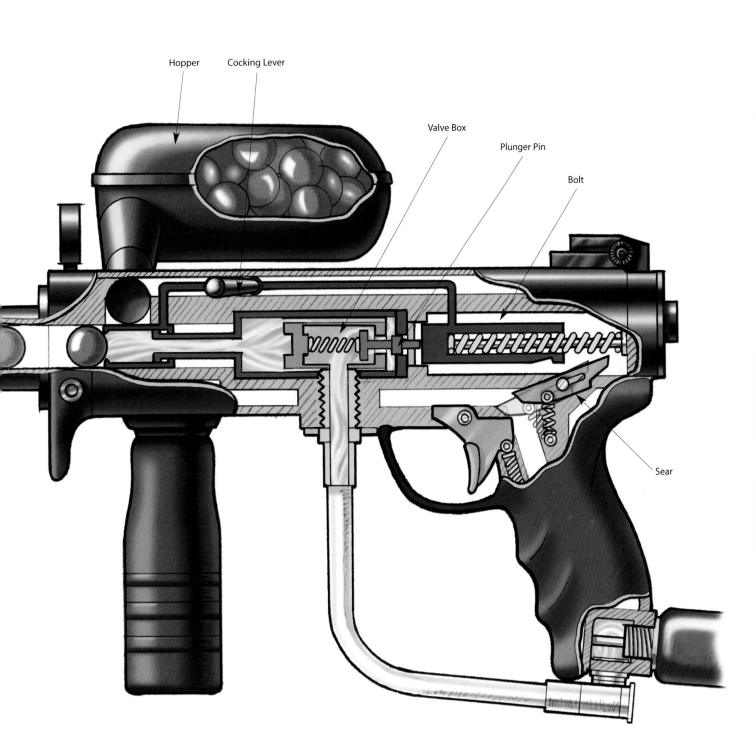

Hopper

Cocking Lever

Valve Box

Plunger Pin

Bolt

Sear

Safety First, Not Last

One of the most important developments in the history of paintball has been safety equipment. When a paintball hits you on the body, you only feel a brief sting. But a speeding paintball could actually "knock your eye out," as moms everywhere might say. In the early days, many players wore no eye protection at all, and others wore only basic safety goggles. These days, paintball players wear full face masks and helmets. This protects them from damage to the eyes, ears, nose, and mouth. Just as in football and hockey, safety equipment is a necessity in paintball.

How **YO-YOs** Work

On the surface, the yo-yo is an incredibly simple toy—it's really nothing but a spool attached to a length of string. But in the right hands, it can be something extraordinary: An accomplished yo-yoist can send the toy flying out in all directions, make it hover in mid air, then snap it back. Ordinary string and plastic are brought to life!

HSW Web Links

www.howstuffworks.com

How Gyroscopes Work
How Boomerangs Work
How Clutches Work
How Bearings Work
More Toy Articles

Name Dropping

The word *yo-yo* and the modern yo-yo design come from the Philippines. Unlike the original Chinese and Greek yo-yos, Philippine yo-yos (the word means *come come* or *come back* in the native Tagalog language) had the ability to sleep. This unique toy may have been an adaptation of the Chinese yo-yo, or it may have developed out of a Philippine hunting weapon. In any case, it apparently dates back a few hundred years in the region.

In the 1920s, a Philippine immigrant named Pedro Flores decided to bring this yo-yo design to the United States. He achieved some success right away, and in 1929, he sold his company to a businessman named Donald Duncan. Duncan trademarked *yo-yo* and, over the next few decades, built his company up into the premier yo-yo manufacturer. In 1965, the Federal Court of Appeals ruled that the term *yo-yo* had become generic, and so could be used by anyone.

A yo-yo may seem like magic, but it's actually just physics at work. Both the classic yo-yo and the sophisticated automatic yo-yos that have popped up in the past few years are remarkable demonstrations of fundamental scientific principles.

One Good Turn

The yo-yo is one of the most popular and enduring toys of all time. The ancient Greeks were playing with terra cotta versions more than 2,500 years ago, and there's some evidence that the Chinese had developed similar toys before that. In any case, the yo-yo has demonstrated phenomenal longevity—it's older than any other toy except the doll.

In early yo-yos, the string was tied securely to the axle. In the modern yo-yo, brought to the United States from the Philippines in the 1920s, the string is looped around the axle, not tied to it. This change gives the yo-yo the ability to "sleep"—to spin on the end of its string so that it doesn't immediately return.

The physical principles at work are very simple. Sitting in the yo-yoist's palm, the yo-yo has a certain amount of potential energy (energy of position). This potential energy takes two different forms:

- The yo-yo is held up in the air, giving it the potential to fall to the ground.
- The yo-yo has string wound around its axle, giving it the potential to spin as it unwinds.

Releasing the yo-yo converts both forms of potential energy to kinetic energy. The yo-yo spool falls straight to the ground (and the yo-yoist may help it along with a flick of the wrist as well). As it falls, it builds a certain amount of *linear momentum* (momentum in a straight line). At the same time, the string unwinds, and the spool spins, which builds

up *angular momentum* (momentum of rotation). When the yo-yo reaches the end of the string, it can't fall any farther. But, because it has a good deal of angular momentum, it will keep spinning.

The spinning motion gives the yo-yo gyroscopic stability. A spinning object resists changes to its axis of rotation. This phenomenon keeps a yo-yo's axis perpendicular to the string, as long as the yo-yo is spinning fast enough.

Since the spool isn't tied securely to the string, the yo-yo can spin freely once it unwinds. The yo-yoist jerks the string a little bit to make the yo-yo return. This tug briefly increases the friction between the string and the axle so that the axle starts rewinding the string. Once it starts rewinding, the string around the spool provides friction to reel in more string.

Asleep at the Wheel

The ability to make the yo-yo spool spin on the end of its string—to sleep, as yo-yoists put it—makes yo-yoing a much more interesting challenge. Yo-yoists try to keep the spool sleeping while making shapes with the string and swinging the yo-yo around the shapes. Another trick is to "walk the dog"—to let the spinning spool roll along the ground before pulling it back in.

Over the years, manufacturers have come up with a number of mechanisms to make doing these sorts of tricks easier. One of the simplest improvements was to redistribute the weight in the yo-yo to alter its moment of inertia. An object's *moment of inertia* is a measure of how resistant it is to changes in rotation. This is determined by two factors: how much mass the object has and how far that mass is from the object's axis of rotation. Increased mass makes an object harder to rotate and harder to stop rotating, as does increased distance between the mass and the axis of rotation. By concentrating mass around the edges of the discs, manufacturers can improve yo-yo performance.

Another approach is to further reduce friction between the yo-yo string and the axis. One popular method is to configure a ball bearing assembly around the yo-yo axle so that the axle itself is separated from the string. If the bearings are properly lubricated, they will significantly reduce friction. This reduced friction lets the axle spin more easily, which increases sleeping time.

Pop the Clutch

A new type of automatic yo-yo hit the scene in the 1990s. Yomega, the leading manufacturer of these yo-yos, advertises its model as "the yo-yo with a brain." It does seem like these yo-yos have some intelligence, since they know exactly when to sleep and wake up, but the "brain" is actually a centrifugal clutch.

As in the ball-bearing yo-yo, the string in an automatic yo-yo does not touch the axle directly. Instead, the string winds around a spindle piece. The axle, which is mounted to the two halves of the yo-yo, runs through the middle of the spindle, but the axle and the spindle are not actually connected.

The spindle and axle move in unison when the yo-yo spins slowly. The clutch locks the axle and spindle together. The clutch mechanism, which is housed inside one of the yo-yo discs, consists of two metal spring-loaded arms. These arms are weighted at one end and connected to the body of the yo-yo at the other end. When the yo-yo is stationary or spinning slowly, the springs press the arms up against the spindle so that the spindle's rotation turns the entire yo-yo. But as the yo-yo speeds up, centrifugal force pushes the weighted ends of the arms outward, against the springs. The arms release the spindle so that the spindle and the rest of the yo-yo move independently.

When you throw the yo-yo, it starts out spinning slowly. The clutch is locked, so the unwinding spindle spins the discs. But just before the yo-yo reaches the end of its string it starts spinning fast enough for the clutch to release the spindle. The disc's angular momentum keeps the yo-yo spinning, but the spindle is free to stop. Eventually, the discs slow down too, and the centrifugal force acting on the arms decreases. When the outward centrifugal force on the weights dips below the inward force of the springs, the arms clamp shut on the spindle. This transfers the spinning motion of the discs back to the spindle, which causes the spindle to rewind the string and return to your palm.

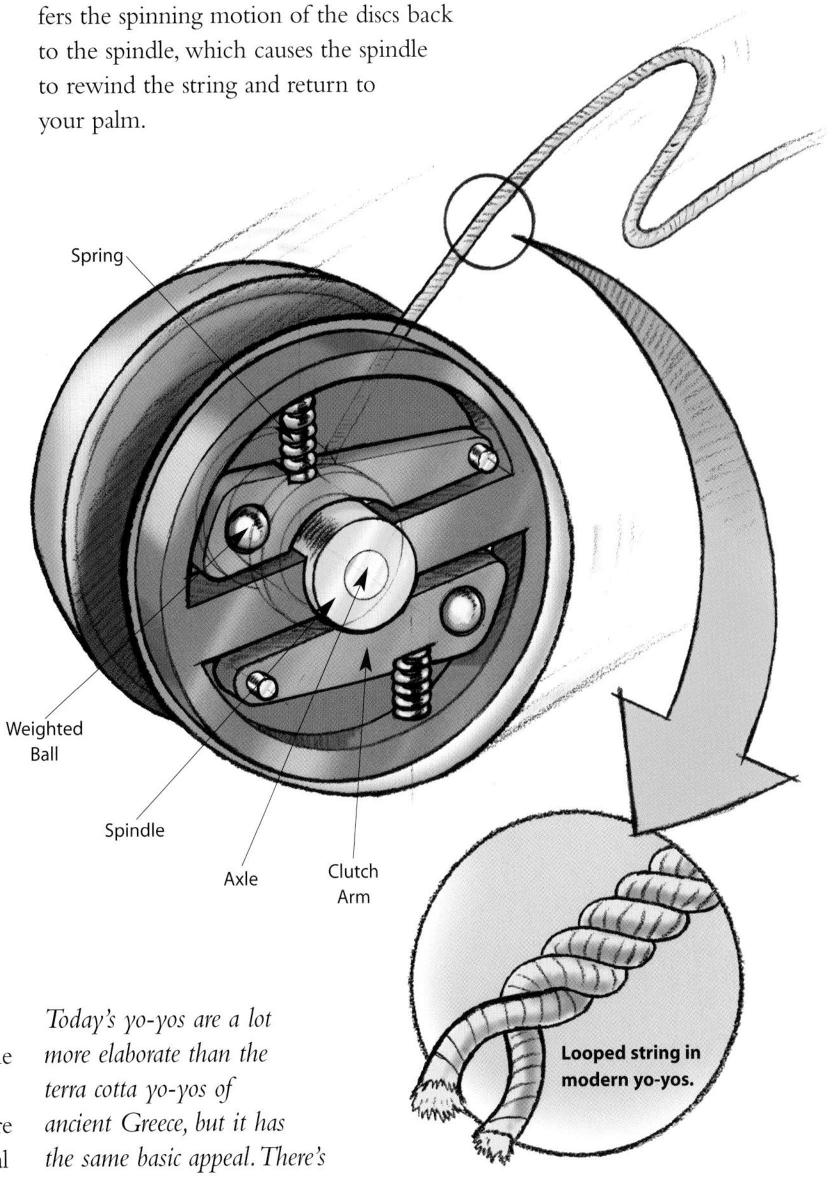

Spring

Weighted Ball

Spindle

Axle

Clutch Arm

Looped string in modern yo-yos.

Today's yo-yos are a lot more elaborate than the terra cotta yo-yos of ancient Greece, but it has the same basic appeal. There's something magical about taking an ordinary spool and, with nothing but a flick of the wrist, turning it into an active, spinning top!

How **WAVE POOLS** Work

Modern civilization is filled with artificial re-creations of nature. People decorate houses with artificial Christmas trees, wear artificial hair, play games on artificial grass, and build zoo cages with artificial rocks. Scientists have even developed artificial hearts! One of mankind's oddest re-creations of nature is the artificial ocean shore, also known as the wave pool. These popular water park attractions are sanitized, regimented revisions of the organic surfs created by nature. In wave pools, the water is chlorinated, the beach is concrete, and the waves arrive like clockwork, once every few minutes.

HSW Web Links

www.howstuffworks.com

How Water Blasters Work
How Toilets Work
How Hydraulic Machines Work
How Floods Work
How Washing Machines Work
How Water Towers Work
How Roller Coasters Work

In the ocean, most waves are created by the wind. The rushing air pushes some water molecules together, producing a swell of water—a disturbance in the ocean's surface—at a particular point. These molecules push on the molecules next to them, which push on the molecules next to them, and so on. In this way, the disturbance is passed along the surface of the ocean, while the individual water molecules stay in roughly the same area.

Makin' Waves

You could replicate this type of wave action in a number of ways. All you need is a basin of water and some way to create a periodic disturbance. You could use a strong blast of air along the surface, a rotating paddle wheel (like the ones used on steamboats), or an oscillating plunger. Basically, you push on the water at one point and this energy travels outward through the surrounding water. It's the same thing that happens when you drop a rock into a pond.

It's simple to make small waves with this sort of system, but it's a lot harder to form large, surfable waves. You would need an absurdly intense blast of air or a large, awfully strong plunger. Such devices would likely be inefficient, cumbersome, and dangerous, so they wouldn't make for particularly good water park attractions. Instead, water parks use water-pumping wave systems.

Just Add Water

A typical wave machine forms a wave by dumping a huge volume of water into the deep end of the pool. The water level in the pool wants to balance out again, so the surge travels all the way to the beach. Because water is fairly heavy, it pushes pretty hard to find its own level. If you dump more water in, you increase the size and strength of the wave.

The idea is simple, but there's a lot of powerful equipment involved in this process. The typical wave pool has five basic parts:

- A water-pumping system
- A water-collection reservoir
- A series of release valves at the bottom of the reservoir
- A giant, slanted swimming pool
- A return canal, leading from the beach area to the pumping system

The pumping system consists of a powerful motor, a long drive shaft, and a propeller—all mounted underwater in the return canal. The spinning propeller drives water from the canal up a pipe to the water collection reservoir.

Keep It Clean

On a busy summer day, thousands and thousands of people will take a dip in a wave pool, and even the cleanest among them leave a certain amount of dirt and oil behind. Water parks need a 24-hour filtering system just to keep the water sanitary. In a typical filter setup, a powerful pump sucks water in from the canal, sends it through the filter system, and shoots clean water back out. Chlorine levels are monitored constantly to make sure that the water is safe.

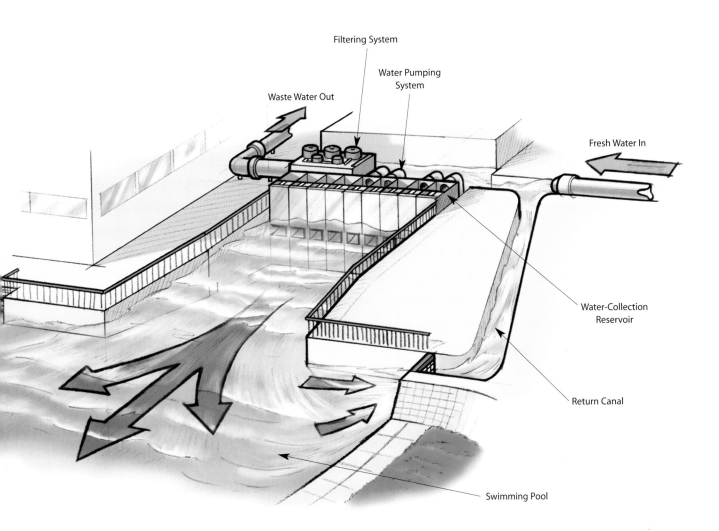

Filtering System

Water Pumping System

Waste Water Out

Fresh Water In

Water-Collection Reservoir

Return Canal

Swimming Pool

The water reservoir is broken up into connected chambers, each with its own release valve. The release valve has three major elements:

- **The valve seat**—The opening that leads down to the pool
- **The valve plate**—A wide piece of metal that fits snugly on top of the valve seat
- **The hydraulic piston**—Housed inside an oil-filled cylinder, this piston attaches to the plate.

The plate is fairly heavy, so it naturally will stay fixed over the valve seat if nothing is pushing on it. The plate plugs up the reservoir so water can't escape. (A rubber gasket around the valve seat keeps the valve from leaking too much.)

The hydraulic piston opens the valve plate, unplugging the reservoir. The water in the reservoir drops into a curved passageway leading to the pool. As it rushes into the pool, the water runs into a sort of reef in the concrete floor. This concentrates the water into a good-sized, surfable wave.

As you can see, the entire system is incredibly simple: It's basically a giant toilet. The reservoir is like the toilet tank that fills up with water, and the release valve is like the flushing mechanism. At the most basic level, that's all there is to it.

How **CHESS COMPUTERS** Work

How can a computer play chess? For many people a computer that can play chess is a mind-boggling concept. Chess seems like a distinctly human activity requiring intelligence and thought, so how can a computer possibly do it?

HSW Web Links

www.howstuffworks.com

How Computer Programs Work

How Video Game Systems Work

How Encryption Works

It turns out that computers don't really "play" chess like people do. A computer that is playing chess isn't thinking like a person does when a person plays chess. Instead, it's calculating through a set of formulas that cause it to make good moves. As computers have gotten faster and faster, the quality of these calculated moves has gotten better and better. Computer chess calculators are now the best chess players on the planet.

People and Chess

If you have ever watched a person first learning to play chess, you know that a human chess player starts with very limited abilities. Once a player understands the basic rules that control each piece, he or she can play chess. However, the new player is not very good. Each early defeat comes as something of a surprise—"Oh, I didn't think about that!" or "I didn't see that coming!" are common exclamations.

The human mind absorbs these experiences, stores away different board configurations, discovers certain tricks and ploys, and generally soaks up the nuances of the game one move at a time. As the level of skill develops, the player will often read books to discover patterns of play used by the best players. The player develops strategies and tactics that he or she uses to guide his or her play through each game.

Chess Computer

For a human being, therefore, the game of chess involves a great deal of high-level abstract reasoning—visual-pattern matching to recall board positions, adherence to rules and guidelines, conscious thought, and even psychology!

Computers do none of this. . . .

Computers and Chess

The current state-of-the-art in computer chess is fairly intricate, but all of it involves blind computation that is very simple at the core.

Let's say you start with a chess board set up for the start of a game. Each player has 16 pieces. Let's say that white starts. White has 20 possible moves:

- The white player can move any pawn forward one or two positions.
- The white player can move either knight in two different ways.

The white player chooses one of those 20 moves and makes it.

For the black player, the options are the same: 20 possible moves. So black chooses a move.

Now white can move again. This next move depends on the first move that white chose to make, but there are about 20 or so moves white can make given the current board position, and then black has 20 or so moves it can make, and so on.

This is how a computer looks at chess. It thinks about it in a world of all possible moves, and it makes a big tree for all of those moves.

In this tree, there are 20 possible moves for white to start the game. There are $20 \times 20 = 400$ possible configurations after black has gone for the first time. Then there are $400 \times 20 = 8,000$ possible configurations after white has gone the second time. Then there are $8,000 \times 20 = 160,000$ configurations after black has gone for a

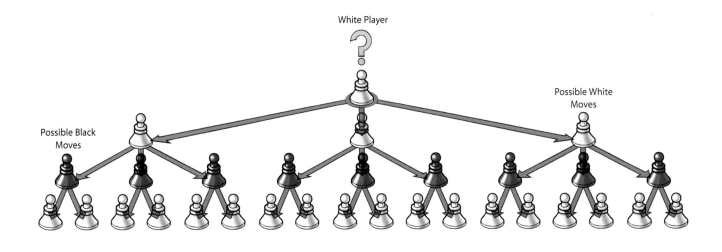

White Player

Possible Black Moves

Possible White Moves

second time, and so on. If you were to fully develop the entire tree for all possible chess moves, the total number of board positions is about 1,000,000,000,000,000,000,000,000, 000,000,000,000,000,000,000,000,000, 000,000,000,000,000,000,000,000,000, 000,000,000,000,000,000,000,000,000, 000,000,000,000,000,000,000,000,000, 000,000, or 10^{120}, give or take a few. That's a very big number. For example, there have only been 10^{26} nanoseconds since the big bang. There are thought to be only 10^{75} atoms in the entire universe. When you consider that the Milky Way galaxy contains billions of suns, and there are billions of galaxies, you can see that that's a whole lot of atoms. That number is dwarfed by the number of possible chess moves. Chess is a pretty intricate game!

No computer is ever going to calculate the entire tree. What a chess computer tries to do is generate the board-position tree 5 or 10 or 20 moves into the future. Assuming that there are about 20 possible moves for any board position, a 5-level tree contains 3,200,000 board positions. A 10-level tree contains about 10,000,000,000,000 (10 trillion) positions. The depth of the tree that a computer can calculate is controlled by the speed of the computer that's playing the game. The fastest chess computers can generate and evaluate millions of board positions per second.

After it generates the tree, then the computer needs to evaluate the board positions. That is, the computer has to look at the

pieces on the board and decide whether that arrangement of pieces is good or bad. The way it does this is by using an evaluation function. The simplest possible function might just count the number of pieces each side has. If the computer is playing white and a certain board position has 11 white pieces and 9 black pieces, the simplest evaluation function might be:

$$11 - 9 = 2$$

Obviously, for chess that formula is way too simple. Some pieces are more valuable than others, so the formula might apply a weight to each type of piece. But the position of the various pieces is also important. As the programmer thinks about it, he or she makes the evaluation function more and more complicated by adding things like control of the center, vulnerability of the king to check, vulnerability of the opponent's queen, and tons of other parameters. No matter how complicated the function gets, however, it is condensed down to a single number that represents the "goodness" of that board position.

The computer looks at millions of board arrangements using its evaluation function, and picks the best ones—the ones that return the maximum value from the function. That is all that the computer does. What's amazing is that, using such a simple technique, a computer can play an excellent game!

How **VIDEO GAME SYSTEMS** Work

Home video game systems, also known as consoles, are an incredibly popular form of entertainment. People spend billions of dollars a year on game consoles and games to run on them. Sony PlayStation, Nintendo 64, and Microsoft XBox systems can be found in houses around the globe. Children and adults alike spend hours on end manning the controls of the popular consoles.

HSW Web Links

www.howstuffworks.com

How Computer Memory
  Works
How Removable Storage
  Works
How PCs Work
How Microprocessors
  Work

At its core, a video game system is a highly specialized computer. In fact, most systems are based on the same central processing units (CPUs) used in many desktop computers. To keep the cost of the video game system low, most manufacturers use a CPU that has been widely available for a long time, so the price has fallen to reasonable levels.

That in mind, a natural question would be, "Why buy a game console instead of a using a normal computer?" There are several reasons:

- A game console is usually much cheaper.
- There's no long wait for the game to load.
- Video game systems are designed to be part of your entertainment system. This means that they are easy to connect to your normal TV and stereo.
- There are no compatibility issues, such as operating system, DirectX drivers, correct audio card, supported game controller, resolution, and so on.
- Game developers know exactly what components are in each system, so games are written to take full advantage of the hardware.
- The degree of technical knowledge required to set up and use it is much lower. Most game consoles are truly plug-and-play machines.
- Most video game systems have games that allow multiple players. Setting up multiple-player games is a difficult process with a typical home computer.

A Short History

Video games have been around since the early 1970s. The first true removable game system to gain commercial success was produced by Atari. Introduced in 1977 as

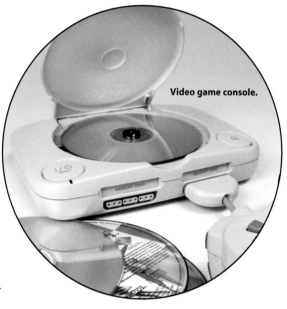

Video game console.

the Atari Video Computer System (VCS), the 2600 stored games on removable cartridges, allowing a multitude of games to be played using the same hardware.

The hardware in the 2600 was quite sophisticated at the time, although it seems incredibly simple now. It consisted of:

- A MOS 6502 microprocessor
- Stella, a custom graphics chip that controlled the synchronization to the TV and all other video processing tasks
- 128 bytes of RAM
- 4-KB ROM-based game cartridges

The chips were attached to a small printed circuit board (PCB) that also connected to the joystick ports, cartridge connector, power supply, and video output. Games were encoded on ROM chips and housed in plastic cartridges. The ROM was wired on a PCB that had a series of metal contacts along one edge. These contacts seated into a plug on the console's main board when a cartridge was plugged into the system. When power was supplied to the system, it would sense the presence of the ROM and load the game software into memory.

Systems like the Atari 2600; its descendant, the 5200; Coleco's ColecoVision; and Mattel's IntelliVision helped to generate interest in home video games for a few years. But interest began to wane because the quality of the home product lagged far behind arcade standards. But in 1985, Nintendo introduced the Nintendo Entertainment System (NES), and everything changed.

The NES introduced three very important concepts to the video game system industry:

- It used a pad controller instead of a joystick.
- It created authentic reproductions of arcade video games for the home system.
- It made the hardware a loss leader by aggressively pricing it, then making a profit on the games themselves.

Game System Basics

The basic pieces really haven't changed that much since the birth of the Atari 2600. Here's a list of the core components that all video game systems have in common:

- Graphics processor
- CPU
- RAM
- Software kernel
- Storage medium for games
- Video output
- Audio output
- Power supply
- Game controllers

The most important part—the part that controls the graphical realism on the screen and the speed and feel of the game—is the graphics processor. Processors are rated by the number of polygons that they can draw per second. The more polygons the processor can draw, the more detailed and realistic the games can look.

The user control interface allows the player to interact with the video game. Early game systems used paddles or joysticks, but most systems today use sophisticated controllers with a variety of buttons and special features.

The software kernel is the console's operating system. It controls the various pieces of hardware, allowing video game programmers to write code using common software libraries and tools.

The two most common storage technologies used for video games today are CDs and ROM-based cartridges. Current systems also offer some type of solid-state memory cards for storing saved games and personal information. Newer systems, like the PlayStation 2, have DVD drives.

Just like computers, video game systems are constantly getting better. New technology developed specifically for video game systems is being coupled with other new technologies, such as DVD discs, to produce new, more, and more-realistic gaming experiences.

Video game controllers.

How **3-D GLASSES** Work

Whether you've used them in the big screen theater or at home in front of your television, 3-D glasses are incredibly cool! They make the movie or television show you're watching look like a 3-D scene that's happening right in front of you. With objects flying off the screen and careening in your direction and characters reaching out toward you, wearing 3-D glasses makes you feel like you've become a part of the action in the movie. Considering they have such high entertainment value, you'll be surprised at how amazingly simple 3-D glasses are.

HSW Web Links

www.howstuffworks.com

How 3-D Graphics Work
How Lasers Work
How Shockwave 3-D
 Technology Works
How Stereolithography
 (3-D Layering) Works

Red-and-blue lens 3-D glasses.

Human beings come equipped with two eyes and an absolutely amazing binocular vision system. For objects up to about 20 feet (6 to 7 meters) away, the binocular vision system lets you easily tell with good accuracy how far away an object is. For example, if there are multiple objects in your field of view, you can automatically tell which ones are farther and which are nearer, and how far away they all are. If you look at the world with one eye closed, you can still perceive distance, but your accuracy goes down and you have to rely on visual cues or focusing distances, both of which take more time for your mind to figure out.

Your Brain as Data Processor

The binocular vision system relies on the fact that our two eyes are spaced about 2 inches (5 cm) apart. Each eye therefore sees the world from a slightly different perspective, and the binocular vision system in your brain uses the difference to calculate distance. Your brain has the ability to correlate the images it sees in its two eyes even though they are slightly different. The correlator can pick out objects in the two scenes it sees and calculate how far apart an object is between the two images. Objects that are farther apart in the two images are closer to you than objects that are not so far apart.

Double Vision

If you have ever used a View-Master viewer or a stereoscopic viewer, you have seen your binocular vision system in action. In a View-Master viewer, each eye is presented with an image. The images are created by two cameras that photograph the same image from slightly different positions. Your eyes can

correlate these images automatically because each eye sees only one of the images.

It turns out that 3-D glasses work a lot like a View-Master viewer. If you go to see a 3-D movie in a movie theater, you wear 3-D glasses to feed different images into your eyes, just like the View-Master viewer does. The screen actually displays two images, and the glasses cause one of the images to enter one eye and the other to enter the other. There are two systems commonly used for doing this:

- **Polarization**—At Disney World, Universal Studios, and other locations that offer 3-D movies, the preferred method uses polarized lenses because they allow color viewing. Two synchronized projectors project two views onto the screen, each with a different polarization. The glasses allow only one of the images into each eye because they contain lenses with different polarization.

- **Red/green or red/blue**—Polarization cannot work on an ordinary TV screen, so the red-green system is used. Two images are displayed on the screen, one in red and the other in green or blue. The filters on the glasses allow only one image to enter each eye, and your brain does the rest. You can't really have a color movie when you're using color to provide the separation, so the image quality is not nearly as good as with the polarized system.

There are some more complicated systems as well, but because they are expensive they're not widely used. For example, in one system a TV screen displays the two images by alternating one right after the other. Special LCD glasses block the view of one eye and then the other in rapid succession. This system allows color viewing on a normal TV, but requires you to buy special equipment.

How **SLOT MACHINES** Work

The technology of slot machines has changed a lot over the years—the classic mechanical designs have given way to sophisticated computer-operated models. But the game has remained the same. The player pulls a handle to rotate a series of reels (typically three) that have pictures printed on them. Winning or losing depends on which pictures line up with the pay line, a line in the middle of a viewing window. If each reel shows the same winning picture along the pay line, you win (certain single images are sometimes winners as well). The amount you win—the payout—depends on which pictures land along the pay line.

The classic mechanical slot machine (often called the one-armed bandit) design works on an elaborate configuration of gears and levers. A coin detector initially registers that a coin has been inserted and unlocks a brake. The central element is a metal shaft that supports the reels. This shaft is connected to a handle mechanism that gets things moving. A braking system brings the spinning reels to a stop, and sensors communicate the position of the reels to the payout system.

There are any number of ways to arrange these elements, and manufacturers have tried dozens of approaches over the years. The diagram shows a representative model.

This design includes three reels mounted on a central shaft. The central shaft also supports three notched discs, which are connected to the three reels. A second shaft below the central shaft supports a kicker, a piece of metal with three paddles. The kicker paddles line up so they can push against the notches on the three discs. The second shaft also supports a series of connected stoppers—teeth that lock into the notches on the discs to stop each reel.

The kicker and the stoppers are both connected to springs, which hold them in a standby position. The kicker is held in place behind the discs, while the stoppers are held up against the discs, locking them into place.

Here's what happens when a player pulls the handle:

1) The handle rotates a hook mechanism, which grabs hold of the kicker, pulling it forward.

2) A catch on the opposite end of the kicker grabs a control cam piece and pivots it forward. This rotates a series of gears connected to the control cam. A spring pulls the control cam back to its original position, but the gear assembly slows it down considerably—the gears act as a mechanical delay.

3) The control cam pulls the stoppers away from the notched discs. As the kicker keeps moving, it pushes the stoppers against several catches on a cam plate. These notches hold the stoppers in place, so the discs and reels can rotate freely.

4) As the handle continues to move the kicker, the kicker paddles push the discs forward briefly. When the handle is all the way back, it releases the kicker.

5) The kicker spring jerks the kicker backward at a good speed. The kicker paddles hit the notches on the discs, spinning the reels rapidly.

6) While all of this is happening, the control cam is slowly returning to its original position. When it does return, it pushes the cam plate back, which releases the stoppers. The different catches holding on to the different stoppers are positioned so that the cam plate will release the stoppers one at a time. Each stopper springs forward and locks into a notch, holding the reel in position.

7) For the sake of simplicity, let's say there is only one winning position for each reel. The "winning notch" in this design is deeper than all the "losing notches." When a reel stops in a winning position, the stoppers engage the winning notch. In other words, the stoppers move farther forward when the winning image comes up. If all three stoppers engage winning notches,

HSW Web Links

www.howstuffworks.com

How Lotteries Work
How Billiard Tables Work
How Chess Computers Work
How Pinball Machines Work

15

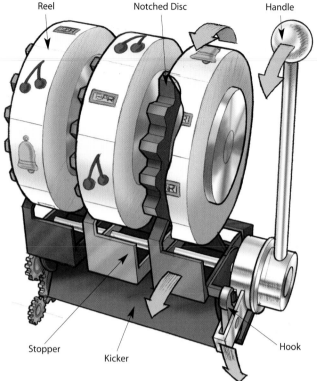

Reel Notched Disc Handle

Stopper Kicker Hook

Front view.

they move a mechanical linkage which releases a coin shutter to release the payout money. (This is only one simplified payout scheme—machines can use a number of methods to determine reel positions.)

From the player's point of view, here's how it looks:

1) The player pulls the handle.
2) There is a clunk, and the three reels start spinning.
3) The three reels stop abruptly one at a time, followed by the payout (if the symbols have lined up).

The stopping-one-at-a-time part builds suspense. If the first reel stops on the jackpot symbol, then you have to wait for the next reel to stop to see if it's a jackpot, and then finally the third. If all three display the right symbol, the player wins. The payout system determines the location of the reels and spits out the appropriate amount of money.

Conventional mechanical slot machines eventually gave rise to electrical machines that worked on similar principles. In an electrical machine, motors spin the reels and electromagnets activate the stoppers. In the past decade, these models have been

almost completely replaced by computer-operated machines.

Modern Machines

Most modern slot machines are designed to look and feel like the old mechanical models, but they work on a completely different principle. The outcome of each pull is actually controlled by a central computer inside the machine, not by the motion of the reels.

The computer uses step motors to turn each reel and stop it at the predetermined point. Step motors are driven by short digital pulses of electricity. These pulses move the motor a set increment, or step, with great precision.

Even though the computer tells the reels where to stop, the games are not preprogrammed to pay out at a certain time. A random number generator at the heart of the computer ensures that each pull has an equal shot at hitting the jackpot.

Whenever the slot machine is turned on, the random number generator spits out whole numbers (typically between 1 and several billion) hundreds of times a second. The instant you pull the arm back (or press the button), the computer records the next few numbers from the random number generator. Then it feeds these numbers through a simple program to determine where the reels should stop.

Here's how the complete process plays out in a typical three-reel machine.

1) You pull the handle, and the computer records the next three numbers from the random number generator. The first number is used to determine the position of the first reel, the second number is used for the second reel, and the third number is used for the third reel. For illustration purposes, let's say the first number is 312,769,458.
2) To determine the position of the first reel, the computer divides the first random number by a set value. Typically, slot machines divide by 32, 64, 128, 256, or 512. In this example, we'll say the computer divides by 64.
3) When the computer divides the random number by the set value, it records the remainder of the quotient. In our example, it finds that 64 goes into 312,769,458 a total of 4,887,022 times, with a remainder of 50.

4) Obviously, the remainder can't be more than 63 or less than 0, so there are only 64 possible end results of this calculation. The 64 possible values act as stops on a large virtual reel. The actual reel that the player sees might have only 11 pictures on it and 22 stops. The computer now has to map the virtual reel to the actual reel.

5) Each of the 64 stops on the virtual reel corresponds to one of the 22 stops on the actual reel. The computer consults a table that tells it how far to move the actual reel for a particular value on the virtual reel. Since there are far more virtual stops than actual stops, some of the actual stops will be linked to more than one virtual stop.

What are the Odds?

The odds of hitting a particular symbol or combination of symbols depends on how the virtual reel is set up. Simply put, the odds of hitting a particular image on the actual reel depend on how many virtual stops correspond to the actual stop.

In a typical weighted slot machine, the top jackpot stop for each reel corresponds to only one virtual stop. This means that the chance of hitting the jackpot image on one reel is 1 in 64. If all of the reels are set up the same way, the chances of hitting the jackpot image on all three reels is 1 in 64^3, or 262,144. For machines with a bigger jackpot, the virtual reel may have many more stops. This decreases the odds of winning that jackpot considerably.

The losing images above and below the jackpot image may correspond to more virtual stops than other images. Consequently, a player is most likely to hit the non-winning images right next to the winning image. This creates the impression that they just missed the jackpot, which encourages them to keep gambling, even though the proximity of the physical images is meaningless.

Show Me the Money

The program is carefully designed and tested to achieve a certain payback percentage. The payback percentage is the percentage of the money put in that is eventually paid out to the player. With a payback percentage of 90, for example, the casino would take about 10% of all money put into the slot machine and give

away the other 90%. The lower the payback percentage, the more money the casino makes.

In most gambling jurisdictions, the law requires that payback percentages be above a certain level (usually somewhere around 75%). The payback percentage in most casino machines is much higher than the minimum—often in the 90% to 97% range. Casinos don't want their machines to be a lot tighter than their competitors' machines or the players will take their business elsewhere.

The odds for a particular slot machine are built into the program on the machine's computer chip. In most cases, the casino cannot change the odds on a machine without replacing this chip. Despite popular opinion, there is no way for the casino to instantly "tighten up" a machine.

Machines don't loosen up by themselves either. That is, they aren't more likely to pay the longer you play. The computer always pulls up new random numbers, so you have exactly the same chance of hitting the jackpot every single time you pull the handle. The idea that a machine can be "ready to pay" is all in the player's head, at least in the standard system.

Originally, casinos installed slot machines as a diversion for casual gamers. Unlike traditional table games (such as blackjack or craps), slot machines don't require any gambling knowledge, and anyone can get in the game with a very small bet.

Slot machines proved to be a monstrous success. They eventually moved off the sidelines to become the most popular and the most profitable game in town, bringing in more than 60% of the annual gaming profits in the United States.

How Much Loot?

There are several different pay-out schemes in modern slot machines. A standard flat top or straight slot machine has a set payout amount that never changes. The jackpot payout in a progressive machine, on the other hand, steadily increases as players put more money into it, until somebody wins it all.

In one common progressive setup, multiple machines are linked together in one computer system. The money put into each machine contributes to the central jackpot. In giant progressive games, machines are linked up from different casinos all across a city, or even an entire state.

Back view.

Kicker

Cam Plate

Stopper

How **BILLIARD TABLES** Work

Billiards, commonly known as pool, is an indoor sport that is played throughout the world. Many restaurants, bars, and pubs have billiard tables, and the popularity of billiard tables in private homes is increasing.

Billiards means any game played on a table with a cue and balls. Billiards relies on the fundamentals of physics and geometry, and becoming an expert in billiards requires skillful mastery of the game's equipment. This equipment includes:

• **Balls**—Although each variation of billiards has different rules, the goal is always to strike the ball and move it in some fashion.

• **Cue**—A long, tapered rod that has a cushioned tip on the narrow end for striking the balls.

• **Table**—The playing surface that the balls travel on. Depending on the game it's designed for, the table may or may not have pockets (holes) for the balls to fall into, but the vast majority of tables have pockets in each corner and also in the middle of the longest side.

It turns out that the most interesting pieces of equipment are the balls and the table, especially in the case of the coin-operated machines that you often find in public.

A Clean Slate

The playing surface of a billiard table has traditionally been made using a large slab of slate covered with cloth. Slate is a bluish-gray rock that cleaves (splits naturally) in broad, flat segments. Slate is formed when layers of clay sediment (the soil and debris that settles at the bottom of a body of water) that have large concentrations of chlorite, mica, and quartz are compressed into sedimentary rock. The sediment hardens in thin layers as it compresses, creating a very hard rock with hundreds of naturally flat layers.

Slate can be ground and polished into a perfectly flat surface fairly easily, which is why it is sought after for billiard tables.

The slate is carved out around the edges where the pockets will go. Also, holes are drilled along the edge so that the slate can be bolted to the top of the table frame.

Inexpensive recreational billiard tables use non-slate playing surfaces, including:

• **Slatron and Permaslate**—Hard, synthetic materials that are basically sheets of plastic layered over particle board

• **Honeycomb**—A stiff plastic honeycomb structure between two sheets of plastic

• **Medium density fiberboard (MDF)**—A flat piece of material made from compressing tiny pieces of wood together, also called *pressed wood* or *particle board*

The biggest problem with non-slate surfaces is that they warp easily and don't have the durability to maintain a perfectly flat playing surface for any length of time. But even the highest-quality playing surface doesn't guarantee a good table unless it has a solid foundation to support it.

A Solid Foundation

The table cabinet starts with a large, rectangular, wooden frame, typically made of thick hardwood planks. Usually, there are one or more cross beams, along with a center beam, to provide additional support to the slate. The frame is connected at the corners either with metal brackets or wooden blocks. The metal brackets or wooden blocks are placed in each corner and bolted to the planks, forming a very solid frame.

Depending on the size of the table and the thickness and weight of the slate, there will be four, six, or eight legs supporting the table. Some designer tables replace the legs with a large pedestal base. Table legs can be hollow or solid, although solid legs are preferable. While the legs may just go to the bottom of the frame, most experts agree that solid legs that extend to the underside of the slate provide the best support.

The Path of Least Resistance

If you could look inside a commercial table, you would see a system of chutes that start at the table's six pockets. Each chute is angled slightly downward from the pocket to the ball return. When a ball falls into that pocket, gravity causes it to roll along in the chute until it reaches a collection chamber where the balls line up single file in a trough. These balls remain locked in the chamber, which you can see behind a piece of clear acrylic, until someone wants to play a game and inserts some coins. By placing coins in a slot and pushing the coin arm in, you trip a lever that allows the balls to roll out of the trough into a large open access area at the foot end of the table.

Cue Balls

On a commercial table, the balls go to the collection chamber when they go in a pocket. But what about the cue ball? If a player accidentally pockets the cue ball (an act known as a scratch), the cue ball needs to come back out from the access area at the foot of the table. For the most part, coin-operated tables use two types of cue balls:

- An oversized ball that is separated by a radius gauging device
- A magnetic cue ball that triggers a magnetic detector

The oversized ball is approximately $2^3/8$ inches (6 cm) in diameter, which is about $1/8$ of an inch (2 mm) larger than a normal ball. This slight difference in size allows the cue ball to be separated before it gets to the storage compartment. The smaller, numbered balls are able to pass through a gauging mechanism, while the larger cue ball is directed through a second chute, where it falls out into an opening on the side of the table.

For players who dislike using the slightly larger cue ball, there are coin-operated machines that can use a magnetic ball. The magnetic cue ball has a magnet built into its core. Magnetic cue balls that go into a pocket are separated by a magnetic detector. As the magnetic ball passes this detector, the magnet triggers a deflecting device that sends the cue ball on a different path than the other balls and, again, sends it into the opening on the side of the table.

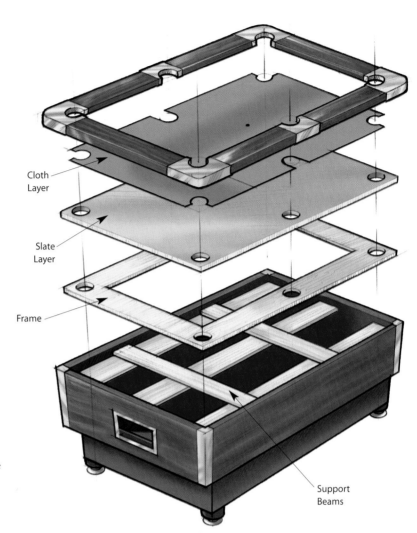

Cloth Layer

Slate Layer

Frame

Support Beams

Both oversized and magnetic cue balls can be used interchangeably on most of today's coin-operated tables, but each has its shortcomings. If you're a beginning pool player the larger ball might not affect your play, but it can disrupt the play of some advanced players who are used to playing with the normal $2^1/4$-inch diameter (5.7 cm) cue ball. Likewise, some players will notice a difference in the properties of a magnetic ball, which sometimes lacks a true roll. Also, because the magnetic ball has the magnetic material inside of it, it has a greater tendency to shatter if dropped on a hard surface.

Billiards seems so simple—balls rolling on a flat table. As with most things, however, it gets more involved if you want to do things right. A full-size slate table will last for centuries and will stay just as flat as it was the day you bought it. However, it also weighs 2,000 pounds!

And Another Thing...

Billiard cloth is often referred to as felt, but it actually is nothing like real felt. Real felt is not a woven material but is formed from compressed and matted fibers and would not work well at all as a smooth playing surface. Billiard tables normally use tightly woven cloth made primarily of wool with a synthetic such as nylon added for durability. The cloth provides a consistent and smooth playing surface.

19

How PINBALL MACHINES Work

Since Gottlieb released the first true pinball game, Humpty Dumpty, in 1947, people have had an obsession with pinball machines. In the past half-decade, countless technologies have been added to pinball tables to attract players, but the goal of the game remains the same: Score points and keep the pinball from going down the drain.

HSW Web Links

www.howstuffworks.com

How Video Game Systems Work

How Sony's PlayStation 2 Works

How Sega's Dreamcast Works

How Nintendo's Game Boy Advance Works

Jackpot!

Scoring in pinball games is a mystery to most amateur pinball players, whose basic goal is simply to keep the ball from going down the drain. The pinball expert however, is after much more.

Your current score is kept on the dot matrix display located at the base on the backglass. This display is the center point of all scoring options and profiles. In most newer games, this display guides you, through animations or words, by telling you which ramps to shoot or what targets to hit in order to score the most points.

The experienced pinball player is able to score the most points through *combination shots*. These shots consist of a specific sequence of moves that activates a jackpot or some other scoring mechanism—the specifics vary with the different themes of the pinball machines.

The original pinball machines were purely mechanical, but modern pinball machines are a strange intermingling of electronic and mechanical parts. Machines have lights and LCD panels, sometimes even a TV screen, but they also have mechanical flippers, bumpers, and switches.

Start the Ball Rolling

The key components of the game are the flippers and the pinball. The flippers are usually located at the bottom of the playfield, directly above the drain. The flippers propel the ball toward the bumpers and ramps at the other end of the table so that you can score points. Sometimes extra flippers are further up on the table to give you more options. Two buttons, one on either side of the machine, control the flippers. The button sends electricity to a powerful solenoid (an electromagnetic switch controlled by an electrical current) that moves the flipper.

A pinball is a 1 1/16-inch diameter (about 3 cm) steel sphere that weighs 2.8 ounces (80 g). On a normal (unwaxed) table, the ball can reach speeds of up to 90 miles per hour (145 kph). With this traditional steel pinball, the ball's magnetic properties sometimes come into play, as some machines use magnets to trap the ball in certain places on the playfield. In a few machines, a ceramic pinball called a powerball is used. This ball weighs only 2.28 ounces (65 g), so it moves faster around the table and is immune to the magnets used on some games. Certain

multistart pinball games mix up the action of the play by using both magnetic and ceramic pinballs simultaneously.

The Backbox

The backbox portion of the table—the horizontal part that is typically pushed against a wall—does two things: it holds the main electronics of the game and the points display, and it attracts players with lights and art. The electromechanical bumpers and flippers found on the table are all tied into the main controller board located behind the backglass. In a modern game, the controller is a microprocessor. A ROM chip contains all the information needed to play the game.

The wiring that runs from the controller board to the rest of the machine is massive, usually consisting of over a half-mile (0.8 km) of wire. These wires carry commands back and forth between the main board and the flippers, bumpers, targets, and ramps.

There are usually two other pieces of electronics contained in the backbox. A dot-matrix display board, usually either 128 x 32 or 192 x 64 pixels in dimension, is located at the base of the backglass. This display is used to relay information to the player, such as the score and hints about how to increase the score and possibly get a free game. Also, since the early 1990s, there are speakers in the backbox that have replaced the traditional bell. Pinball sounds are now digital and machines can play songs or sampled sounds when the player scores points.

The underside of a pinball playfield.

The Playfield

The pinball playfield itself is usually made of a wood base that's been coated with several layers of paint and finish. The playfield inclines at a 6- to 7-degree angle toward the player. The bumpers, ramps, and flippers are all mounted onto the playfield with screws and glue. All of these obstacles and targets wire into the main controller board in the backglass area so that the computer can tell where the ball is and react by giving points or activating special features.

Once you press the start button, a solenoid shoots the ball into the launch lane in front of the plunger. On some machines, you still have to pull the plunger back and let it go to launch the ball into play. On many newer machines, there is simply a button or some kind of themed device that you activate to launch the ball. When activated, a solenoid behind the ball kicks the ball into play. The solenoid is also wired into the computer so that if a ball gets lost on the table or the player receives a multiball award, another ball can be launched onto the playfield with no action required by the player.

Tilt!

One component of a pinball machine that most players have experience with is the tilt sensor. The tilt sensor is there to make sure players don't cheat—at least not too much, anyway. By shaking the machine, a player is able to influence how the ball travels down the playfield and score more points in the process. A skilled player knows exactly how much he or she is able to shake the machine without setting off the tilt sensor. The sensor consists of a metal ring with a cone-shaped pendulum bob hanging though the center of it. Normally, the bob hangs so that none of it is touching the ring.

As the machine is shaken, the bob comes closer to the edges of the conductive

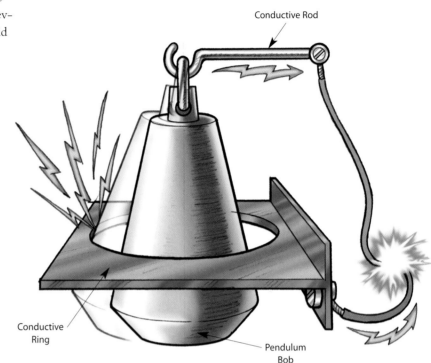

Conductive Rod

Conductive Ring

Pendulum Bob

Basic tilt mechanism in a typical pinball machine.

ring. Once the bob touches the ring, a current flows through the ring and a tilt is registered. Depending on the machine, you might immediately lose your ball, or you might only get warned. Most new machines give you two warnings before all the flippers stop working and your ball goes down the drain.

There are also devices in the machine that look for slam tilts. A slam tilt is a heavy abuse of the game, usually in the form of someone picking the machine up or kicking the front end very hard. Registering one of these immediately ends the game.

There was once a time when an arcade had nothing but pinball machines. Video games changed all that, and will probably one day make the pinball machine extinct.

Did You Know?

One great pinball idea that ended up being shelved was called Pinball 2000. It involved the merging together of video games and pinball machines to form a pinball machine that interacted with virtual targets on a video monitor. While it was billed as the future of pinball, it never had a chance to completely flourish. Two games (Revenge from Mars and Star Wars: Episode 1) made it through production, and a third was in development when Williams/Bally shut the doors of its pinball division for good.

How **KARATE** Works

To the untrained observer, karate skills can seem like magical superpowers. Using only her body, a 5-foot 5-inch, 120-pound karate master can take down a 6-foot, 200-pound man in a matter of seconds. In strength and sheer bulk, the man has the upper hand by a considerable margin. But somehow the karate master prevails with a few elegant punches and kicks. Using the same set of skills, advanced karate students can break thick bricks and boards with their bare hands and feet. How is all this possible?

HSW Web Links

www.howstuffworks.com

How Exercise Works
How Force, Power, Torque
   and Energy Work
How Muscles Work
How Verbal Self-Defense
   Works
How Hypnosis Works
How Performance-
   Enhancing Drugs Work
How the Physics of
   Football Works

The word *karate* is Japanese for *open hand*. *Te* is Japanese for *hand*, and signifies that the body is the primary weapon. *Kara* is Japanese for *open,* and means that karatekas (practitioners of karate) are open to the world around them, making them better equipped to handle any attack.

On the most basic level, karate is simply a method of using physics to your advantage. In any fight between two people, both fighters bring a certain amount of energy to the situation. The total amount of potential energy depends on the fighters' size, muscle strength, and physical health. The object of karate is to use your body to channel this energy.

them rely on this same basic idea. The point of impact is reduced to some small, usually bony area of your hand or foot, and the force of your attack is focused on this point. Karatekas strengthen their hands and feet so

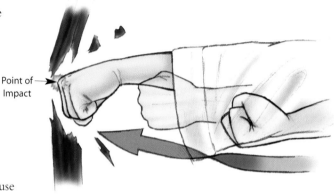

Point of Impact

Smart Hitting

A karateka concentrates his or her strength into a relatively small area. If you open your hands wide and shove somebody, the force of your attack spreads out across your palm and fingers. This dissipates the force of your attack over a fairly wide area; your opponent feels a relatively blunt force. But if you hold all of your fingers tightly together and hit the person with only the side of your hand, or with only your fingertips, that same amount of force is applied to a much smaller area. In that area, the impact is much more intense.

In karate, there are a number of punching and kicking techniques, but most of

Karatekas in action.

that they can execute these punches and kicks effectively.

Karatekas maximize the force of the impact by putting their whole body into the punch or kick. If you watch karatekas fight, you'll see that they often pivot their torso and shift their weight from one leg to the other when they throw a punch. In this way, the energy of their moving body goes into each strike along with the energy of the arm muscles. Karatekas also practice hitting with great speed, as this increases the force of each blow.

One of the most important elements in karate is following through on punches and kicks. When you hit something, say a piece of board, your natural instinct is to slow down your swing just before impact; you hesitate because you don't want to hurt your hand. Karatekas deprogram this hesitation instinct; they visualize pushing their fist to some point past their target (the other side of the board, for example). To maximize the force of each movement, it's essential that the karateka follows through. To help their concentration before each attack, karatekas take a deep breath. As they release the punch or kick, they let this breath out.

Deflection Defense

If somebody punches you squarely in the chest, you feel the brunt of the fist's force. You would prefer to miss the punch altogether, or to have the blow glance off. In karate, the object is to sweep your opponent's arm (or leg, if your opponent is kicking) away from you with your own arm so that you avoid being struck.

When you sweep a blow away to the side, your opponents' own momentum upsets their balance. This leaves them vulnerable to attack, and you can land a successful hit or pin them to the ground. You might also grab attackers and pull them forward, increasing their forward momentum. Using this defense, a karateka can throw attackers to the ground. Throwing is not a central element in karate, but it does play an important part in other martial arts forms, notably judo and aikido.

Stand and Deliver

To protect against attacks, karatekas take on particular fighting stances. Generally, karatekas stand with one leg in front of them and one leg behind them. This effectively turns their bodies to the side and shields the front (center, vital areas) of the body from attack, and it also gives the karateka better balance. Karatekas hold themselves with their center of gravity relatively low to the ground, so it is more difficult for an opponent to knock them down.

In a karate competition, both karatekas concentrate on guarding themselves against attack while waiting for an opening in their opponent's defenses. Often, a karateka can land a

successful hit immediately after deflecting the opponent's attack, as this is when the opponent is most vulnerable.

A lot of karate is based on paying attention to what's going on around you so you can recognize an opportunity when it arises.

The Ki is Key

To advance through the higher levels of karate, it is essential that a karateka cultivate spiritual power, whatever his or her religious beliefs. The basic element of this power is ki.

Ki is generally described as the energy of life itself. It binds all living things together and gives each person his or her spiritual, physical, and mental power. As karatekas develop heightened physical control, they become more aware of the seat of ki in their body.

Karatekas may apply the lessons of karate to all aspects of their life. In the same way that a karateka can deflect a physical punch, he or she might deflect a less physical blow (losing a job, for example, or taking an insult). Karatekas leave themselves open to opportunities and redirect any attacks.

You could practice these principles without undertaking karate, of course, but karatekas claim that exercising this philosophy physically as well as mentally strengthens their resolve and character. If you know to automatically deflect physical attacks and land effective kicks and punches, you will also know how to handle everyday threats and opportunities.

Show of Skill

To demonstrate their mastery of punches, kicks and blocks, karatekas simulate various combat scenarios. In one exercise, called kata, karatekas carry out a predetermined sequence of movements against an imaginary group of attackers. Kata is extremely important to beginning karate students, as it helps them perfect their technique.

More advanced karatekas may engage in kumite, a sort of freestyle sparring. Karatekas also demonstrate strength through tameshiwari, also known as "the breaking demonstration." With a great deal of practice and concentration, karatekas can break boards and bricks with only their feet and hands. Basically, they turn their appendages into natural chisels, breaking the structural integrity of an object by focusing the force of their entire body into a small area.

How **BICYCLES** Work

Bicycles are simple and beautifully elegant machines that attract just about every kid at an early age. The coolest thing about riding a bicycle is that it is so efficient. Riding a bike is a lot faster and expends less energy than walking or running—especially if you are riding downhill.

HSW Web Links

www.howstuffworks.com

How Mountain Bikes Work
How Gear Ratios Work
How Exercise Works
How Tire Pressure Gauges
 Work
How Sweat Works

The other neat thing about bicycles—for anyone interested in machines and mechanics—is that all the working parts are completely exposed. There are no covers or pieces of sheet metal hiding anything—on a bicycle it's all out in the open. Many kids with mechanical tendencies can't resist the desire to take their bikes apart. And why shouldn't they? It's a great way to learn about lots of things. By examining all of the different parts of a bicycle, anyone can completely understand how a bike works.

Bicycle Parts

A good way to start talking about bicycles is to name all of the parts. When you look at a bike, the parts that you can immediately see and identify are:

- **The frame**—A bicycle's frame is made of metal tubes welded together.
- **The wheels**—The wheels are made of the hub, the spokes, the metal rim, and the rubber tire.
- **The seat and seat post**—The seat post supports the seat and connects it to the frame.

Bicycle Chain

Spokes

Brakes

Gears

Wheel Axle

Pedal

- **The handlebars and the handlebar stem**—The handlebar stem supports the handlebars and connects the handlebars to the frame.
- **The cranks and the pedals**—The pedals attach to the cranks, which attach to the axle.
- **The brakes**—The brake system includes the actuators on the handlebars, the brake cable, the brake calipers, and the brake pads.
- **The chain and gears**—This includes the front chain wheels, the rear freewheel, the front and rear derailleur, the shift levers on the handlebars, and the cables.

Bicycles really are simple, and that is one of the things that make them so beautiful.

Bicycle Bearings

Bicycles use ball bearings to reduce friction. You can find ball bearings in:

- The front and rear hubs for the wheels
- The bottom bracket, where an axle connects the two pedal cranks together
- The fork tube, where they allow the handlebars to turn
- The pedals
- The freewheel, where they do double-duty (in the freewheel they also help provide the one-directional feature)

The ball bearings ride in a cup. The cones screw into the cup and ride on the bearings. The cones are adjusted to be tight enough so there is no play, but not so tight that they squeeze the ball bearings and cause them to bind. The fork, wheel hubs and pedals work this way, with the cones providing the adjustment. In the crank axle, one of the cups provides the adjustment instead of the cones. A little bit of grease in the bearings makes them even smoother.

Periodically you have to disassemble the bearings on a bicycle to clean out the dirt

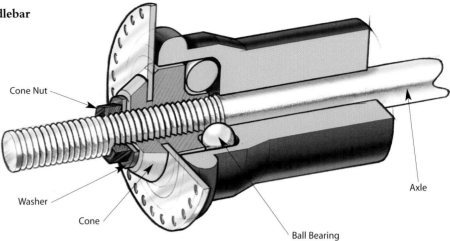

Cone Nut
Washer
Cone
Ball Bearing
Axle

and put in fresh grease. Some more-expensive bicycles have sealed bearing cartridges that never need adjustment or lubrication.

Bicycle Gears

The idea behind having multiple gears on a bicycle—whether it's an older 10-speed bike or a modern mountain bike with 24 gears—is that you can change the amount the bike moves forward with each pedal stroke. The less distance the bike travels with each stroke, the easier it is to pedal.

The gears at the front are called the chain wheels. Most bikes have two or three chain wheels. Attached to the rear wheel is the freewheel, which has between 5 and 9 gears on it, depending on the bike.

Bicycle Power

Have you ever wondered if, while riding your bike, you could generate enough power to run a small appliance?

It turns out that a little pedaling prowess can really pay off.

Because laptops are designed to run off batteries, they are very efficient. A laptop might consume 15 watts. It would be extremely easy to generate 15 watts (0.02 horsepower) on a bicycle.

How many calories would you burn doing this? To generate 1 watt for an hour, you burn about 0.85 calories. Rounding up, that's about 1 calorie per watt-hour. So you would burn about 15 calories per hour using your bike to power your laptop. At that rate, a single 60-calorie chocolate-chip cookie could power a laptop for four hours!

A freewheel spins freely in one direction and locks in the other. That allows the rider to pedal or not—when the rider isn't pedaling, the bike coasts.

To change the gears, a bicycle has front and rear derailleurs. The rear derailleur has two small cogs on it that both spin freely. The purpose of the arm and lower cog of the derailleur is to tension the chain. The cog and arm are connected to a spring so that the cog pulls backwards at all times. As you change gears you will notice that the angle of the arm changes to take up or let out slack.

The top cog is very close to the freewheel. When you adjust the gears with the lever on the handlebar, this cog moves to a different position on the freewheel and drags the chain with it. The chain naturally slips from one gear to the next as you turn the pedals.

The lowest gear ratio on the bike might be when the front chain wheel has 22 teeth and the rear gear has 30 teeth. That means that the gear ratio is 0.73 to 1. For each pedal stroke, the rear wheel turns 0.73 times.

Let's say that the bike has wheels that are 26 inches in diameter. The circumference of the wheels would be about 81.5 inches. So

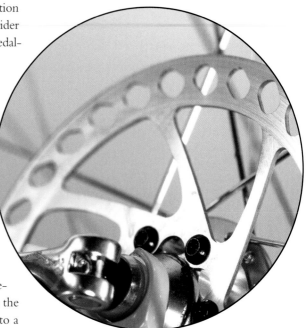

for each pedal stroke, the bicycle moves 0.73×81.5 inches, or about 60 inches—which is 5 feet.

If the rider completes a pedal stroke each second, the bike travels 300 feet in a minute. This works out to about 3.4 miles per hour (5.4 kph), which is about the slowest that you can go on a bike without falling off.

The highest gear ratio on the bike might be when the front chain wheel has 44 teeth and the rear gear has 11 teeth. That creates a 4 to 1 gear ratio. For each pedal stroke, the rear wheel turns 4 times.

Still assuming that the wheels are 26 inches in diameter, the bike moves forward 326 inches with each pedal stroke. At a 60 rpm pedaling rate, the speed of the bike is 18.5 mph (30 kph). By doubling the pedaling rate to 120 rpm, you can get the bike up to a speed of 37 mph (60 kph), which is about as fast as any racer can make a bike go on flat ground.

Everything about a bicycle is simple, elegant, and lightweight. That's what makes bikes such great machines to ride, and also great machines to look at as works of art!

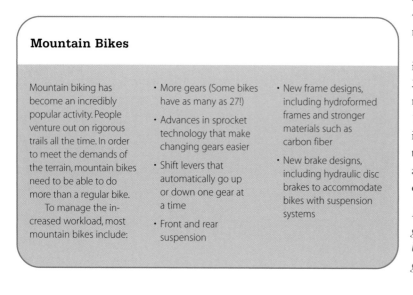

Mountain Bikes

Mountain biking has become an incredibly popular activity. People venture out on rigorous trails all the time. In order to meet the demands of the terrain, mountain bikes need to be able to do more than a regular bike.

To manage the increased workload, most mountain bikes include:

- More gears (Some bikes have as many as 27!)
- Advances in sprocket technology that make changing gears easier
- Shift levers that automatically go up or down one gear at a time
- Front and rear suspension

- New frame designs, including hydroformed frames and stronger materials such as carbon fiber
- New brake designs, including hydraulic disc brakes to accommodate bikes with suspension systems

chapter two

IN PUBLIC

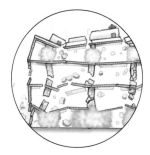

How **WATER TOWERS** Work

Have you ever experienced a water failure? That is, have you ever turned on your faucet and found no water? If you get your water from a municipal water system, the answer is probably no.

HSW Web Links

www.howstuffworks.com

How Water Heaters Work
How Car Cooling Systems Work
How Ice Rinks Work

We have power failures all the time. Cable TV goes out fairly frequently. Although less common, the phone system goes down every so often, and it's now common to get an all-circuits-busy message when making long-distance calls. But the water in any city or suburb is always there. Water pressure is very reliable.

A big reason for that level of reliability is the water tower. You see water towers everywhere, especially if you live in a flat area full of small towns. Each water system has one or more towers.

In most towns, the water people drink comes from a well, a river, or a reservoir (normally a local lake). The water is treated in a water treatment plant to remove sediment and bacteria. The output from the water treatment plant is clear, germ-free water. A high-lift pump pressurizes the water and sends it to the water system's primary feeder pipes. If the pump is producing more water than the water system needs, the excess flows automatically into the water tower's tank. If the community is demanding more water than the pump can supply, then water flows out of the tank to meet the need.

Tower, Tank, and Pump

A water tower is an incredibly simple device. Although water towers come in all shapes and sizes, they are simply large, elevated tanks of water. Water towers are tall to provide pressure. Each foot of height provides 0.43 psi (pounds per square inch) of pressure. A typical municipal water supply runs at between 50 and 100 psi (major appliances, such as a dishwasher or washing machine, require at least 20 to 30 psi). The water tower must be tall enough to supply that level of pressure to all of the houses and businesses in the area of the tower. So water towers are typically located on high ground, and they are tall enough to provide the necessary pressure. In hilly regions, a simple tank located on the highest hill in the area can sometimes replace a tower.

A water tower's tank is normally quite large. A normal in-ground swimming pool in someone's backyard might hold something like 20,000 or 30,000 gallons, and a typical water tower might hold 50 times that amount. Typically, a water tower's tank is sized to hold about a day's worth of water for the community served by the tower. If the pumps fail—for example, during a power failure—the water tower holds enough water to keep things flowing for about a day.

One of the big advantages of a water tower is that it lets a municipality size its pumps for average consumption rather than peak demand. That can save a community a lot of money. Say that the water consumption for a pumping station averages 500 gallons of water per minute (or 720,000 gallons over the course of a day). There will be times during the day when water consumption is much greater than 500 gallons per minute. For example, lots of people wake up at about the same time on weekdays (say 7:00 a.m.). They go to the bathroom, take a shower, brush their teeth, and so on. Water demand, therefore, might peak at 2,000 gallons per minute around 7 a.m. Mondays through

Fridays. There is a big cost difference between a 500-gallon-per-minute pump and a 2,000-gallon-per-minute pump. Because of the water tower, the municipality can purchase a 500-gallon-per-minute pump and let the water tower handle the peak demand. At night, when demand normally falls to practically zero, the pump can make up the difference and refill the water tower.

Form and Function

Water towers come in all shapes and sizes. For example, there's a giant peach-shaped water tower along I-85 in Gaffney, South Carolina! Complete with stem and leaf, this amazing structure holds one million gallons of water. The folks in Clanton, Alabama liked this design so much, that they now have a smaller version (it holds half a million gallons of water) just outside of town.

In a city, tall buildings often need to solve their own water pressure problems. Because the buildings are so tall, they often exceed the height that the city's water pressure can handle. Therefore, a tall building will have its own pumps and its own water towers. If you were to look around the skyline from atop the Empire State Building in New York City,

you would see dozens of small water towers on the rooftops of the surrounding buildings.

The next time you are out driving around, especially if you are driving through a series of small towns, take the time to notice the water towers. Now that you know how they work, you will be amazed by how many you see and by all the different forms they take!

Cool Facts

Sunset Beach, California refused to completely retire its water tower after it was deactivated in 1974. Instead of housing water, this tower now houses people! This 3-level home has offered an incredible view of the surrounding beach to its many owners since the mid 1980s.

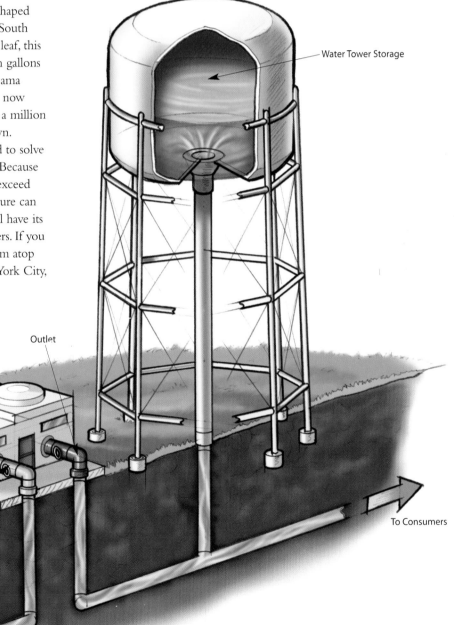

Water Tower Storage

Outlet

Pump House

Intake Pump

From Treatment Plant

To Consumers

How **BRIDGES** Work

A bridge provides passage over some sort of obstacle. You see bridges all the time—they're a part of everyday life. In fact, if you've ever laid a plank or log down over a stream to keep from getting wet, you've constructed a bridge. The mere mention of the word might even cause you to picture a certain bridge in your mind—such as the Golden Gate or Brooklyn bridge.

HSW Web Links

www.howstuffworks.com

How House Construction
  Works
How Floating Cities
  Will Work
How Smart Structures
  Will Work
How Skyscrapers Work

Cool Facts

When an army marches across a bridge, the soldiers are often told to "break step." This is to avoid the possibility that their rhythmic marching will start resonating throughout the bridge. An army that is large enough and marching at the right cadence could start a bridge swaying and undulating until it breaks apart.

There are three major types of bridges:

- Beam bridges
- Arch bridges
- Suspension bridges

The biggest difference between these three types of bridges is the distances they can cross in a single span. A *span* is the distance between two bridge supports, whether they are columns, towers, or the wall of a canyon. A modern beam bridge, for instance, is likely to span a distance of up to 200 feet (60 meters), while a modern arch can safely span up to 800 or 1,000 feet (240 to 300 meters). A suspension bridge, the pinnacle of bridge technology, is capable of spanning up to 7,000 feet (2,100 meters).

Span This

What is it that lets an arch bridge span greater distances than a beam bridge, or a suspension bridge span a distance seven times that of an arch bridge? The answer lies in how each bridge type deals with two important forces:

- **Compression**—When a force is compressing or shortening the thing it's acting on
- **Tension**—When a force is expanding or lengthening the thing it's acting on

A simple, everyday example of compression and tension is a spring. When you press down, or push the two ends of the spring together, the force of compression shortens the spring. When you pull up, or pull apart the two ends, you're creating tension in the spring.

Compression and tension are present in all bridges, and it's the job of the bridge

design to handle these forces without the bridge buckling or snapping. *Buckling* is what happens when the force of compression overcomes an object's ability to handle compression, and *snapping* is what happens when tension overcomes an object's ability to handle tension.

The best way to deal with these forces is to either dissipate them or transfer them. When you *dissipate* force you're spreading it out over a larger area so that no one spot has to bear the brunt of the concentrated force. If you want to *transfer* force, you need to move it from an area of weakness to a stronger area that's designed to handle that force. An arch bridge is a good example of dissipation, while a suspension bridge is a good example of transference.

The Beam Bridge

A beam bridge is basically a rigid horizontal structure that is resting on two piers, one at each end. The piers directly support the weight of the bridge and any traffic traveling on it. This weight is pressing directly downward. The force of compression manifests itself on the top side of the beam bridge's deck (or roadway). This causes the upper portion of the deck to shorten. Compression on the upper portion of the deck causes tension in the lower portion of the deck. This tension causes the lower portion of the beam to lengthen. To keep this kind of bridge from buckling or snapping, the forces acting on the bridge need to be dissipated.

Many beam bridges that you find on highway overpasses use concrete or steel beams to handle the load. The size of the beam, and in particular the height of the beam, controls the distance that the beam can span. By increasing the height of the beam, the beam has more material to dissipate the tension. To create very tall beams, bridge designers add supporting latticework, a *truss,* to the bridge's beam. This support truss adds rigidity to the existing beam, greatly increasing its ability to dissipate the compression and tension. Once the beam begins to compress, the force is dissipated through the truss.

Despite the ingenious addition of a truss, the beam bridge is still limited in the distance it can span. As the distance increases, the size of the truss must also increase, until it reaches a point where the bridge's own weight is so large that the truss cannot support it.

Force

Beam bridge tension diagram.

Tension

Force

Arch bridge tension diagram.

Tension

The Arch Bridge

An arch bridge is a semicircular structure with abutments on each end. Arch bridges are always under compression. The force of compression is pushed outward along the curve of the arch toward the abutments. The design of the arch—the semicircle—naturally diverts the weight from the bridge deck to the abutments, dissipating the force of compression. The natural curve of the arch and its ability to dissipate the force outward also greatly reduces the effects of tension on the underside of the arch. However, the larger the semicircle of the arch, the greater the effects of tension are on the underside. As with the beam bridge, the limits of size will eventually overtake the natural strength of the arch.

The Suspension Bridge

A suspension bridge has cables (or ropes or chains) strung across the gap. The deck is suspended from these cables. Modern

suspension bridges have two tall towers through which the cables are strung, so the towers are supporting the majority of the roadway's weight.

The force of compression pushes down on the suspension bridge's deck, but because it is a suspended roadway, the cables transfer the compression to the towers, which dissipate the compression directly into the earth where they are firmly entrenched.

The supporting cables, running between the two anchorages, handle the tension forces. The cables are literally stretched from the weight of the bridge and its traffic as they run

And Another Thing...

Arches are fascinating in that they are a truly natural form of bridge. It is the shape of the structure that gives it its strength. An arch bridge doesn't need additional supports or cables. In fact, an arch made of stone doesn't even need mortar. Ancient Romans built arch bridges (and aqueducts) that are still standing, and structurally sound, today. These bridges and aqueducts are real testaments to the natural effectiveness of an arch as a bridge structure.

Buckling and Snapping

A good example of buckling and snapping is easy to see. Take a two-by-four and place each end on top of two chairs, making a sort of bridge. Then place a 50-pound weight on top of the board exactly in the middle. The two-by-four will start to bend because the top side of the board is under compression and the bottom side is under tension. If you keep adding weight, eventually the two-by-four will break because the top side will buckle and the bottom side will snap.

from anchorage to anchorage. The anchorages are also under tension, but since they, like the towers, are held firmly to the earth, the tension they experience is dissipated.

In addition to cables, almost all suspension bridges have a supporting truss system beneath the bridge deck (a *deck truss*). This structure helps to stiffen the deck and reduce the tendency of the roadway to sway and ripple.

Additional Bridge Forces

In addition to compression and tension, there are other forces that also must be considered when designing a bridge. These forces are usually specific to a particular location or bridge design.

Torsion is a rotational or twisting force. The natural shape of the arch and the additional truss structure of the beam bridge have eliminated the destructive effects of torsion on these bridges. However, because they hang from a pair of cables, suspension bridges are somewhat more susceptible to torsion, especially in high winds. Innovative designs with special attention to truss structures help handle this problem.

Resonance is a vibration caused by an external force that is in harmony with the natural vibration of the original object. If left unchecked, this force can be fatal to a bridge. Resonant vibrations will travel through a bridge in the form of waves.

To lessen the resonance effect in a bridge, designers build dampeners into the bridge design to interrupt the resonant waves. Interrupting them is an effective way to prevent the growth of the waves regardless of the duration or source of the vibrations.

Dampening techniques generally involve inertia. Say a bridge has a solid roadway. A resonant wave can easily travel the length of this bridge. However, if a bridge roadway is made up of different sections that have overlapping plates, it's not that easy. The movement of one section is transferred to another via the plates, which, since they are overlapping, creates a certain amount of friction. The trick is to create enough friction to change the frequency of the resonant wave. Changing the frequency prevents the wave from building. This change can create two different waves, neither of which can build off the other into a destructive force.

The force of nature, specifically weather, is by far the hardest to combat. Rain, ice, wind, and salt can each bring down a bridge on its own, and in combination they can be hard on a bridge. Bridge designers have learned their craft by studying the failures of the past. Iron has replaced wood, and steel has replaced iron. Each new material or design technique builds off the lessons of the past. The problems of weather, however, have yet to be completely conquered.

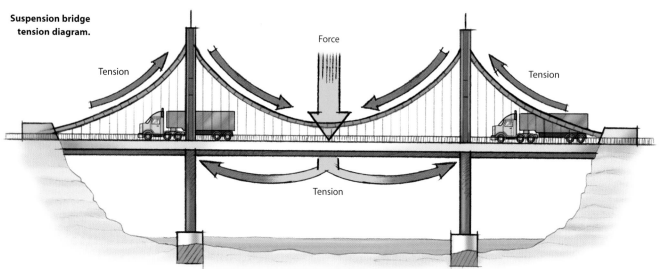

Suspension bridge tension diagram.

Force

Tension

Tension

Tension

How SKYSCRAPERS Work

Throughout the history of architecture, there has been a continual quest for height. Thousands of workers toiled on the pyramids of ancient Egypt, the cathedrals of Europe, and countless other towers, all striving to create something awe inspiring. People build skyscrapers primarily because they are convenient—you can create a lot of real estate out of a relatively small ground area. But ego and grandeur do sometimes play a significant role in deciding the scale of the construction, just as they did in earlier civilizations.

The main obstacle in building upward is the downward pull of gravity. Imagine carrying a friend on your shoulders. If the friend is light, you can support him or her by yourself. But if you were to put another person on your friend's shoulders (build your tower higher), the weight would probably be too much for you to carry alone. To make a tower that is multiple-people high, you need more people on the bottom to support the weight of everybody above.

This is how cheerleader pyramids work, and it's also how real pyramids and other stone buildings work. There has to be more material at the bottom to support the combined weight of all the material above. In normal buildings made of bricks and mortar, you have to keep thickening the lower walls as you build new upper floors. After you reach a certain height, this is highly impractical. If there's almost no room on the lower floors, what's the point in making a tall building?

In the late 1800s, new manufacturing processes made it possible to produce long beams of solid iron. This gave architects a whole new set of building blocks to work with. Narrow, relatively lightweight metal beams could support much more weight than the solid brick walls in older buildings, while taking up a fraction of the space. Steel, which is lighter and stronger than iron, made it possible to build even taller buildings.

Giant Girder Grids

The vertical columns in a skyscraper consist of dozens of steel beams riveted end to end. At each floor level, these vertical columns are connected to horizontal girder beams. Many buildings also have diagonal beams running between the girders for extra structural support.

In this giant three-dimensional grid—called the *superstructure*—all the weight in the building is transferred directly to the vertical columns, which concentrate the force into small areas at the building's base. The *substructure* under the building spreads this force out again.

In a typical substructure, each vertical column sits on a spread footing. The column rests directly on a cast-iron plate, which sits on top of a *grillage*—a stack of horizontal steel beams, lined side-by-side in multiple layers. The grillage rests on a thick concrete pad underground. The entire structure is covered with concrete.

The structure expands out, lower in the ground, distributing the concentrated weight from the columns over a wide area. Ultimately, the entire weight of the building rests directly on the hard clay material under the earth's surface. In very heavy buildings, the base of the spread footings rest on massive concrete or steel piers that may extend all the way to the earth's bedrock layer.

In this design, the outer walls need only support their own weight. Architects can open the building up as much as they want; they can even make the outer walls entirely out of glass.

HSW Web Links

www.howstuffworks.com

How Iron and Steel Work
How Building Implosions Work
How House Construction Works
How Bridges Work
How Earthquakes Work
How Tower Cranes Work

What's the Tallest Skyscraper?

It depends on the rules of the competition. Traditionally, the architectural community defines a building as an enclosed structure built primarily for occupancy. This excludes a lot of extremely tall freestanding structures, such as Toronto's 1,815-foot (550 meter) CN Tower, from the running.

If you include rooftop antennas in the total height measure, the Sears Tower takes first prize at 1,730 feet. Without including antenna height, the Petronas Towers in Malaysia, built in 1997, win with 1,483 feet each. The top part of this structure is only decorative, however, and it just barely creeps into the record books by the tips of its thin spires. Many Chicagoans point out that their Sears Tower has the highest occupied floor, at 1,431 feet, and the highest traditional roof, at 1,454 feet.

Conventionally, decorative structures count toward height, but antennas do not, giving the Petronas Towers the official lead.

33

Functionality

Skyscrapers would never have worked without the coincident emergence of elevators. Designing skyscraper elevator systems is a balancing act of sorts. As you add more floors to a building, you increase the building's occupancy. When you have more people, you need more elevators. But elevator shafts take up a lot of room, so you lose floor space for every elevator you add. To make more room for people, you have to add more floors. Deciding on the right number of floors and elevators is one of the most important parts of designing a building.

Building safety is also a major consideration in design. Skyscrapers wouldn't have worked so well without the advent of new fire-resistant building materials in the 1800s. These days, skyscrapers are also outfitted with sophisticated sprinkler equipment that puts out most fires before they can spread very far.

Architects also pay careful attention to the comfort of the building's occupants. The Empire State Building, for example, was designed so its occupants would always be within 30 feet of a window. A building is only successful when the architects have focused not only on structural stability, but also usability and occupant satisfaction.

Wind Resistance

In addition to the vertical force of gravity, skyscrapers also have to deal with the horizontal force of wind. Most skyscrapers can easily move several feet in either direction, like a swaying tree, without damaging their structural integrity. But if the building moves a substantial horizontal distance, the occupants will definitely feel it.

To keep bigger skyscrapers from swaying too much, engineers have to construct especially strong cores through the center of the building. In the older super skyscrapers, a sturdy steel truss fortifies the area around the central elevator shafts. Newer buildings have one or more concrete cores built into the center of the building.

Making buildings more rigid also braces them against earthquake damage. The entire building moves with the horizontal vibrations of the earth, so the steel skeleton isn't twisted and strained.

Some buildings use advanced wind-compensating dampers. In one method, a hydraulic system pushes a 400-ton concrete weight back and forth on one of the top floors, shifting the weight of the entire building from side to side. A sophisticated computer system carefully monitors how the wind is shifting the building and moves the weight accordingly.

How High?

Experts are divided about how high we can build skyscrapers. Some say we could build a mile-high (5,280 ft, or 1,609 m) building with existing technology, while others say we would need to develop lighter, stronger materials; faster elevators; and advanced sway dampers first. Future technology advances could conceivably lead to sky-high cities, giant buildings that house a million people or more.

We might be compelled to build farther upward in the future simply to conserve land. When you build upward, you can concentrate much more development into one area, instead of spreading out into untapped natural areas. Skyscraper cities would also be very convenient: More businesses could be clustered together in a city, reducing commuting time.

The main force behind the skyscraper race might turn out to be vanity, rather than necessity. Where monumental height once honored gods and kings, it now glorifies corporations and cities. These structures come from a fundamental desire—everybody wants to have the biggest building on the block. This drive has been a major factor in skyscraper development over the past 120 years, and it's a good bet it will continue to push buildings up in centuries to come.

How **BUILDING IMPLOSIONS** Work

You can demolish a stone wall with a sledgehammer, and it's fairly easy to level a five-story building using excavators and wrecking balls. But when you need to bring down a massive structure, say a 20-story skyscraper, you have to haul out the big guns. Explosive demolition is the preferred method for safely and efficiently demolishing larger structures. When a building is surrounded by other buildings, it may be necessary to implode *the building—to make it collapse in on itself.*

The basic idea of explosive demolition is quite simple: If you remove the support structure of a building at a certain point, the section of the building above that point will fall on the structure below. If this upper section is heavy enough, it will crush the lower part. The explosives are just the trigger—gravity actually brings the building down.

Demolition blasters load explosives onto support columns at several different levels of the building so that the building structure falls down on itself at multiple points. The main challenge is controlling which way it falls.

Ideally, a blasting crew will be able to tumble the building over on one side, into a parking lot or other open area. Tipping a building is something like felling a tree. To topple the building to the north, the blasters detonate explosives on the north side of the building first, in the same way you would chop into a tree from the north side to tip it that way. Blasters may also secure steel cables to support columns in the building and pull the support columns in the direction the blasters want the building to fall.

When other buildings surround the building marked for demolition, the blasters proceed with a true implosion: demolishing the building so that it collapses straight down into its own *footprint* (the total area at the base of the building).

Implosion Planning

Blasters approach each project a little differently, but the basic idea of an implosion is to treat the building as a collection of separate towers. The blasters set the explosives so that each "tower" falls toward the center of the building, in roughly the same way that they would set the explosives to topple a single structure to the side. When the explosives are detonated in the right order, the toppling towers crash against each other, and all of the rubble collects at the center of the building. Another option is to detonate the columns at the center of the building before the other columns so that the building's sides fall inward, toward the unsupported center.

Generally speaking, blasters will explode the major support columns on the lower floors first, and then on a few upper stories. In a 20-story

HSW Web Links

www.howstuffworks.com

How Land Mines Work
How Skyscrapers Work
How Nuclear Bombs Work
How Earthquakes Work
How Smart Structures
  Work
How Force, Power, Torque
  and Energy Work

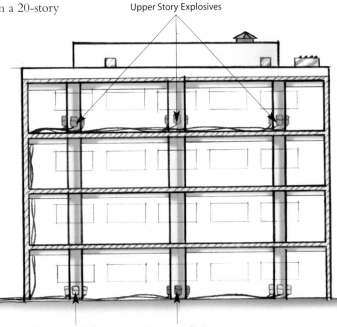

Upper Story Explosives

Secondary Explosive   Primary Explosive

building, for example, the blasters might blow the columns on the 1st and 2nd floor, as well as the 12th and 15th floors. In most cases, blowing the support structures on the lower floors is sufficient to collapse the building; exploding columns on a few upper floors helps break the building material into

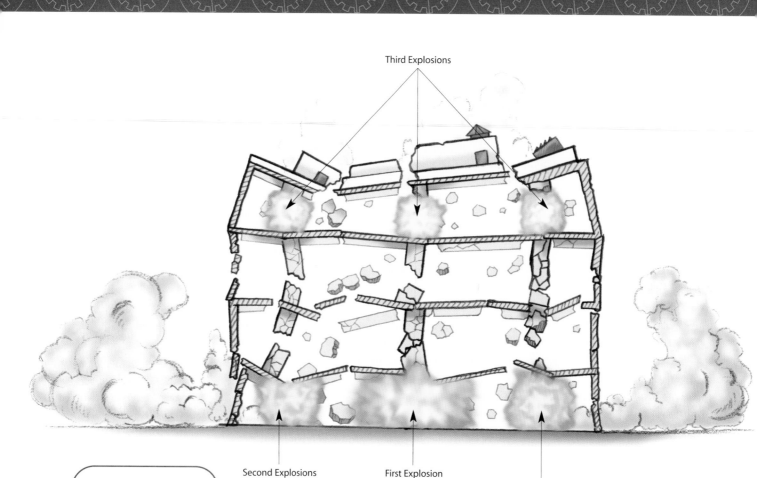

Third Explosions

Second Explosions

First Explosion

There is no "blaster school" anywhere in the world; the only way to become a demolition expert is to learn on the job. Prospective blasters work at an established blasting company until they know the field inside and out. When they have some experience, they can either stay on with their boss or venture out on their own and compete with the blasters who trained them.

Clients are understandably cautious about building implosion, and they tend to hire a demolition company based on the jobs it has done in the past. Consequently, it's very difficult for a young demolition firm to land major implosion jobs. About 20 well-established companies handle almost all major implosion jobs in the world. In many of these companies, blasting is passed on from generation to generation. Parents teach their children the skills, and the children then raise little blasters of their own.

smaller pieces as it falls. This means easier cleanup following the blast.

Detonators and Dynamite

Once the blasters have a clear idea of how the structure should fall, it's time to prepare the building. The first step is to clear out any furniture and debris. Next, destruction crews take out non-load-bearing walls within the building. This makes for a cleaner break at each floor: If these walls were left intact, they would stiffen the building, hindering its collapse. Destruction crews may also weaken the supporting columns with sledgehammers or steel cutters so that they give way more easily.

Then it's time to load the appropriate columns with explosives. Blasters determine the necessary type and quantity of explosives, based on the columns' composition and thickness. For concrete columns, blasters use traditional dynamite or a similar explosive material. Blasters drill narrow bore holes in the columns and insert the explosive material. The explosion generates

a shock wave, shattering the concrete into tiny chunks.

For buildings with a steel support structure, blasters typically use a specialized explosive material called cyclotrimethylen-etrinitramine—RDX for short. RDX-based explosive compounds expand up to 27,000 feet per second (8,230 meters per second). Instead of disintegrating the entire column, the concentrated, high-velocity pressure slices right through the steel, splitting it in half. Blasters may also ignite dynamite on one side of a steel column to push it over in a particular direction.

Blasters ignite dynamite or RDX with blasting caps. Many modern explosives need a sharp shock, rather than heat, to set them off. The blasting cap provides the shock by exploding.

These days, blasters typically use an electrical detonator to set off the blasting cap. An electrical detonator fuse, called a lead line, is just a long length of electrical wire. At the detonator end, the wire is surrounded by a layer of explosive material.

This detonator is attached directly to the primer charge, which is connected to the main explosives. When you send current through the wire (by hooking it up to a battery, for example), electrical resistance causes the wire to heat up. This heat ignites the flammable substance on the detonator end, which in turn sets off the primer charge, triggering the main explosives.

To control how the explosives go off, blasters configure the blasting caps with simple delays—sections of slow-burning material between the fuse and the primer charge. By using a longer or shorter length of delay material, the blasters can adjust how long it takes the explosives to go off.

To reduce flying debris, blasters may wrap chain-link fencing and geotextile fabric around each column. Blasters may also wrap fabric around the outside of each floor that's been rigged with explosives.

The Big Bang

After the structure has been pre-weakened and all the explosives have been loaded, it's time to make the final preparations. Blasters perform a last check of the explosives and make sure the building and the area surrounding it are completely clear. Surprisingly, implosion enthusiasts sometimes sneak past barriers for a closer view of the blast, despite the obvious risks. With the level of destruction involved, it is imperative that all spectators are a good distance away.

When the area is clear, the blasters retreat to the detonator controls and begin the countdown. Electrical detonators have a detonator controller with two buttons, one labeled "charge" and one labeled "fire." Toward the end of the countdown, a blaster presses and holds the charge button until an indicator light comes on. This builds up the intense electrical charge needed to activate the detonators. After the detonator-control machine is charged, and the countdown is completed, the blaster presses the fire button (while still holding down the charge button), releasing the charge into the wires so it can set off the blasting caps.

In most cases, the actual implosion only takes a few seconds. To many onlookers, the speed of destruction is the most incredible aspect of an implosion. How can a building that took months and months to build and stood up to the elements for a hundred years or more collapse into a pile of rubble as if it were a sand castle?

Following the blast, a cloud of dust billows out around the wreckage, enveloping nearby spectators. This cloud can be a nuisance to anyone living near the blast site, but blasters point out that it is actually less intrusive than the dust kicked up by non-explosive demolition. When workers take down buildings using sledgehammers and wrecking balls, the demolition process may take weeks or months. When the building is leveled in one moment, all the dust is concentrated in one cloud, which lingers for a relatively short period of time.

After the cloud has cleared, the blasters survey the scene and review videotapes to see if everything went according to plan. At this stage, it is crucial to confirm that all of the explosives were detonated and to remove any explosives that did not go off.

Most of the time, experienced blasters bring buildings down exactly as planned. Damage to nearby structures, even ones immediately adjacent to the blast site, is usually limited to a few broken windows. And if something doesn't work out quite right, the blasters log it in their mental catalog and make sure it doesn't happen on the next job. In this way, job by job, the science and art of implosion continues to evolve.

The Right Balance

Blasters determine how much explosive material to use based largely on their own experience and the information provided by the architects and engineers who originally built the building. But most of the time, they won't rely on this data alone. To make sure they don't overload or underload the support structure, the blasters perform a test blast on a few of the columns, which are wrapped in a shield for safety. The blasters try out varying degrees of explosive material, and based on the effectiveness of each explosion, determine the minimum explosive charge needed to demolish the columns. By using only the necessary amount of explosive material, the blasters minimize flying debris, reducing the likelihood of damaging nearby structures.

How **AUTOMATED TELLER MACHINES** Work

You're short on cash, so you walk over to the automated teller machine (ATM), insert your card into the card reader, respond to the prompts on the screen, and within a minute you walk away with your money and a receipt. These machines can now be found at most supermarkets, convenience stores, and travel centers, all over the country, from coast to coast. But have you ever wondered about the process that makes your bank funds available to you at any of the thousands of ATMs?

HSW Web Links
www.howstuffworks.com

How Stocks and the Stock
  Market Work
How MusicTellers Will
  Work
How Credit Cards Work
How the Euro Works

An ATM is simply a data terminal with two input and three output devices. A normal data terminal has some sort of keyboard for input, some sort of screen for output, and a network connection that lets it talk to a server somewhere on the network. An ATM adds a card reader as an input device, along with a printer and an amazing money dispenser as output devices, to create a complete package.

Settlement Funds

Let's say you want to get some money from an ATM at a convenience store. Chances are that the merchant who owns the store either owns or rents the ATM. So the merchant fills the ATM with cash each day, and it is the merchant's cash that you receive when you get money from the ATM.

You walk up to the ATM, insert your card, and type your password. The card tells the machine your bank and account information. The ATM forwards this information to the host processor, which routes the transaction request to your bank.

If you're requesting cash, the host processor causes an electronic funds transfer to take place from your checking account to the host processor's account. Once the funds are transferred to the host processor's bank account, the processor sends an approval code to the ATM authorizing the machine to dispense the cash. The host processor then sends your funds into the merchant's bank account by automated clearing house, usually the next bank business day. So when you request cash, the money moves electronically from your account to the host's account to the merchant's account, and you get the merchant's cash.

Now you know what the virtual process is, but what's actually going on inside the machine?

Parts of the Machine

An ATM has two input devices, the card reader and the keypad. The card reader captures the account information stored on the magnetic stripe on the back of an ATM card. The keypad lets the cardholder tell the bank what kind of transaction is required (cash withdrawal, balance inquiry, or whatever) and for what amount. Also, the bank requires the cardholder's personal identification number (PIN) for verification.

The most important output device is the heart of an ATM—the safe and cash-dispensing mechanism. The entire bottom portion of most small ATMs is a safe that contains the cash. The cash is stored in a series of cassettes—a big ATM in a high-traffic area can hold up to $100,000. The bill count and all of the information pertaining to a particular transaction is recorded in a journal. The journal information is printed out periodically and the machine owner maintains a hard copy for two years.

Besides the electric eye that counts each bill, the cash-dispensing mechanism also has a sensor that evaluates the thickness of each bill. If two bills are stuck together, then instead of being dispensed to the cardholder they are diverted to a reject bin. The same thing happens with a bill that is excessively worn or torn, or is folded. So, while it's not likely you're going to get an extra 20-dollar bill with your next withdrawal, you'll be happy to know you won't get half of a bill either.

How **PAWNSHOPS** Work

Have you ever been to a pawnshop? For most people, there seems to be something, well, shady about these places. But if you haven't been to a pawnshop, you may be missing out on some great bargains. A pawnshop is a lot like a dozen garage sales and a flea market all rolled into one. Pawnshops also play an important role in many communities by providing people with an easy, fast way to borrow small amounts of money.

Pawnshops and pawnbroking have been around for thousands of years. The basic idea behind any pawnshop is to loan people money. You bring in something you own and give it to the pawnbroker as collateral for a loan. This act is called *pawning*. The pawnbroker loans you money against that collateral. When you repay the loan plus the interest, you get your collateral back. If you don't repay the loan, the pawnbroker keeps the collateral.

Here's an example. Say that you go to a pawnshop and present your wedding band as collateral for a loan. It's a nice wedding ring—bought new, the ring cost about $140.

When you ask the pawnbroker how much you can borrow for the ring, he replies, "$10." You decide to do it and the pawnbroker gives you $10 and a pawn ticket.

I've Got a Golden Ticket!

A pawn ticket includes some important information. It lists the item pawned, the amount of money loaned for that item, and the amount of money due in 30 days to get the item back. So, your ticket for your wedding ring says that you received $10 and that you need to pay $12.20 (that's 2% interest plus 20% in fees on the $10) in 30 days to get the ring back. Within 30 days, you have three options:

- You can return to the pawnshop, pay the full amount ($12.20 in this case) and retrieve your wedding ring.
- You can return to the shop and pay the monthly fee ($2.20 in this case) to extend the loan for another 30 days. At this point, you'd have to enter into a new contract for the next 30 days. Here's where some pawnshops differ on the second-month contract: Some would make you

pay 22% on the new principal, $12.20, for a total of $14.88 due after another 30 days, while other pawnshops will allow you to continue paying 22% on the original principal of $10.

- Your last option is to do nothing, in which case the pawnshop keeps your ring and sells it.

The process seems pretty straightforward. But you're probably wondering how aboveboard it all is. It seems like a system that could be taken advantage of by thieves. Just how legitimate is this business?

Licensed to Sell

Pawnshops are regulated at the state level in the United States, and every state has different rules. Some of the details are absolutely fascinating! As an example, let's take a look at how a pawnshop works in the state of North Carolina.

Pawnshops are a business just like any other. But unlike many other businesses, pawnshops have a special set of laws that keep them on the straight and narrow. Pawnshops are specially licensed, and it turns out that they have to cooperate with police on a daily basis to prevent the movement of stolen merchandise.

When you pawn an item, the pawnbroker takes your name and address, verifies it with your valid driver's license, and then inspects the item carefully. Most pawnshops have the ability to test diamonds and gold for authenticity. If you're bringing in something like a TV or VCR, the pawnbroker tests it to make sure it works properly. If the item has a serial number, it's also recorded on the pawn ticket.

Every day, the pawnbroker must submit a list of all merchandise received, including

- First month—$100.00
- Second month—$75.00
- Third month—$75.00
- Fourth to sixth months—$50.00

In practical terms, this means that the maximum loan that someone can get at a North Carolina pawnshop is about $500. For a bigger loan, the fees would exceed the state-mandated maximum. Even with this regulated fee structure, the interest rate still works out to well over 100% annually. In North Carolina, after six months you either have to pick up the collateral and make a new loan on it or the item becomes the property of the pawnshop and the shop can sell it.

serial numbers, to the police. The police compare the serial numbers against records of stolen merchandise. Anything stolen is recovered this way and returned to the owner. Why do pawnshops bother giving this information to the police? If a stolen item is found in a pawnshop and the item was not reported to the police by the pawnshop when it came in, the pawnbroker can be charged with receiving stolen merchandise.

In Their Best Interest

What about the interest rate? In North Carolina, the maximum interest rate that a pawnshop can charge is 2% per month, or 24% per year. That's about the same as some credit cards. However, a pawnshop can also tack on other charges, such as fees for handling, appraisal, storage, and insurance. The maximum allowed charge for these additional fees is 20% per month. In addition to the 20% cap (which does not include the 2% interest cap), there are the following total limits for interest + fees:

Bargain of the Century

Because many of the people never pick up the items they pawn, a pawnshop is a consumer store as much as it is a lending institution. People also come in and sell used items outright to the pawnshop. This makes any pawnshop a huge, daily garage sale! Some of the things you typically find in pawnshops include: electronics, tools, lawn equipment, sporting goods, musical instruments, and jewelry. In fact, pawnshops are a great place to purchase jewelry, because you generally pay about half the retail value. If you're worried about authenticity, you can even bring an appraiser with you. Since jewelry is often the most valuable thing a person owns, it's one of the most commonly pawned items.

You will also find random things you might not expect to see at a pawnshop—such as saddles, wheel chairs, motorcycle leathers, and bicycles. Anything of value that you can find at a garage sale is probably for sale at a pawnshop near you. Be sure to haggle!

How **UPC BAR CODES** Work

If you look in your refrigerator or pantry right now, you will find that just about every package you see has a UPC bar code printed on it. In fact, nearly every item that you purchase from a grocery store, department store, or mass merchandiser has a UPC bar code on it somewhere. Have you ever wondered where these codes come from and what they mean?

UPC stands for *Universal Product Code*. UPC bar codes were originally created to help grocery stores speed up the checkout process and keep better track of inventory, but the system quickly spread to other retail products because it was so successful.

What's in a Number . . . ?

UPCs originate with an organization called the Uniform Code Council (UCC). A manufacturer applies to the UCC for permission to enter the UPC system. The manufacturer pays an annual fee for the privilege. In return, the UCC issues the manufacturer a six-digit manufacturer identification number and provides guidelines on how to use it. The UPC symbol printed on a package has two parts:

- The machine-readable bar code
- The human-readable 12-digit UPC number

The first six digits of the UPC number represent the manufacturer's identification number. The next five digits are the item number. A person employed by the manufacturer, called the UPC coordinator, assigns item numbers to products, making sure the same code is not used on more than one product, retiring codes as products are removed from the product line, and so on. In general, every item the manufacturer sells, as well as every size package and every repackaging of the item, needs a different item code.

The last digit of the UPC code is called a check digit. This digit lets the scanner determine if it scanned the number correctly or not. Here is how the check digit is calculated for the other 11 digits, using the code number 639382000393 as an example:

1) Add together the value of all of the digits in odd positions (digits 1, 3, 5, 7, 9, and 11).
   6 + 9 + 8 + 0 + 0 + 9 = 32

2) Multiply that number by 3.
   32 × 3 = 96

3) Add together the value of all of the digits in even positions (digits 2, 4, 6, 8, and 10).
   3 + 3 + 2 + 0 + 3 = 11

4) Add this sum to the value in Step 2.
   96 + 11 = 107

5) To create the check digit, determine the number that, when added to the number in Step 4, creates a multiple of 10.
   107 + 3 = 110

6) The check digit is therefore 3.

Each time the scanner scans an item it performs this calculation. If the check digit it calculates is different from the check digit it reads, the scanner knows that something went wrong and the item needs to be rescanned.

The Price is Right

There's no price information encoded in a bar code. When the scanner at the checkout line scans a product, the cash register sends the UPC number to the store's central POS (point of sale) computer to look up the UPC number. The central computer sends back the actual price of the item at that moment.

This approach allows the store to change the price whenever it wants—to reflect sale prices, for example. If the price were encoded in the bar code, prices could never change.

On the other hand, not encoding a fixed price gives the store an easy way to rip off customers. When you hear about scanner fraud in the news that's what the newsperson is talking about. It's incredibly easy for a store to mistakenly or purposefully overprice an item.

HSW Web Links

www.howstuffworks.com

How Scanners Work
How Anti-shoplifting Devices Work
How Smart Labels Will Work

How **OIL REFINING** Works

In movies and television shows—Giant, Oklahoma Crude, Armageddon, and The Beverly Hillbillies—we have seen images of thick, black crude oil gushing out of the ground or a drilling platform. But when you pump the gasoline for your car, you've probably noticed that it is clear. And there are many other products besides gasoline that come from oil, including crayons, plastics, heating oil, jet fuel, kerosene, synthetic fibers, and tires. How is it possible to start with crude oil and end up with gasoline and all of these other products?

HSW Web Links

www.howstuffworks.com

How Grills Work
How Car Engines Work
How Diesel Engines Work
How Gas Turbine Engines Work
How Gas Prices Work

Crude oil is unprocessed oil—the stuff that comes out of the ground. It's also known as *petroleum*. Crude oil is a fossil fuel, meaning that it was made naturally from decaying plants and animals living in ancient seas millions of years ago—anywhere you find crude oil was once a sea bed. Crude oils vary in color, from clear to tar-black, and in viscosity, from water-like to almost solid.

Crude oils are a useful starting point for many different substances because they contain hydrocarbons. Hydrocarbons are molecules that contain hydrogen and carbon and come in various lengths and structures, from straight chains to branching chains to rings.

There are two things that make hydrocarbons exciting to chemists:

- Hydrocarbons contain a lot of energy.
- Hydrocarbons can take on many different forms.

From Crude Oil

The problem with crude oil is that it contains hundreds of different types of hydrocarbons all mixed together. You have to separate the different types of hydrocarbons to have anything useful. Fortunately, there's an easy way to separate things, and this is what oil refining is all about. Different hydrocarbon chain lengths all have progressively higher boiling points, so they can all be separated by distillation. This is what happens in an oil refinery—in one part of the process, crude oil is heated and the different chains are pulled out by their vaporization temperatures. Each different chain length has a different property that makes it useful in a different way.

The Refining Process

Oil refining separates everything into useful substances. Chemists use the following steps to do this:

1) The oldest and most common way to separate things into various components (called *fractions*) is to use the differences in boiling temperature. This process is called *fractional distillation*. You basically heat crude oil up, let it vaporize, and then condense the vapor.

2) To produce marketable products, many of the components separated by fractional distillation require chemical processing on some of the fractions to make other fractions, in a process called *conversion*. Chemical processing, for example, can break longer chains into shorter ones. This allows a refinery to turn diesel fuel into gasoline depending on the demand for gasoline.

3) Treat the fractions to remove impurities.

4) Combine the various fractions (processed, unprocessed) into mixtures to make desired products. For example, different mixtures of chains can create gasoline with different octane ratings.

Where Does It All Go?

To understand the diversity contained in crude oil and to understand why refining crude oil is so important in our society, look through the following list of products that come from crude oil:

- **Petroleum gas**—Used for heating, cooking, making plastics

- **Naphtha or ligroin**—Intermediate product that will be further processed to make gasoline

- **Gasoline**—Motor fuel

- **Kerosene**—Fuel for jet engines and tractors; starting material for making other products

- **Gas oil or diesel distillate**—Used for diesel fuel and heating oil; starting material for making other products

- **Lubricating oil**—Used for motor oil, grease, other lubricants

- **Heavy gas or fuel oil**—Used for industrial fuel; starting material for making other products

The products are stored on-site until they can be delivered to various markets such as gas stations, airports, and chemical plants. In addition to making the oil-based products, refineries must also treat the wastes involved in the processes to minimize air and water pollution.

Fractional Distillation and Chemical Processing

Here are the steps in fractional distillation:

1) You heat the mixture of two or more substances (liquids) with different boiling points to a high temperature. Heating is usually done with high pressure steam to temperatures of about 1112°F (600°C).

2) The mixture boils, forming vapor (gases); most substances go into the vapor phase.

3) The vapor enters the bottom of a long column (*fractional distillation column*) that is filled with trays or plates. The trays have many holes or bubble caps (like a loosened cap on a soda bottle) in them to allow the vapor to pass through. The trays increase the contact time between the vapor and the liquids in the column. The trays also help to collect liquids that form at various heights in the column. Note: There is a temperature difference across the column (hot at the bottom, cool at the top).

4) The vapor rises in the column.

5) As the vapor rises through the trays in the column, it cools.

6) When a substance in the vapor reaches a height where the temperature of the column is equal to that substance's boiling point, it will condense to form a liquid. (The substance with the lowest boiling point will condense at the highest point in the column; substances with higher boiling points will condense lower in the column.)

7) The trays collect the various liquid fractions.

The collected liquid fractions may:

• Pass to condensers, which cool them further, and then go to storage tanks.

• Go to other areas for further chemical processing.

Fractional distillation is useful for separating a mixture of substances with narrow differences in boiling points, and is the most important step in the refining process. But very few of the components come out of the fractional distillation column ready for market. Many of them must be chemically processed to make other fractions.

You can change one fraction into another by using one of three chemical processing methods:

• Breaking large hydrocarbons into smaller pieces (*cracking*)
• Combining smaller pieces to make larger ones (*unification*)
• Rearranging various pieces to make desired hydrocarbons (*alteration*)

Treating and Blending the Fractions

Distilled and chemically processed fractions are treated to remove impurities, such as organic compounds containing sulfur, nitrogen, oxygen, water, dissolved metals, and inorganic salts. Treating is usually done by passing the fractions through the following:

1) A column of sulfuric acid to remove unsaturated hydrocarbons (those with carbon-carbon double bonds), nitrogen compounds, oxygen compounds, and residual solids (tars, asphalt)

2) An absorption column filled with drying agents to remove water

3) Sulfur treatment and hydrogen-sulfide scrubbers to remove sulfur and sulfur compounds

After the fractions have been treated, they are cooled and then blended together to make various products. The next time you change the oil in your car, reach for the soap bottle by the kitchen sink, turn on the furnace, or take an airplane ride, remember all of these things were, in some way, made possible through oil refining.

Did You Know?

Only 40% of distilled crude oil is gasoline; however, gasoline is one of the major products made by oil companies. Rather than continually distilling large quantities of crude oil, oil companies chemically process some other fractions from the distillation column to make gasoline; this processing increases the yield of gasoline from each barrel of crude oil.

How **LANDFILLS** Work

You've just finished your meal at a fast food restaurant and you throw the uneaten food, food wrappers, drinking cup, utensils, and napkins into the trashcan. On trash pickup day in your neighborhood, you push your can out to the curb, and workers dump the contents into a big truck and haul it away. Americans generate trash at an astonishing rate of four pounds per day per person, which translates to 600,000 tons per day, or 210 million tons per year! This is almost twice as much trash per person as most other developed countries. What happens to this trash?

HSW Web Links

www.howstuffworks.com

How Sewer and Septic
  Systems Work
How Composting Works

Some of this trash gets recycled or recovered and some is burned, but you might be surprised to learn that the majority of this trash is actually buried—in landfills.

Waste Not, Want Not

Landfills are carefully designed structures built into or on top of the ground. The goal of a landfill is to isolate trash from the surrounding environment (groundwater, air, rain). This isolation is accomplished with a bottom liner and daily covering of soil. There are two types of landfills:

- **Sanitary landfill**—A landfill that uses a clay liner to isolate the trash from the environment
- **Municipal solid waste (MSW) landfill**—A landfill that uses a synthetic (plastic) liner to isolate the trash from the environment

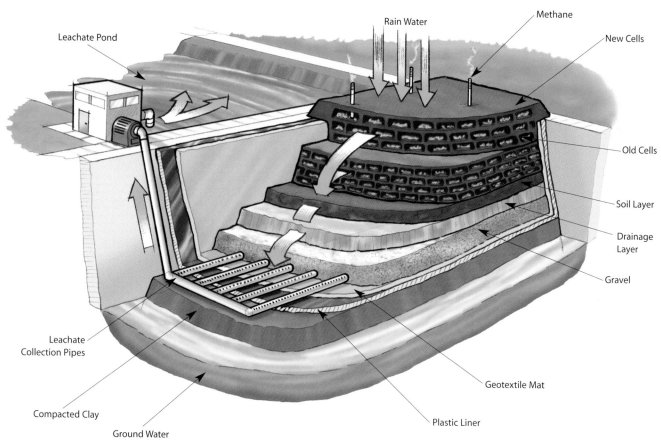

Leachate Pond

Rain Water

Methane

New Cells

Old Cells

Soil Layer

Drainage Layer

Gravel

Geotextile Mat

Plastic Liner

Leachate Collection Pipes

Compacted Clay

Ground Water

The purpose of a landfill is to bury the trash in such a way that it will be isolated from groundwater, will be kept dry, and will not be in contact with air. Under these conditions, trash will not decompose much.

Landfill Composition

The basic parts of a landfill are:

- Bottom liner system
- Cells (old and new)
- Storm-water drainage system
- Leachate collection system
- Methane collection system
- Covering or cap
- Groundwater monitoring stations

Each of these parts is designed to address specific problems that are encountered in a landfill.

Bottom Liner System

A landfill's major purpose and one of its biggest challenges is to contain the trash so that the trash doesn't cause problems in the environment. The bottom liner prevents the trash from coming in contact with the outside soil, particularly the groundwater. In MSW landfills, the liner is usually some type of durable, puncture-resistant synthetic plastic. It is usually about 1 to 4 inches (30 to 100 millimeters) thick. The plastic liner may be combined with compacted clay soils as an additional liner. The plastic liner may also be surrounded on either side by a fabric mat that helps to keep the plastic liner from being torn or punctured by the nearby rock and gravel layers.

Cells (Old and New)

Perhaps the most precious commodity and overriding problem in a landfill is space. The amount of space is directly related to the capacity and usable life of the landfill. If you can increase the space, then you can extend the usable life of the landfill. To preserve space, trash is compacted into areas called *cells* that contain only one day's trash. Once the cell is made, it is covered with six inches of soil and compacted further. Cells are arranged in rows and in layers of adjoining cells called *lifts*.

Storm Water Drainage

It's important to keep the landfill as dry as possible to reduce the amount of *leachate*—water leaching through the landfill from top to bottom. Leachate can be reduced in two major ways:

- **Excluding liquids from the solid waste.** Solid waste must be tested for liquids before entering the landfill. This is done by passing samples of the waste through standard paint filters. If no liquid comes through the sample after 10 minutes, then the trash is accepted into the landfill.
- **Keeping rainwater out of the landfill.** To exclude rainwater, the landfill has a storm drainage system. Plastic drainage pipes and storm liners collect water from areas of the landfill and channel it to drainage ditches surrounding the landfill's base.

The ditches are either concrete or gravel-lined and carry water to collection ponds at the side of the landfill. In the collection ponds, suspended soil particles are allowed to settle and the water is tested for leachate chemicals. Once settling has occurred and the water has passed tests, it is then pumped or allowed to flow off-site.

Leachate Collection System

No system meant to exclude water from the landfill is perfect—water does get into the landfill. The water percolates through the cells and soil in the landfill much like water percolates through ground coffee in a drip coffee maker. As the water percolates through the trash, it picks up contaminants, just like water picks up coffee in the coffee maker. This water with the dissolved contaminants is leachate and is typically acidic.

To collect leachate, perforated pipes run throughout the landfill. These pipes drain into a leachate pipe, which carries leachate to a leachate collection pond. Leachate can

In most parts of the world, there are regulations that govern where a landfill can be placed and how it can operate. The whole process begins with someone proposing the landfill. In the United States, taking care of trash and building landfills are local government responsibilities. Before a city or other authority can build a landfill, an environmental impact study must be done on the proposed site to determine:

- The area of land necessary for the landfill
- The composition of the underlying soil and bedrock
- The flow of surface water over the site
- The impact of the proposed landfill on the local environment and wildlife
- The historical or archaeological value of the proposed site

Once the environmental impact study has been completed, permits must be obtained from the local, state, and federal governments. In addition, money will have to be raised from taxes or municipal bonds to build and operate the landfill. Because funding usually comes from some public source, public approval must be obtained through local governments or a referendum.

be pumped to the collection pond or can flow to it by gravity.

The leachate in the pond is tested for its levels of various chemicals and allowed to settle. After testing, the leachate must be treated like any other sewage; the treatment may occur on-site or off-site. Some landfills recirculate the leachate and later treat it. This method reduces the volume of leachate from the landfill, but increases the concentrations of contaminants in the leachate.

Methane Collection System

Because the landfill is airtight, bacteria in the landfill break down the trash in the absence of oxygen. A byproduct of this breakdown is landfill gas, which contains approximately 50% methane and 50% carbon dioxide, with small amounts of nitrogen and oxygen. This gas presents a hazard because the methane can explode and/or burn. So, the landfill gas must be removed. To do this, a series of pipes are embedded within the landfill to collect the gas. In some landfills, this gas is vented or burned.

More recently, this landfill gas has been recognized as a usable energy source. The methane can be extracted from the gas and used as fuel.

Covering or Cap

Each cell is covered daily with six inches of compacted soil. This covering seals the compacted trash from the air and prevents pests such as rats and flying insects from getting into the trash. This soil takes up quite a bit of space. Because space is a precious commodity, many landfills are experimenting with tarps or spray coverings of paper or cement/paper emulsions. These emulsions can effectively cover the trash, but take up only a quarter of an inch instead of 6 inches.

When a section of the landfill is finished, it is covered permanently with a polyethylene cap that's about 1.5 inches (40 mm) deep. The cap is then covered with a 2-foot layer of compacted soil. The soil is then planted with vegetation to prevent erosion

of the soil by rainfall and wind. The vegetation usually is grass. No trees, shrubs, or plants with deep penetrating roots are used so that the plant roots do not contact the underlying trash and allow leachate out of the landfill.

Occasionally, leachate may seep through a weak point in the covering and come out on to the surface. It appears black and bubbly. Later, it will stain the ground red. Leachate seepages are promptly repaired by excavating the area around the seepage and filling it with well-compacted soil to divert the flow of leachate back into the landfill.

Groundwater Monitoring

Groundwater monitoring stations can be found at many points surrounding a landfill. Pipes are sunk into the groundwater so water can be sampled and tested for the presence of leachate chemicals. Because the temperature rises when solid waste decomposes, the temperature of the groundwater is measured. An increase in groundwater temperature could indicate that leachate is seeping into the groundwater. Also, if the pH of the groundwater becomes acidic, that could indicate seeping leachate.

The Long Haul

Trash put in a landfill will stay there for a very long time. Inside a landfill, there is little oxygen and little moisture. Under these conditions, trash does not break down very rapidly. In fact, when old landfills have been excavated or sampled, 40-year-old newspapers have been found with easily readable print. Landfills are not designed to break down trash, merely to bury it. When a landfill closes, the site, especially the groundwater, must be monitored and maintained for up to 30 years.

Landfills are complicated structures that, when properly designed and managed, serve an important purpose. In the future, new technologies called bioreactors *will be used to speed the breakdown of trash in landfills and produce more methane.*

How **HYDROPOWER PLANTS** Work

Worldwide, hydropower plants produce about 24% of the world's electricity and supply more than 1 billion people with power. According to the National Renewable Energy Laboratory, the world's hydropower plants output a combined total of 675,000 megawatts, the energy equivalent of 3.6 billion barrels of oil. There are more than 2,000 hydropower plants operating in the United States, making hydropower the country's largest renewable energy source. How is it that falling water can create this much energy?

When you're watching a river roll by, it's hard to imagine the force it's carrying. If you've ever been white-water rafting, then you've felt a small part of the river's power. White-water rapids are created as a river that's carrying a large amount of water downhill bottlenecks through a narrow passageway. As the river is forced through this opening, its flow quickens. Floods are another example of how much force a tremendous volume of water can have.

Hydropower plants harness water's energy and use simple mechanics to convert that energy into electricity.

Harnessing the Power of Water

Hydropower plants are actually based on a pretty simple concept—water flowing through a dam turns a turbine, which turns a generator.

HSW Web Links

www.howstuffworks.com

How Nuclear Power Works
How Power Distribution
 Grids Work
How Electric Motors Work
How Gas Turbine Engines
 Work
How Floods Work

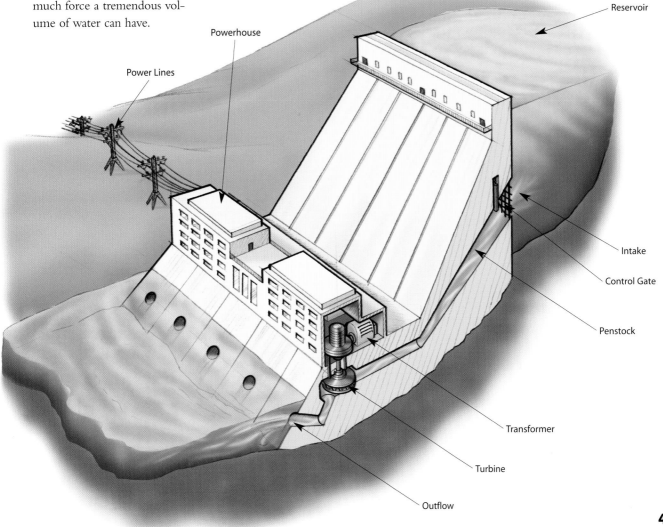

Reservoir

Powerhouse

Power Lines

Intake

Control Gate

Penstock

Transformer

Turbine

Outflow

The basic idea of hydropower is to use the power of a moving liquid to turn a turbine blade. Typically, a large dam has to be built in the middle of a river to capitalize on this power. A new invention is taking advantage of hydropower on a much smaller scale to provide electricity for portable electronic devices.

Inventor Robert Komarechka of Ontario, Canada, has come up with the idea of placing small hydropower generators into the soles of shoes. He believes these micro-turbines will generate enough electricity to power almost any portable gadget.

As a person walks, the compression of electrically conductive fluid in a sac located in the shoe's heel will force fluid through the conduit and into the hydroelectric generator module. As the user continues to walk, the heel will be lifted and downward pressure will be exerted on the sac under the ball of the person's foot. The movement of the fluid will rotate the rotor and shaft to produce electricity.

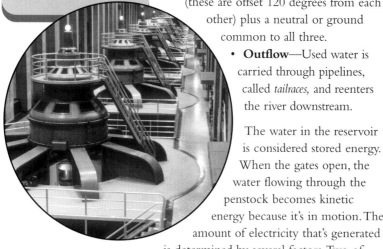

Here are the basic components of a conventional hydropower plant:

- **Dam**—Most hydropower plants rely on a dam that holds back water, creating a large reservoir.
- **Intake**—Gates on the dam open and gravity pulls the water through the *penstock,* a pipeline that leads to the turbine. Water builds up pressure as it flows through this pipe.
- **Turbine**—The water strikes and turns the large blades of a turbine, which is attached to a generator above it by way of a shaft. The most common type of turbine for hydropower plants is the Francis turbine, which looks like a big disc with curved blades.
- **Generators**—As the turbine blades turn, so do a series of magnets inside the generator. Giant magnets rotate past copper coils, producing alternating current (AC) by moving electrons.
- **Transformer**—The transformer inside the powerhouse takes the AC and converts it to higher-voltage current.
- **Power lines**—Out of every power plant come four wires: the three phases of power being produced simultaneously (these are offset 120 degrees from each other) plus a neutral or ground common to all three.
- **Outflow**—Used water is carried through pipelines, called *tailraces,* and reenters the river downstream.

The water in the reservoir is considered stored energy. When the gates open, the water flowing through the penstock becomes kinetic energy because it's in motion. The amount of electricity that's generated is determined by several factors. Two of those factors are the volume of water flow and the amount of hydraulic head. The head refers to the distance between the water surface and the turbines. As the head and flow increase, so does the amount of electricity that's generated. The head is usually dependent upon the amount of water in the reservoir.

Inside the Generator

The heart of the hydroelectric power plant is the generator. Most hydropower plants have several of these generators. The generator, as you might have guessed, generates the electricity. The basic process of generating electricity in this manner is to rotate a series of magnets inside coils of wire. This process moves electrons, which produces electrical current.

The Hoover Dam has a total of 17 generators, each of which can generate up to 133 megawatts. The total capacity of the Hoover Dam hydropower plant is 2,074 megawatts. Each generator is made of a shaft, an exciter, a rotor, and a stator.

As the turbine turns, the exciter sends an electrical current to the rotor. The rotor is a series of large electromagnets that spin inside a tightly wound coil of copper wire, which is the stator. The magnetic field between the coil and the magnets creates an electric current.

In the Hoover Dam, the generators produce 16,500 volts. The voltage moves from the generator to the transformer, where the current ramps up to 230,000 volts before being transmitted.

Throw It In Reverse

The majority of hydropower plants work in the manner described above. However, there's another type of hydropower plant—a pumped-storage plant. In a conventional hydropower plant, the water from the reservoir flows through the plant, exits, and is carried downstream. A pumped-storage plant has two reservoirs:

- **Upper reservoir**—Like a conventional hydropower plant, a dam creates a reservoir. The water in this reservoir flows through the hydropower plant to create electricity.
- **Lower reservoir**—Water exiting the hydropower plant flows into a lower reservoir rather than reentering the river and flowing downstream.

Using a reversible turbine, the plant can pump water back to the upper reservoir. This is done in off-peak hours. Essentially, the second reservoir refills the upper reservoir. By pumping water back to the upper reservoir, the plant has more water to generate electricity during periods of peak consumption.

How **RED LIGHT CAMERAS** Work

Red light cameras collect all of the evidence authorities need to prosecute people who run traffic lights. If a camera catches you speeding through the intersection, you can expect a ticket (along with a photograph of the violation) to arrive in your mailbox a month or two later. These systems rely on some sophisticated technology, but conceptually they are very simple.

In a typical set-up, digital cameras mounted on poles observe an intersection from multiple angles. One or more triggers detect when a car has moved past a particular point in the road. A computer monitors the triggers and the traffic light cycle.

Trigger Power

The main trigger technology used in red-light systems is the *induction loop,* a coiled length of electrical wire buried just under the asphalt. The loop is hooked up to an electrical power source and a meter. Sending electrical current through the wire generates a magnetic field and creates a giant inductor.

The induction's intensity depends on the structure and composition of the loop. When a car drives over the loop, the huge mass of metal changes the inductance of the loop.

The electrical meter constantly monitors the total inductance level of the circuit. When the inductance changes significantly, the computer knows that a car has passed over the loop.

Caught Red-Handed

When the light is green or yellow for incoming traffic, the computer ignores the triggers and does not activate the cameras. The system doesn't turn on until it receives a signal that the light is red.

In most systems, the computer will not activate the cameras unless a car moves over the loops at a particular speed.

These systems have two loop triggers for each lane. When both triggers are activated in quick succession, the computer knows a car has moved into the intersection at high speed. If there is more of a delay, the computer knows the car is moving more slowly.

When a car activates both triggers after the light is red, the computer automatically takes a picture. This first shot shows the car just as it's entering the intersection. The computer then hesitates briefly and takes another shot when the car is in the middle of the intersection. The computer calculates the length of the delay based on the measured speed of the car. It's important to get two pictures of the car to show that it entered the intersection when the light was red and then proceeded through the intersection.

The computer superimposes relevant information over the image, including the date, the time, the location, the car's speed, and the elapsed time between when the light turned red and the car entered the intersection.

HSW Web Links

www.howstuffworks.com

How Brakes Work
How Air Bags Work
How Crash Testing Works
How NASCAR Safety
  Works

Second Camera

First Camera

Second Induction Loop

First Induction Loop

With all this information, the police have everything they need to charge the driver. In most areas, the police, or a private firm hired to maintain the system, look up the license plate and send the ticket in the mail. The driver (or car owner) can pay the fine through the mail and be done with it, or he or she can try to contest the ticket in court. Of course, the police send the photos along with the ticket— so most drivers end up just paying the fine.

How **JUMBO TV SCREENS** Work

If you've ever been to a sporting event that has a large-screen TV in the stadium, then you've witnessed the gigantic and amazing displays that make games so much easier to follow. The jumbo TV can display instant replays, close-ups, and player profiles. You also see these large-screen TVs at racetracks, concerts, and in large public areas like Times Square in New York City. Have you ever wondered how they can create a television that is 30 to 60 feet (10 to 20 meters) high?

HSW Web Links

www.howstuffworks.com

How Projection Television
  Works
How Television Works
How HDTV Works
How Light Emitting
  Diodes Work
How Digital Television
  Works

Cool Facts

A typical 60-foot jumbo TV can consume up to 1.2 watts per pixel, or approximately 300,000 watts for the full display.

A jumbo TV that is 60 feet high has to do the same thing that a normal television set does—it has to take a video signal and convert it into points of light. A normal TV uses a CRT (cathode ray tube) to do that, but CRTs can't get much bigger than 3 feet high. A jumbo TV uses a totally different technology.

Decoding the Signal

Let's take a look at how a black-and-white TV works in order to understand how a normal television set converts a video signal to light:

- The electron beam in a CRT paints across the screen one line at a time. As it moves across the screen, the beam energizes small dots of phosphor, which then produce light that we can see.
- The video signal tells the CRT beam what its intensity should be as it moves across the screen.
- A five-microsecond pulse at zero volts (the horizontal retrace signal) tells the electron beam that it is time to start a new line. The beam starts painting on the left side of the screen, and zips across the screen in 42 microseconds. The varying voltage following the horizontal retrace signal adjusts the electron beam to be bright or dark as it shoots across.
- The electron beam paints lines down the face of the CRT and then receives a vertical retrace signal that tells it to start again at the upper left-hand corner.

A color screen works the same way, but it uses three separate electron beams and three dots of phosphor (red, green, and blue) for each pixel on the screen. A separate color signal indicates the color of each pixel as the electron beam moves across the display.

As the electron beam paints across the screen, it's hitting the phosphor on the screen with electrons. The electrons in the electron beam excite small dots of phosphor and the screen lights up. By rapidly painting 480 lines on the screen at a rate of 30 frames per second, the TV screen allows the eye to integrate everything into a smooth moving image.

CRT technology works great indoors, but as soon as you put a CRT-based TV set outside in bright sunlight, you cannot see the display anymore. The phosphor on the CRT simply is not bright enough to compete with sunlight. Also, CRT displays are limited to about a 36-inch screen. You need a different technology to create a large, outdoor screen that is bright enough to compete with sunlight.

Jumbo Screens

There are two big differences between a jumbo TV screen that you see at a stadium and the TV in your home:

- Obviously, it's gigantic compared to your TV. It might be 60 feet (20 meters) high instead of 18 inches (0.5 meters) high.
- It's incredibly bright, so that people can see it in sunlight.

To be so big and so bright, almost all large-screen outdoor displays use light emitting diodes (LEDs) to create the image. LEDs are, essentially, little colored light bulbs. Modern LEDs are small, extremely bright, and use relatively little power for the light that they produce.

On a color CRT television set, all of the colors are produced using red, green, and blue phosphor dots for each pixel on the screen. In a jumbo TV, red, green, and blue LEDs are used instead of phosphor. A "pixel"

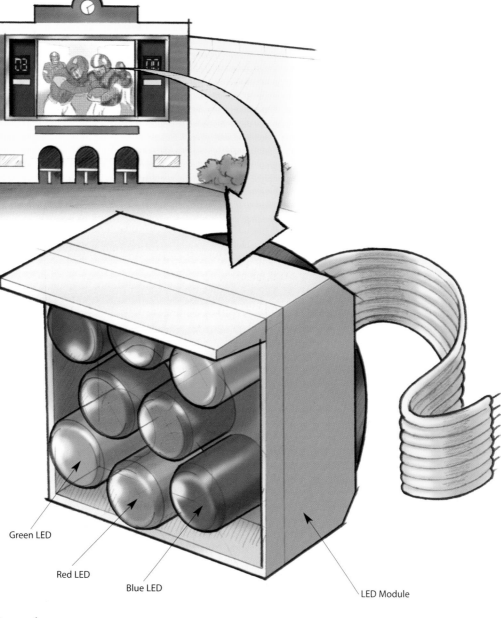

Green LED

Red LED

Blue LED

LED Module

on a jumbo TV is a small module that can have as few as three or four LEDs in it (one red, one green, and one blue). In the biggest jumbo TVs, each module could have dozens of LEDs. Modules typically range from about 0.2 to 1.5 inches (4 mm to 4 cm) in size.

To build a jumbo TV, you take thousands of these LED modules and arrange them in a rectangular grid. For example, the grid might contain 640 by 480 LED modules, or 307,200 modules. The size of the ultimate screen depends on the size of the LED modules.

To control a huge LED screen like this, you use a computer system, a power control system, and a lot of wiring. The computer system looks at the incoming TV signal and decides which LEDs it will turn on and how bright they should be. The computer samples the intensity and color signals and translates them into intensity information

for the three different LED colors at each pixel module. The power system provides power to all of the LED modules, and modulates the power so that every LED has the right brightness.

As LED prices have dropped, jumbo TV screens have started to pop up in all sorts of places and in all sorts of sizes. You now find jumbo TVs indoors (in places like shopping malls and office buildings) and in all sorts of outdoor environments—especially in areas that attract lots of tourists.

How **CAR WASHES** Work

Car washes have been popular ever since two Detroit men opened the first one, the Automated Laundry, in 1914. A lot of people wash their own cars at home, but the convenience of an automated car wash and its relatively low cost can be hard to beat.

⚙ **HSW Web Links**

www.howstuffworks.com

How Hydraulic Machines
 Work
How Electric Motors Work
How Gas Prices Work
How Oscillating Sprinklers
 Work
How Washing Machines
 Work

Car washes fall into five categories:

- **Self-service**—A coin or bill operated, open-bay car wash that has a pressure sprayer and, sometimes, a foaming brush.
- **Exterior rollover**—An automated system where you drive your car inside the bay and a signal informs you to stop. At that point, the car wash equipment moves back and forth over your car on a track—washing, rinsing, and waxing it.

- **Detail shop**—A detail shop may hand wash or use an automated system to wash the car. Then, attendants completely clean and polish the car—inside and out.

The conveyor-driven systems used in exterior-only and full-service systems use some amazing machinery to make a vehicle squeaky clean.

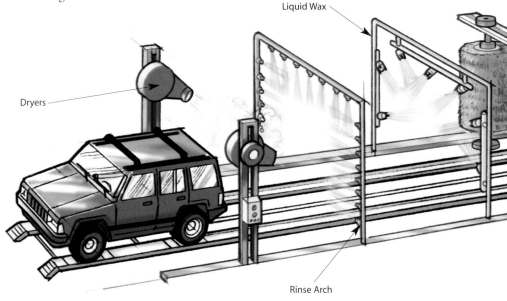

Liquid Wax

Dryers

Rinse Arch

- **Exterior only**—An automated system where you drive your car into the entrance of a long, tunnel-like bay and position your front tire (usually the driver's side tire) on a special conveyor belt and put the car in neutral. The conveyor belt guides the car through the bay.
- **Full service**—A modification of the exterior-only system, full service uses the same conveyor-belt-based automated system. The difference is that attendants manually clean the interior of your car.

Drive in Slowly

Car washes are normally either touchless or cloth friction wash. A touchless car wash relies on high-powered jets of water and strong detergents to clean the car. Only the water and cleaning solutions actually come in physical contact with the vehicle. Cloth friction wash systems use soft cloth that moves around against the surface of the car. The system that we write about uses cloth friction wash technology, but touchless car washes use quite a few of the same components.

The first part of the system is the conveyor track. At the beginning of the conveyor is a device called a *correlator*. This is simply a series of rollers that allow the wheel of the car to slide sideways until it is aligned with the conveyor. The driver turns the car off and places it in neutral. Most conveyor systems have small rollers that pop up behind the wheel once it is on the conveyor. The roller pushes the wheel forward, causing the car to roll along through the tunnel—the long bay used to house all the equipment.

the car down and begins loosening the dirt on the car.

A lot of car washes also have a set of nozzles arranged near the ground that are called *tire applicators*. These nozzles spray the tires with a solution designed specifically for removing brake dust and brightening the black rubber of the tire.

On many car washes, a *mitter curtain* is used to spread the presoak over the car. This is a series of long strips of cloth that hang from a frame near the top of the tunnel. The frame is connected to a motorized shaft that

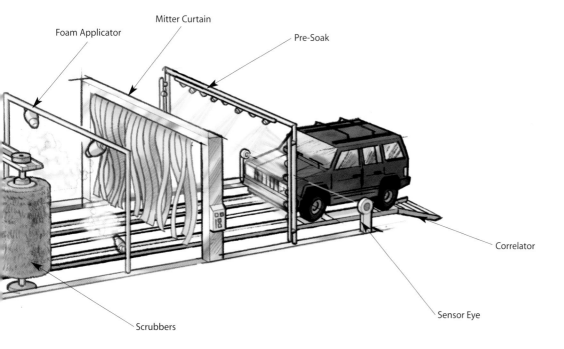

Foam Applicator
Mitter Curtain
Pre-Soak
Correlator
Sensor Eye
Scrubbers

Once the car enters the tunnel, it passes through an infrared beam and a sensor, called an *eye*. As soon as the beam is interrupted, the eye sends a signal to the digital control system (DCS), the computer that runs the automated portion of the car wash. By measuring the amount of time that the signal is interrupted, the DCS determines the length of the vehicle and adjusts the system accordingly.

It's Suds Time

Immediately after the eye, most car washes have a presoak. This is an arch that contains several small nozzles that spray a special solution all over the car. This solution wets

moves the frame up and down in a circular pattern. This motion makes the cloth strips rub back and forth across the horizontal surfaces of the car.

Next, a foam applicator applies a detergent to the car that becomes a deep-cleaning foam on contact. The nozzles on the foam applicator, as well as most other spray systems in a car wash, can be adjusted to change the angle of the spray and the size of the opening. The foam is created by mixing a chemical cleaner with water and air. There are usually separate adjustment controls for determining the exact mix of the three components. The chemical typically contains some coloring agent as well to make the foam more eye-pleasing and obvious.

Scrub-a-Dub

Scrubbers are large vertical cylinders with hundreds of small cloth strips attached to them. The scrubbers rotate rapidly, anywhere from 100 to 500 times a minute, spinning the cloth strips until they are perpendicular to the cylinder. Although the cloth strips are quite soft, they would feel like a whip if they hit you because they're moving so fast.

Scrubbers normally have hydraulic motors that spin them—a hydraulic motor avoids problems with electrical shock that electric motors would create. There is at least one scrubber on each side of the car wash, and there may be two or more. As the car moves past the scrubbers, the cloth strips brush along the vertical surfaces of the car.

Some car washes also have wrap-around washers. These are scrubbers on short booms that can move around to the front and rear of the vehicle, scrubbing those vertical surfaces. Like most of the mechanical equipment in the car wash, the washers are run by hydraulics. Normally, a single, large hydraulic power unit is connected to all of the various hydraulic pumps throughout the car wash.

Wash, Rinse, Repeat

Next, the car goes through a rinse arch. This is a series of nozzles arranged on an arch that use clean water to remove whatever residue is left after the high-pressure washer, scrubbers, and mitter curtain have done their respective jobs. In an average car wash, there are multiple rinse arches, usually placed after each major cleaning station.

The last rinse arch in the tunnel, aptly called the final rinse, should always use clean, non-recycled water to ensure that all residue is removed from the surface of the car.

Wax On, Wax Off

A standard feature of the car wash is the wax arch. The wax that is used in a car wash, which forms a water-resistant coating, is quite different from the wax you would apply by hand. One of the key differences is that car-wash wax is formulated to work on glass, chrome, and rubber, as well as the painted plastic and metal surfaces of the car. Also, it leaves a clear, thin film that does not have to be rubbed. However, car-wash wax does not provide the same level of protection or remove or cover up tiny scratches as standard wax does.

The wax arch uses one of two methods to apply wax. The first type of wax arch uses a system of foam applicators, the most common being a triple-foam applicator, to apply a foam wax. The second type uses nozzles, similar to those of the rinse arch, to apply a liquid wax.

In the case of a liquid-wax system, the next step is usually to go through a rinse arch. But when wax foam has been applied, the car usually goes through another set of scrubbers and another mitter curtain before going through a rinse arch.

Blow Me Over

After the car is completely washed, the final step in the automated process is the dryer. Much like a giant hair dryer, the dryer in a car wash moves large amounts of air and forces it out through a series of nozzles. The blast of air rapidly dries the surface of the car. The dryer has a large, flat, round section just before the nozzle opening. This section is the silencer. Like a muffler or the silencer on a gun, the dryer's silencer deadens the noise created by the air being forced through the system.

Some car washes apply a special chemical after the final rinse and before the dryer that speeds up the drying process. Most full-service car washes set the dryer lower than exterior-only car washes. This is because a full-service car wash usually has attendants who hand-dry the car with towels to remove all of the water.

HEAVY EQUIPMENT

How IRON AND STEEL Work

If you had to pick one technology that has had a tremendous effect on modern society, the refining of iron and steel would have to be somewhere near the top of the list. Iron and steel show up in a huge array of modern products. Cars, tractors, bridges, trains (and their rails), tools, skyscrapers, guns, and ships all depend on iron and steel to make them strong and inexpensive. Iron is so important that primitive societies are measured by the point at which they learn how to refine iron and enter the Iron Age!

⚙ **HSW Web Links**

www.howstuffworks.com

How Metal Detectors Work

How Sword Making Works

How Electromagnets Work

How Billiard Tables Work

How Skyscrapers Work

Have you ever wondered how people refine iron and steel? You probably have heard of iron ore, but how is it that you extract a metal from a rock?

The Advantages of Iron

Iron is an incredibly useful substance for several reasons:

- Relatively speaking, and especially when compared to wood or copper, iron is extremely strong.
- When heated, iron is relatively easy to bend and shape using simple tools.
- Unlike wood, iron can handle heat, so you can build things like engines from it.
- Unlike most substances, you can magnetize iron, making it useful in the creation of electric motors and generators.
- Iron is plentiful—5% of the earth's crust is iron, and in some areas it concentrates in ores that contain as much as 70% iron.
- Refining iron using simple tools is relatively easy.

When you compare iron and steel with something like aluminum, you can see why it was so important historically. To refine aluminum you must have access to huge amounts of electricity. To shape aluminum you must either cast it or extrude it. Iron is much easier to deal with by comparison. Iron has been useful to man for thousands of years, while aluminum really did not exist in any meaningful way until the twentieth century.

An object like the flintlock rifle would be impossible to create without iron. Fortunately, iron can be created relatively easily with tools available to primitive societies.

There will likely come a day when we become so technologically advanced that iron is completely replaced by aluminum, plastics, and things like carbon and glass fibers. But right now the economic equation gives inexpensive iron and steel a huge advantage over these much more expensive alternatives.

The only real problem with iron and steel is rust. Fortunately you can control rust with paint, galvanizing, chrome plating, or sacrificial anodes.

Iron Ore

To make iron you start with iron ore. Iron ore is nothing more than rock that happens to contain a high concentration of iron. One thing that gave certain countries an edge between the fifteenth and twentieth centuries was the availability of iron ore deposits. For example, England, the U.S., France, Germany, Spain, and Russia all have good iron ore deposits. When you think of the historical importance of all of these countries, you can see the correlation!

Common iron ores include:

Hematite	Fe_2O_3	70% iron
Magnetite	Fe_3O_4	72% iron
Limonite	$Fe_2O_3 + H_2O$	50% to 66% iron
Siderite	$FeCO_3$	48% iron

Usually you find these minerals mixed into rocks containing silica.

Creating Iron

All of the iron ores contain iron combined with oxygen. To make iron from iron ore, you need to eliminate the oxygen to create pure iron.

The most primitive facility used to refine iron from iron ore is a *bloomery.* In a bloomery, you burn charcoal with iron ore and a good supply of oxygen (provided by a bellows or blower). Charcoal is essentially pure carbon. The carbon combines with oxygen to create carbon dioxide and carbon monoxide (releasing lots of heat in the process). Carbon and carbon monoxide combine with the oxygen in the iron ore and carry it away, leaving iron metal.

In a bloomery, the fire does not get hot enough to melt the iron completely, so you are left with a spongy mass (the *bloom*) containing iron and silicates from the ore. By heating and hammering the bloom, the glassy silicates mix into the iron metal to create wrought iron. Wrought iron is tough and easy to work, making it perfect for creating tools in a blacksmith shop.

The more advanced way to smelt iron is in a blast furnace. A blast furnace is charged with iron ore, charcoal or coke (coke being charcoal made from coal), and limestone ($CaCO_4$). Huge quantities of air blast in at the bottom of the furnace. The calcium in the limestone combines with the silicates to form slag. At the bottom of the blast furnace, liquid iron collects along with a layer of slag on top. Periodically you let the liquid iron flow out and cool. Typically, the liquid iron flows into a channel and indentations in a bed of sand. Once it cools, this metal is known as *pig iron*.

To create a ton of pig iron you start with 2 tons of ore, 1 ton of coke, and half a ton of limestone, and the fire consumes 5 tons of air. The temperature reaches almost 3000°F (1600°C) at the core of the blast furnace!

Pig iron contains 4% to 5% carbon and is so hard and brittle it is almost useless. You do one of two things with pig iron:

- You melt it, mix it with slag, and hammer it to eliminate most of the carbon (down to 0.3%) and create wrought iron. Wrought iron is the stuff a blacksmith works with to create tools, horseshoes, and so on. When you heat wrought iron, it is malleable, bendable, weld-able, and very easy to work with.
- You create steel.

Creating Steel

Steel is iron that has most of the impurities removed. Steel also has a consistent concentration of carbon (0.5% to 1.5%) throughout. Impurities like silica, phosphorous, and sulfur weaken steel tremendously, so they must be eliminated. The advantage of steel over iron is greatly improved strength.

The open-hearth furnace is one way to create steel from pig iron. Into an open hearth furnace goes pig iron, limestone, and iron ore. It is heated to about 1600°F. The limestone and ore forms a slag that floats on the surface. Impurities, including carbon, are oxidized and float out of the iron into the slag. When the carbon content is right, you have carbon steel.

Most modern steel plants use what's called a basic oxygen furnace to create steel today. The advantage is that using a basic oxygen furnace speeds up the process—it's about 10 times faster than using the open-hearth furnace.

A variety of metals might be alloyed with the steel at this point to create different properties. For example, the addition of 10% to 30% chromium creates stainless steel, which is very resistant to rust. The addition of chromium and molybdenum creates chrome-moly steel, which is strong and light.

When you think about it, there are two accidents of nature that have made it much easier for humans to move forward rapidly. One is the huge availability of something as useful as iron ore. The second is the availability of vast quantities of oil and coal to power the production of iron. This is an amazing coincidence, because without iron and energy, we would not have gotten nearly as far as we have today!

How **HYDRAULIC MACHINES** Work

From backyard log splitters to the huge machines you see on construction sites, hydraulic equipment is amazing in its strength and agility! On any construction site, you are likely to see some form of hydraulically operated machinery, such as bulldozers, backhoes, shovels, loaders, forklifts, and cranes. At a garage, you see hydraulics lifting the cars so that mechanics can work underneath them, and many elevators are hydraulically operated. Even the brakes in your car use hydraulics!

HSW Web Links

www.howstuffworks.com

How Hydraulic Cranes
  Work
How the Jaws of Life Work
How Wave Pools Work
How Force, Power, Torque
  and Energy Work
How a Block and Tackle
  Works

The basic idea behind any hydraulic system is very simple: Force applied at one point is transmitted to another point using an incompressible fluid. The fluid is almost always an oil of some sort. The force is almost always multiplied in the process.

Physics in Action

Let's say you have two pistons that fit into two cylinders filled with oil that are connected to one another with an oil-filled pipe. If you apply a downward force to one piston, then the force is transmitted to the second piston through the oil in the pipe. Since oil is incompressible, the efficiency is very good—almost all of the applied force appears at the second piston. The pipe connecting the two cylinders can be any length and shape, allowing it to snake through all sorts of things separating the two pistons. The pipe can also fork, so that one master cylinder can drive more than one slave cylinder, if desired.

The neat thing about hydraulic systems is that adding force multiplication (or division) to the system is very easy. In a hydraulic system, all you do is change the size of one piston head and cylinder relative to the other. The fluid has the same pressure (pounds per square inch) at every point in the system. Since the pressure at the larger piston is working on a larger area, that piston pushes upward with a greater force.

Direction of Force

Piston

Oil

To determine the multiplication factor, start by looking at the size of the pistons. Assume that the first piston is 2 inches in diameter (meaning it has a 1-inch radius), while the second piston is 6 inches in diameter (a 3 inch radius). The area of the pistons is πr^2. The area of the first piston is therefore 3.14, while the area of the second piston is 28.26. The second piston is nine times larger than the first.

What that means is that any force applied to the first piston will appear nine times greater on the second piston. So if you apply a 100-pound downward force to the first piston, a 900-pound upward force will appear on the second. The only catch is that you will have to depress the first piston 9 inches to raise the second piston 1 inch.

A More Advanced System

Heavy hydraulic machines work on the same principle of force multiplication, but most use a hydraulic pump—rather than a simple master cylinder piston—to drive pistons. The hydraulic pump is just like a water pump. It collects fluid from a reservoir and forces it into the hydraulic system at high pressure.

The simple pistons we've looked at so far can only apply multiplied force in one direction. In order to pull the piston in as well as push it out, you must use a modified system that can drive fluid to either side of the piston head inside the cylinder.

There are only a handful of elements at work in this system. The hydraulic pump collects fluid from the tank and pressurizes it. The spool valve controls the fluid's path. When you shift the spool to the right, the pressurized fluid (shown in orange) goes to the right side of the cylinder. As the piston slides to the left, it pushes the unpressurized fluid (shown in yellow) back to the tank. The piston retracts. When you shift the spool to the left, the pressurized fluid goes to the left side of the cylinder, driving the fluid on the right side into the tank. The piston extends out of the cylinder.

A multifaceted hydraulic machine, such as a backhoe loader, uses a number of different spool valves to drive a collection of hydraulic pistons. The machine's controls simply shift the valves back and forth to move the pistons in and out.

The Log Splitter

Now that you understand the basics of a simple hydraulic system, let's look at a specific example—the log splitter. A typical log splitter includes:

- A 5-horsepower gasoline engine
- A two-stage hydraulic oil pump rated at a maximum of 11 gallons per minute (gpm) at low pressure or 3 gpm at 2,500 psi
- A 4-inch-diameter, 24-inch-long hydraulic cylinder
- A 3.5-gallon hydraulic oil tank that feeds the pump
- A filter to keep the oil clean

The two-stage pump is an ingenious time-saver. The pump actually contains two pumping sections and an internal pressure-sensing valve that cuts over between the two. One section of the pump generates the maximum gallons-per-minute flow rate at a lower pressure. It is used, for example, to draw the piston back out of a log after the log has been split. Drawing the piston back into the cylinder takes very little force and you want it to happen quickly, so you want the highest possible flow rate at low pressure. When pushing the piston into a log, however, you want the highest possible pressure in order to generate the maximum splitting force. The flow rate isn't a big concern, so the pump switches to its high-pressure, lower-volume stage to split the log.

A typical log splitter boasts a 20-ton splitting force. A 4-inch piston has an area of 12.56 square inches. If the pump generates a maximum pressure of 3,000 pounds per square inch (psi), the total pressure available is 37,680 pounds, or about 2,320 pounds shy of 20 tons.

With all this information, you can determine the cycle time of the piston. To move a 4-inch-diameter piston 24 inches, you need

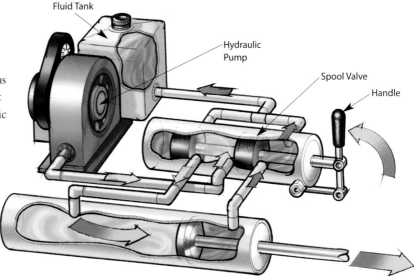

Fluid Tank

Hydraulic Pump

Spool Valve

Handle

Spool valve hydraulic machine.

$3.14 \times 22 \times 24 = 301$ cubic inches of oil. A gallon of oil is about 231 cubic inches, so you have to pump almost 1.5 gallons of oil to move the piston 24 inches in one direction. That's a fair amount of oil to think about that the next time you watch how quickly a hydraulic backhoe or skid/loader is able to move! In our log splitter, the maximum flow rate is 11 gallons per minute. That means that it will take 10 or so seconds to draw the piston back after the log is split, and it may take almost 30 seconds to push the piston through a tough log (because the flow rate is lower at high pressures).

You can see that just to fill the cylinder with oil, you need at least 1.5 gallons of hydraulic oil in the system. You can also see that one side of the cylinder has a larger capacity than the other side, because one side has the piston shaft taking up space and the other doesn't. Therefore, big hydraulic machines usually have:

- Large appetites for hydraulic oil (100 gallons is not uncommon if there are six or eight large hydraulic cylinders used to operate the machine)
- Large external reservoirs to hold the difference in the volume of oil displaced by the two sides of any cylinder

The next time you go to get your tires rotated or pass a construction site, you'll have a new appreciation for the hydraulic technology of the machines that you see all around you!

And Another Thing...

The brakes in your car are a good example of a basic piston-driven hydraulic system. When you depress the brake pedal in your car, it pushes on the piston in the brake's master cylinder. Four slave pistons, one at each wheel, actuate to press the brake pads against the brake rotor to stop the car. (Actually, in almost all cars on the road today, two master cylinders are driving two slave cylinders each. That way, if one of the master cylinders has a problem or springs a leak, you can still stop the car.)

In most other hydraulic systems, hydraulic cylinders and pistons are connected through valves to a pump supplying high-pressure oil.

How the **JAWS OF LIFE** Work

Driving down the interstate, you reach down to grab the road map from the passenger-side floorboard. In an instant, you inadvertently swerve onto the shoulder of the road, and your car flips as you attempt to regain control. When your car comes to rest, you've got a broken leg, your car is upside down, and you're pinned underneath the dashboard. In this type of situation, rescue workers will use a set of tools commonly called the Jaws of Life to cut away the car and get you out.

HSW Web Links

www.howstuffworks.com

How Hydraulic Machines Work

How Backhoe Loaders Work

How Hydraulic Cranes Work

How Brakes Work

How Crash Testing Works

How Air Bags Work

Jaws of Life is actually a brand of tools that is trademarked by the Hurst Jaws of Life company, but the name is often used when talking about other brands of rescue systems. The term actually refers to three different types of tools—cutters, spreaders, and rams.

Simple Hydraulics

If you've read "How Hydraulic Machines Work" in this chapter, you know that hydraulic equipment is based on a simple concept—the transmission of force from point to point using a fluid. Jaws of Life equipment is simple and lightweight, and there are very few moving parts involved. In the cutter and spreader, a portable engine pumps pressurized hydraulic fluid into the hydraulic cylinder through one of two hose ports. A typical Jaws of Life machine uses about 1 quart of hydraulic fluid. An operator-controlled valve switch controls which port the fluid enters through. If it enters one port, the fluid forces the rod up and opens the arms of the spreader or the blades of the cutter. The operator can then toggle the switch and cause the rod to retract, closing the arms or blades.

Spreaders

A spreader consists of pincer-like aluminum alloy arms with tips made of heat-treated steel. The hydraulics open and close the pincers. Rescue workers use the spreader to pry a car or building apart.

Spreaders come in different models, with varying capabilities. For example, the ML-32 Hurst Jaws of Life spreader provides:

- 16,000 pounds (71 kiloNewtons) of spreading force
- 14,400 pounds (64 kN) of pulling force
- 32 inches (81.9 cm) of opening distance

Other spreaders can provide more or less spreading and pulling force. The body of the ML-32 spreader is made out of aluminum alloy and the piston and piston rod are made from forged alloy steel. When the portable engine is started, oil flows through a set of hydraulic hoses into the hydraulic pump inside the machine's housing. A typical power unit might be a 5-horsepower gasoline engine that turns a pump producing 5,000 pounds per square inch (psi), although the pressure differs in different power units. This type of engine can run on 0.5 gallons (2 liters) of gas for about 45 minutes to an hour.

To open the arms of the spreader, the operator slides a valve switch that causes the hydraulic fluid to flow from one hose into the cylinder, pushing the piston and rod up. This rod is attached to linkages attached to the spreader's arms. When the rod pushes up, it causes the linkages to rotate, which opens the arms. To close the arms, the operator moves the valve in the opposite direction, which causes the hydraulic fluid to flow through a second hose.

Cool Facts

Oil is the most commonly used incompressible fluid for hydraulic machines. However, Jaws of Life equipment uses a phosphate-ester fluid, which is fire resistant and electrically noncon-ductive. At a crash scene, this type of synthetic fluid is favored over conventional oil.

To use the spreader, a rescue worker inserts the closed spreader arms into an opening in the vehicle or structure, such as a doorjamb. The spreader can also clamp down on a structure to crush any material between its arms.

Cutters

A cutter is basically a big pair of hydraulically driven scissors. Instead of arms, the cutter has curved, claw-like extensions that come to a point. Just like in the spreader, hydraulic fluid flows into a cylinder, placing pressure on a piston. When the piston rod is raised, the claws open. As the piston rod lowers, the claws of the cutter come together around a structure, such as a car roof, and pinch through it. If you've ever seen this device in action, you know that it can snap a car-door post like a twig in a few seconds.

Cutters, like spreaders, also come in different models, with varying capabilities—the Hurst Jaws of Life ML-40 model gives the operator:

- 12,358 pounds (60 kN) cutting force at the blade center
- 22,455 pounds (99.9 kN) cutting force at the notch
- 4.25-inch (10.8-cm) cuts

Rams

The ram is the most basic type of hydraulic system: It's just a matter of using hydraulic fluid to move a piston head inside a cylinder to extend and retract a piston rod. If you look at some heavy construction equipment, like a backhoe loader (see "How Backhoe Loaders Work" in this chapter), you'll notice that rams are used to control the boom arm.

The ram's function is to push apart sections of a car or building. For instance, a rescue worker can place a ram on the doorframe and extend the piston to push the dashboard up, creating enough space to free a crash victim.

Hydraulics play an important part in many of the machines around us, but none of the machines that use hydraulics may be as vital as the equipment known as the Jaws of Life. These devices have been called upon to save thousands of lives in situations where a few seconds could mean the difference between life and death.

Spreader and cutter.

How **BACKHOE LOADERS** Work

Backhoe loaders are all over the place. You'll find them on most major construction sites, you see them working on the side of the road—you might even see them shoveling snow or lifting heavy pipes. Backhoe loaders are popular because of their power, versatility, and relatively low price. They are the machine for a construction firm just starting out.

HSW Web Links

www.howstuffworks.com

How Skid Steer Loaders Work
How Hydraulic Machines Work
How Car Engines Work
How Diesel Engines Work
How Horsepower Works
How Force, Power, Torque and Energy Work

Backhoe loaders are the Swiss Army knives of construction equipment. Contractors love them because they combine several handy tools in one relatively compact unit. The most basic backhoe loader includes a tractor, a loader, and a backhoe. Some models have attachments for additional tools as well.

Just like the tractors you find on a farm, the backhoe tractor is designed to move easily over all kinds of rough terrain. A typical backhoe loader has a powerful, turbocharged diesel engine, large, rugged tires, and a cab with

These two components serve very different functions, but they work on the same principles. Both are powerful tools driven by hydraulics.

Cab

Hydraulic Piston

Loader

basic steering controls. The tractor's job is to move the backhoe and loader from place to place and to push the loader across the ground.

The loader sits on the front of the tractor, and the backhoe is attached in the back.

Hydraulic systems use pressurized fluid to transmit and multiply force. In a backhoe loader, the engine drives an oil pump that provides a steady supply of pressurized oil to the hydraulic system. A series of spool valves direct the oil to various

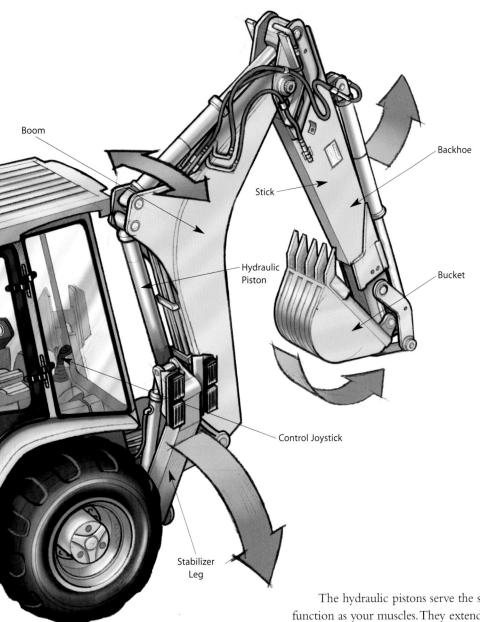

Boom

Stick

Backhoe

Hydraulic
Piston

Bucket

Control Joystick

Stabilizer
Leg

hydraulic pistons to move the backhoe and loader. Inside each cylinder, the pressurized oil can push on either side of the piston head to pull the piston in or push it out. (See "How Hydraulic Machines Work" for more information.)

The Backhoe

Think of the backhoe as an extremely powerful version of your arm. It has three segments—the boom, the stick, and the bucket—that are like your upper arm, forearm, and hand. The segments are joined by hinges.

The hydraulic pistons serve the same function as your muscles. They extend and retract to pivot the various segments back and forth. Two pistons at the base of the boom act like your shoulder muscle—they swing the entire arm left and right. They are synchronized so that when you push with one, the other pulls.

The operator controls all these pistons with two joysticks. The joysticks drive the spool valves in the hydraulic control system, which direct pressurized fluid to the different cylinders. Swinging a joystick from side to side extends and retracts one piston, and tilting it forward and backward moves another piston. An experienced operator can control all the pistons at once to swing the arm with a fluid motion.

63

The backhoe's job is to dig trenches, shovel up hard, compact material, or lift heavy loads. To dig a hole, the operator drives the tractor to the edge of the hole site, swings around in his or her seat and extends the stabilizer legs. The stabilizer legs support the backhoe loader while it's digging. Without the legs, the force of the operation would strain the backhoe loader's wheels and tires, and it would jostle the entire tractor. The backhoe loader might also fall into the hole!

To dig into the earth, the operator angles the bucket so it cuts into the ground, then gradually tilts it so it scoops up the dirt. It's very difficult to master a proper digging technique. The operator has to know how to position the arm segments to achieve optimal leverage, and he or she has to know how to move the bucket to dig and scoop the earth effectively. Learning to do all this is something like learning to drive a car. Eventually, the operator does most of the work unconsciously.

The Loader

If the backhoe is like a high-powered shoveling arm, the loader is like a giant dustpan or scoop. Operators don't generally use it to dig anything; they use it to pick up and carry loose material. It's also used to smooth things over like a butter knife, or to push dirt like a plow. To do all this, the operator has to operate the loader and drive the tractor at the same time.

Like the backhoe, the loader is made up of several jointed segments connected to a series of hydraulic pistons. The illustration shows a loader with two piston pairs and an 8-bar linkage.

The first piston pair, mounted to the tractor body, lifts the bucket in exactly the same way you would lift a heavy box. The hydraulic system extends both pistons together to elevate the entire assembly.

The various bars in the loader are connected in such a way that the bucket doesn't tip as it rises. Basically, the two main sets of parallel bars that hold the bucket move together so that they keep the bucket level

with the ground. Without this parallel lift, the loader would be something like a seesaw with a crate nailed to one end. If you filled the crate with oranges when the seesaw was level, a lot of them would fall out when you tilted the seesaw up. A parallel-lift system allows for more efficient loading because it keeps more of the material in the bucket as it lifts.

The second piston pair is attached to the loader arms and the bucket itself. These rams extend to dump the bucket and retract to tilt it back up.

The operator controls the loader with a single joystick. Pulling the joystick back and forth controls the first set of arms, lifting the bucket. To tilt the bucket, the operator moves the joystick left or right.

Putting It All Together

The loader and backhoe components are a natural combination for all sorts of jobs. When you dig up a lot of dirt to make a ditch, for example, you generally need a loader to either move the dirt out of the area or to fill the dirt back in over pipes or power lines. The most common application for a backhoe loader is digging a trench with the backhoe and then back-filling it with the loader.

The main reason you see backhoes at work all the time is that digging and moving dirt is a big part of a lot of different projects. There are a number of tools that do this sort of work, often more efficiently than a backhoe, but many construction crews use a backhoe instead for the sake of convenience. Backhoe loaders can move around construction sites easier than larger equipment, and operators can drive them on roads for short distances.

Amazingly, experienced operators use backhoe loaders in much the same way you would use a shovel or wheelbarrow at home—they know exactly how to move the controls to dig and load quickly and effectively, and they're always thinking ahead to their next few moves. This extremely powerful piece of heavy equipment becomes an extension of the operator's body!

And Another Thing...

Some backhoe loaders let you connect tools to the backhoe stick or the loader in place of the standard buckets. Loader tools include street sweepers and pallet forks. Backhoe tools include hydraulic hammers for breaking up asphalt, augers for digging circular holes, and grapples for gripping and pulling up rooted material (such as tree stumps).

How **SKID STEER LOADERS** Work

You've probably seen skid steer loaders around commercial construction sites or landscaping projects. Their small size and maneuverability allows them to operate in tight spaces. Their light weight allows them to be towed behind a full-size pickup truck, and their wide array of work-tools makes them very flexible.

Skid steer loaders are used to dig and to move landscaping and building materials. With the right attachments, the machines can also *grade* (make something level or slope evenly), jackhammer cement, and load trucks.

Before we look at how the various systems on a skid steer loader work, let's look at some of the things these loaders can do and learn how to control one.

On the Job

Skid steer loaders are fairly easy to operate. They can turn within their own footprint, just like a tank.

There are several steps to getting started:

1) To get in, you use the grab handles and steps to climb over the bucket. Fasten your seatbelt and lower the wrap-around armrest.
2) Start the engine. If the engine is cold, you might have to wait a few seconds for the glow plugs to warm up.
3) Release the parking brake to unlock all of the hydraulics.
4) The loader has a foot throttle, which, like the gas pedal on your car, makes it go faster. It also makes the loader arms move faster.

Operating the loader is simple. There are two joysticks: The left-hand joystick controls direction, and the right-hand joystick controls the loader. Each of the joysticks controls hydraulic valves that regulate the flow of hydraulic fluid to either the hydraulic motors that power the wheels or the hydraulic cylinders that power the loader.

Control

When you push forward on the left-hand joystick, all four wheels start to spin, or in the case of the multiterrain loader, both tracks start to turn. If you keep the joystick pushed forward and move it to the left, the machine turns left. It does this by slowing down or stopping the two left wheels or the left track. The farther left you push the joystick, the slower the left wheels or left track will move.

The opposite is true when moving in reverse: If you pull the stick all the way back, the machine goes straight backwards, but if you then move the joystick to the left, the right wheels or right track will slow down, causing the machine to turn right. If you center the joystick and then push it to the left, the left wheels or left track will move backward and the right wheels or right track will move forward—this turns the machine around in the smallest possible area.

The right-hand joystick controls the loader arms and bucket. Pulling the joystick back raises the arms, and pushing it forward lowers them. Moving the joystick to the left tilts the bucket up, and moving it to the right causes the bucket to dump.

Now, let's take a look at the drive system.

HSW Web Links

www.howstuffworks.com

How Hydraulic Machines Work
How Hydraulic Cranes Work
How Tower Cranes Work
How House Construction Works
How Backhoe Loaders Work

65

gasoline engines. A skid steer loader may operate 8 or more hours every day. Over the course of a year, a 5% or 10% difference in efficiency can make a real difference in fuel costs.

Although the engine, cooling system, and other accessories are tightly packed into the skid steer loader, the engine compartment is designed to make maintenance easy. A door on the back opens wide, and the radiator and fan tilt up to allow clear access to the engine and all of the maintenance items (such as filters).

The engines in a skid steer loader can range from a naturally aspirated diesel engine to a turbocharged diesel engine. The power is transmitted to a set of hydraulic pumps bolted directly to the output of the engine.

The Pumps

In some skid steer loaders, there are as many as four hydraulic pumps hooked up to the engine:

- Two variable-displacement pumps located in a single housing provide hydraulic power for the two hydraulic drive motors.
- A fixed-displacement pump provides hydraulic power for the loader arms and accessories.
- A smaller fixed-displacement pump provides hydraulic power for circulating the hydraulic fluid through filters and provides pressure to the pilot controls.

This setup allows the skid steer loader to make good utilization of the engine's power without ever stalling it. An engine stalls when the load on it is greater than the power it can produce. On hydraulic machines like these, the power that the engine can produce has to be balanced with the power that the hydraulic system uses. The maximum amount of power that the engine can make depends on its running speed. On a skid steer loader, an engine at full speed can produce its top-rated horsepower.

Between the pumps that power the wheels and the pump that powers the work

Drive System

The drive system on the skid steer loader has no transmission. Instead, it uses pumps and hydraulic motors to provide power to the wheels or tracks.

A hydraulic motor powers each side of the machine. Each of the two motors connects to a sprocket, and each sprocket is connected by two chains to each wheel. The sprockets and chains serve two purposes: They distribute the power from a single hydraulic motor to both wheels, and they provide a gear reduction to increase the torque at the wheels.

In your car, the main parts of the powertrain are the engine and transmission. In a skid steer loader, the powertrain consists of a diesel engine and a set of hydraulic pumps.

The Engine

Why does a skid steer loader use a diesel engine? For the same reason that all construction, mining, and farm equipment does: Diesel engines are more efficient than tools, the hydraulic system can demand more power from the engine than the engine can generate. On most skid steer loaders, it is up to the operator to carefully modulate the controls to keep the engine from stalling.

During an operation like loading a pile of dirt into a truck, the operator uses a lot of the engine power to push the machine into the pile. When the operator lifts a bucket load of dirt out, it takes a lot of force to break the load out of the pile. If the implement pump were to supply the pressure and flow for this operation while the drive pumps were still drawing power, the engine would stall.

To avoid this, some skid steer loaders automatically reduce the displacement of the pumps. This keeps the engine from stalling while still maintaining torque to the wheels or tracks at a reduced speed.

The Loader

The business end of the skid steer loader is the loader arms. These arms and their associated hydraulics can hold a variety of implements, not just buckets.

Many skid steer loaders use a *radial lift* lift-arm design. These lift arms are connected to the machine with a single pin on each side. The pins allow the bucket to follow an arc as it rises. As the bucket starts to rise, it first moves out, away from the machine. After it gets higher than the height of the mounting pin, it moves in closer to the machine.

The bucket sits close to the machine when it is in the down position to make the machine more stable and compact when moving loads around. As the bucket is raised, it moves away from the machine and then straight up. This gives the machine extra reach, making it easier to dump a load into the middle of a truck or place a pallet into a deep shelf.

The Add-Ons

The thing that makes a skid steer loader or multiterrain loader so useful is the wide variety of work tools that are available.

Construction and landscape work sites make use of a wide variety of add-ons, such as:

- Several different buckets that are geared toward load size and type
- Pallet forks are used to lift and carry things such as bales of hay
- Grapple forks are used to pierce or pick up things like paper at a recycling center
- Angle blades are used to move loose material like dirt or gravel
- Augers are used to drill holes for trees, signs or other large objects
- Landscape tillers are used to break up and level soil
- Stump grinders are used to remove tree stumps
- Trenchers are used to cut small, straight trenches

It's amazing how a skid steer loader can do the job of something like 10 different pieces of machinery on any given work site!

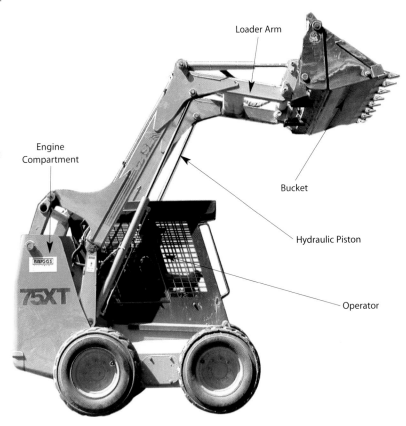

Loader Arm

Engine Compartment

Bucket

Hydraulic Piston

Operator

75XT

How **TOWER CRANES** Work

Tower cranes are a common fixture at any major construction site. They're pretty hard to miss—they often rise hundreds of feet into the air, and can reach out just as far. The construction crew uses the tower crane to lift steel, concrete, large tools like acetylene torches and generators, and a wide variety of other building materials.

 HSW Web Links

www.howstuffworks.com

How Hydraulic Cranes
  Work
How Backhoe Loaders
  Work
How Skid Steer Loaders
  Work
How House Construction
  Works
How Oil Drilling Works

When you look at one of these cranes, what it can do seems nearly impossible: Why doesn't it tip over? How can such a long boom lift so much weight? How is it able to grow taller as the building grows taller?

Parts of a Tower Crane

All tower cranes consist of the same basic parts:

- The **base** is bolted to a large concrete pad that supports the crane
- The base connects to the **mast** (also called the tower), which gives the tower crane its height
- Attached to the top of the mast is the **slewing unit**—the gear and motor—that allows the crane to rotate

On top of the slewing unit are three parts:

- The long horizontal **jib** (also called the working arm), which is the portion of the crane that carries the load
- The shorter horizontal **machinery arm,** which contains the crane's motors and electronics as well as the large concrete counter weights
- The **operator's cab**

How Much Weight Can They Lift?

A typical tower crane has the following specifications:

- **Maximum unsupported height—** 265 feet (80 meters). The crane can have a total height much greater than 265 feet if it is tied into the building as the building rises around the crane.

- **Maximum reach**—230 feet (70 meters)
- **Maximum lifting power**—19.8 tons (18 metric tons), 300 tonne-meters
- **Counterweights**—20 tons (16.3 metric tons)

The maximum load that the crane can lift is 39,690 pounds (18 metric tons), but the crane cannot lift that much weight if the load is positioned at the end of the jib. The closer the load is positioned to the mast, the more weight the crane can lift safely. The 300 tonne-meter rating tells you the relationship. For example, if the operator positions the load 30 meters (100 feet) from the mast, the crane can lift a maximum of 10.1 tonnes.

The crane uses two limit switches to make sure that the operator does not overload the crane:

- The maximum load switch monitors the pull on the cable and makes sure that the load does not exceed 18 tonnes.
- The load moment switch makes sure that the operator does not exceed the tonne-meter rating of the crane as the load moves out on the jib. A cat head assembly in the slewing unit can measure the amount of collapse in the jib and sense when an overload condition occurs.

Why Don't They Fall Over?

When you look at a tall tower crane, the whole thing seems outrageous—why don't these structures fall over, especially since they have no support wires of any kind?

The main element of the tower crane's stability is a large concrete pad that the construction company pours several weeks before the crane arrives. This pad typically measures 30 feet × 30 feet × 4 feet (10 × 10 × 1.3 meters) and weighs 400,000 pounds (182,000 kg). Large anchor bolts are embedded deep into the pad to support the base of the crane. The mast is an extremely strong triangulated truss that is self-supporting.

How Do They Grow?

Tower cranes arrive at the construction site on 10 to 12 tractor-trailer rigs. The crew uses a mobile crane to assemble the jib and the machinery section, and places these

---

Hydraulic Cranes

Hydraulic cranes are very simple by design, but can perform Herculean tasks that would otherwise seem impossible. In a matter of minutes, these machines are able to raise multi-ton bridge beams on highways, heavy equipment in factories, and even lift beachfront houses onto pilings. Hydraulic truck cranes are also used to lift killer whales like Shamu out of water tanks when places like Sea World ship the whales to new destinations.

Using a very simple principle of hydraulics, these machines move thousands of pounds with relative ease, making them an essential component of most construction projects and a great example of the power of basic physics.

When watching a hydraulic truck crane in action, it's hard to believe just how much weight it's moving because it deals with these multi-ton objects with relative ease. Hydraulic truck cranes vary in lifting power. It's

easy to tell how much a particular hydraulic truck crane can lift just by the name of it: A 40-ton crane can lift 40 tons (80,000 lb or 36,287 kg).

Most hydraulic truck cranes use two-gear pumps that have a pair of intermeshing gears to pressurize the hydraulic oil. When pressure needs to increase, the operator pushes the foot throttle to run the pump faster. In a gear pump, the only way to get high pressure is to run the engine at full power.

---

horizontal members on a 40-foot (12-m) mast that consists of two mast sections. The mobile crane then adds the counterweights.

The mast rises from this firm foundation. The mast is a large, triangulated lattice structure, typically 10 feet (3.2 meters) square. The triangulated structure gives the mast the strength to remain upright.

To rise to its maximum height, the crane grows itself one mast section at a time. The crew uses a top climber or climbing frame that fits between the slewing unit and the top of the mast. Here's the process:

1) The crew hangs a weight on the jib to balance the counterweight.
2) The crew detaches the slewing unit from the top of the mast. Large hydraulic rams in the top climber push the slewing unit up 20 feet (6 m).
3) The crane operator uses the crane to lift another 20-foot mast section into the gap opened by the climbing frame. Once bolted in place, the crane is 20 feet taller!

Once the building is finished and it is time for the crane to come down, the process is reversed—the crane disassembles its own mast and then smaller cranes disassemble the rest.

Did You Know?

Most construction companies rent their tower cranes. The rental company ships the crane to the site, assembles it, and charges a monthly fee while the crane is on the site.

The typical fee for installation and disassembly runs around $60,000. This price includes shipping the crane to the site, renting the mobile crane used to assemble the tower crane, the cost of the crew that handles the assembly, and so on. A typical monthly fee for a 150-foot-tall tower crane is approximately $15,000, with an additional charge to rent the climbing frame and extra mast sections.

How **FIRE ENGINES** Work

We see fire engines all the time, but have you ever stopped to think about all of the things that these machines do? Fire engines are amazing pieces of equipment that allow firefighters to perform their jobs and get to fire scenes quickly. The important thing to know about a fire engine is that it is a combination of a personnel carrier, toolbox, and water tanker. All three components are essential to fighting fires.

With different fire departments having varying needs, fire engines come in all shapes, sizes, and colors. They can be custom-made to suit the specific needs of a firehouse. There are pumpers, tankers, aerials (also known as ladder trucks), and combinations of these. There are also models particularly geared to rescue missions. These fire engines come equipped with things like the Jaws of Life (see "How the Jaws of Life Work").

Pump It Up

When firefighters arrive at any house fire, they need a lot of water to put out the blaze. The job of any pumper or tanker fire engine is to move the water around. A tanker arrives at the fire carrying hundreds of gallons of water. A pumper can suck water in from an outside source, such as a fire hydrant, drop tank, swimming pool, or lake. Most of the time, both jobs combine in a single engine.

Lights

Sirens

If the pumper-tanker has its own tank, it will hold something like 1,000 or so gallons of water. The tank is normally located inside the vehicle, with lines that run down the center of the rear of the truck. A 6-inch diameter, hard suction line is used to suck water out of an exterior water source like a pool or a lake. Or the water comes into the pumper from a fire hydrant.

The heart of the pumper-tanker is the *impeller water pump*. An impeller is a rotor-like piece that has curved blades. Driven by its own diesel engine, the impeller spins inside the pump at high speed. When water comes into the pump, it hits the inner part of the impeller and is slung outward. Water pressure is created by centrifugal force from the spinning action of the impeller. A valve

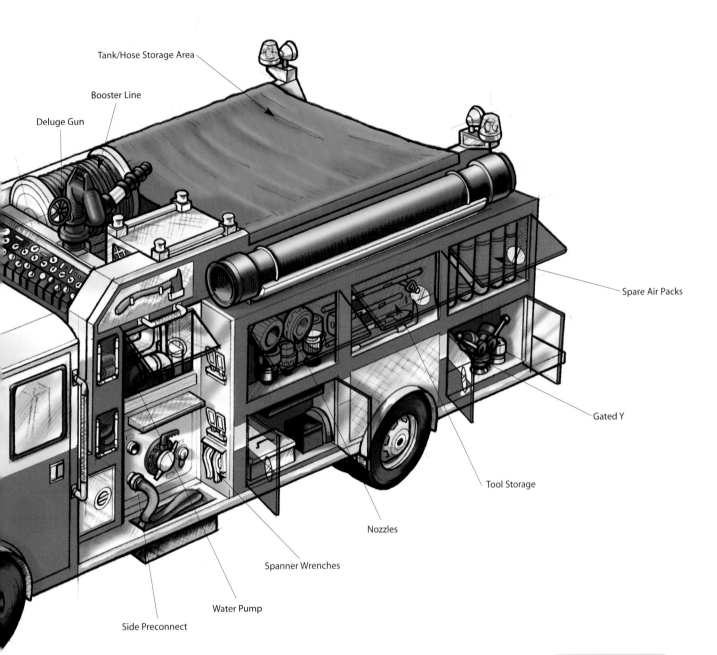

Tank/Hose Storage Area

Booster Line

Deluge Gun

Spare Air Packs

Gated Y

Tool Storage

Nozzles

Spanner Wrenches

Water Pump

Side Preconnect

opens to allow water to hit the center of the rotating impeller.

Firefighters control the hoses using the truck's pump panel. The pump panel is a series of levers and switches that control how much water is flowing and which hoses are getting it; a big pumper can handle many different hoses at the same time. When the firefighters arrive at the scene of a fire, the driver will jump out and rush to the pump panel to begin pump operation. A big indicator—either a dial or a series of red lights on the pump panel—lets the operator know how much water is left in the tank. Other firefighters will hook the pumper up to a source of water.

The first thing the pump operator is going to do is make sure that the valve between the tank and pump is open. An electric switch beside the pump will open that valve, and ensure that water is flowing into the pump. Next, the operator will check to see which lines (hoses) have been pulled off the fire engine by the firefighters, and the operator will discharge those lines. A hose is *discharged* when water flows out of the pump and into the hose. The lines are all color-coded to make it easy for the operator to know which lines to discharge. The color

Priming

Before water can be sucked in from exterior sources, the line has to be primed. Priming the hard suction line involves pumping all of the air out of it. To prime the pump, the operator flips an electronic switch on the pump panel.

of the line corresponds to a plate below each lever on the pump panel.

Most of the discharging is controlled by a built-in electronic device, called a *mastermind*. It automatically controls the pump, and adjusts the pressure up or down. It also has a built-in relief valve so that if one person suddenly cuts off a line, the pressure from that line doesn't automatically get fed into another line.

Many trucks also have a foam system. The foam tank is embedded in the main water tank. Pumper-tankers carry different types of foam. For example, class A foam can be used to saturate materials inside a structure to keep those materials from reigniting. Class B foam is used to fight car fires and other fires where flammable liquids might be present.

Hose It Down

There are many different hoses on the fire engine, and each has its own specific role in putting out a blaze. Hoses can handle different amounts of water depending on the hose length and diameter and the amount of pressure in the pump.

When responding to a house fire, the firefighters will immediately pull off the *crosslay hoses*. These lines are located directly below the pump panel. They lay out in the open and are light, so they are easy to get off the fire engine for attacking a fire. Crosslays are 200 feet (61 m) long, have a diameter of 1.5 inches, and can gush water at 95 gallons (360 L) per minute. For smaller fires, such as small wood fires or chimney fires, the small *booster line* is adequate. A booster line is the smallest hose on the truck and has a diameter of about 1 inch.

There's also the *deluge gun,* sometimes called a deck gun or master stream. Just by looking at it, you know why this water cannon carries those names. The deluge gun is used to put a lot of water on large fires. It can put out in excess of 1,000 gallons per minute.

Trucks might also have a few lines called *preconnects*. These lines are preconnected to the truck in order to save time at the fire scene.

One of the many compartments on the fire engine can be used to store extra sections of hose. For example there might be two extra sections of the 5-inch hose: a 25-foot and a 50-foot section. These two sections are called *curb jumpers,* because they typically lay on the curb. These sections give firefighters just a little bit more line to connect to a fire hydrant without having to get another 100-foot section down.

You might find a hose pack stored alongside the curb jumpers. A *hose pack* is a small, bundled hose that can be taken to the higher levels of a building. It is banded to make it easier to carry up a ladder. A firefighter can just throw it over his or her shoulder and take it up and through a window. Usually, a hose pack is used if the other lines can't reach inside. This hose will connect to the hose that runs up the ladder of a ladder truck.

Going Up!

When a fire breaks out in a multistory building, a ladder truck gets firefighters up to the higher floors quickly.

The ladder on the truck is raised and lowered using a hydraulic piston. As hydraulic fluid enters this piston, the pressure of the fluid will either cause the rod to extend or retract. If the piston rod extends, the ladder will go up. If it retracts, the ladder will come down.

Another set of hydraulic pistons allows the sections of the ladder to telescope up and down. A hydraulic motor is used to rotate the gear that moves the ladder from left to right. While the ladder is in use, four outriggers are extended to stabilize the truck.

On some ladder trucks, the ladder also has a pipe that runs the length of the ladder. This is an extra water line that is sometimes used to spray water on fires that are in a high spot or to spray water down on a fire

Inside the Toolbox

Firefighters have to take dozens of tools and lots of other equipment when responding to a fire or medical call. All of this equipment is stored in compartments that line the sides and the back of the fire engine. Some of the tools you might find on a fire engine are:

Barrel strainer—This is an attachment put on a hard suction hose when sucking water out of a lake or pond. This tool keeps debris out of the water supply.

Nozzles—Different nozzles are needed for different situations. Fog nozzles emit a strong mist of water. Other nozzles direct water in a solid stream. There's also a piercing nozzle that can be used to punch through walls and spray areas that can't be reached otherwise.

Foam inductor—This is a special nozzle used to mix water and foam.

Haligan tool—This tool looks similar to a crowbar.

Sheet rock puller—This tool is used to peel back the sheet rock on walls so that water can be sprayed inside the wall.

Pike poles—These spear-like tools are about 10 to 12 feet long and are thrust into the ceiling to pull sheet rock down.

EMS equipment—Most fire engines carry a defibrillator, an emergency oxygen tank, and a trauma jump kit, which includes all of the first aid equipment needed for emergencies.

Gated Y—This special hose adapter can be attached to a line to allow two smaller lines to run off of the same water source.

Spanner wrenches—These unique tools are used to tighten the lines to the fire engine or to a hydrant.

Hydrant wrench—This is the wrench used to turn the hydrant on.

Jaws of Life—This extrication equipment is used to free victims from car or building accidents. Read "How the Jaws of Life Work" in this chapter to learn more about these hydraulic machines.

Exhaust fan—This fan is placed in the doorway to suck smoke out of the house. Fire engines may also carry a positive-pressure exhaust fan, which blows air through the house and out the other side.

Salvage covers—These are used for covering furniture on a lower floor while firefighters attack a fire on a floor above.

In addition, fire engines also carry bolt cutters, a sledgehammer, a fire extinguisher, a water cooler, a 24-foot (7-m) extension ladder and a 16-foot (5-m) roof ladder. Some trucks may also carry chain saws, rappelling rope, and backboards, which are used to transport injured people. As you can see, a fire engine is a giant toolbox! The design of the fire engine maximizes all possible storage space.

---

from above. The pipe can spray out hundreds, if not a thousand, gallons per minute, depending on the diameter of the pipe.

The ladder is controlled by a series of joysticks at the base of the ladder. The outriggers are controlled in the back of the truck. Each outrigger has four control levers: two for extending the beam out and two for lowering the leg to the ground. Metal pads are placed under the legs to prevent the force of the truck from cracking asphalt surfaces.

Grab a Seat

The unique design of a fire engine allows it to carry a whole team of firefighters to the fire scene. The cabin of the fire engine is divided into two sections: the front seat, where the driver and captain sit, and the jumpseat area, where other firefighters sit.

As mentioned before, the driver is responsible for controlling the pump panel. For this reason, there are some basic controls on the driver's dashboard that are related to that task.

The driver might also have another switch within reach that activates automatic tire chains, which are sometimes needed during the winter to drive through ice and snow. Automatic tire chains save the time and hassle of jacking the truck up and putting tire chains on manually.

The captain sits in the passenger seat next to the driver in the front section of the cab. The front section of the cab has firecoms, which are radio headsets that allow the captain and driver to communicate with the firefighters sitting in the jumpseat area. The captain will often give instructions to the firefighters on the way to the fire scene.

The jumpseat area can hold four to six firefighters on the way to the fire. There is one row of four seats that sit back-to-back with the captain and driver. There are also two fold-down seats directly across from the row of four seats. In between the fold-down seats, there are several yellow pouches that contain the firefighters' masks.

Air packs are located in the back of the four main seats. By already having the air packs on the truck, all the firefighters have to do is put them on their shoulders. Each air pack has 30 minutes of air.

How ESCALATORS Work

You probably ride escalators all the time, but do you know how they move, and flatten, and keep the handrail in sync with the steps? Escalators are one of the largest, most expensive machines people use on a regular basis, but they're also one of the simplest.

HSW Web Links

www.howstuffworks.com

How Electric Motors Work
How a Block and Tackle Works
How Roller Coasters Work
How Bicycles Work
How Gear Ratios Work
How Skyscrapers Work

Did You Know?

Escalator speeds vary from about 90 feet per minute to 180 feet per minute (27 to 55 meters per minute). An escalator moving 145 feet (44 meters) per minute can carry more than 10,000 people an hour—many more people than a standard elevator.

At its most basic level, an escalator is just a simple variation on the conveyer belt. A pair of rotating chain loops pulls a series of stairs in a constant cycle, moving a lot of people a short distance at a good speed.

The core of an escalator is a pair of chains that are looped around two pairs of gears. An electric motor turns the drive gears at the top, which rotate the chain loops. A typical escalator uses a 100-horsepower motor to rotate the gears. The motor and chain system are housed inside the truss, a metal structure extending between two floors.

Instead of moving a flat surface, as in a conveyer belt, the chain loops move a series of steps. Each step in the escalator has two sets of wheels, which roll along two separate tracks. The upper set (the wheels near the top of the step) are connected to the rotating chains, and so are pulled by the drive

gear at the top of the escalator. The other set of wheels simply glides along its track, following behind the first set.

The tracks are spaced apart in such a way that each step will always remain level. At the top and bottom of the escalator, the tracks level off to a horizontal position, flattening the stairway. Each step has a series of grooves in it, so it fits together with the steps behind it and in front of it.

In addition to rotating the main chain loops, the electric motor in an escalator also moves the handrails. A handrail is simply a rubber conveyer belt that is looped around a series of wheels. This belt is precisely configured so that it moves at exactly the same speed as the steps to give riders some stability.

The escalator system isn't nearly as good as an elevator at lifting people dozens of stories, but it is much better at moving people a short distance. This is because of the escalator's high loading rate. Once an elevator is filled up, you have to wait for it to reach its floor and return before anybody else can get on. On an escalator, as soon as you load one person on, there's space for another.

Handrail Drive

Handrail

Step

Electric Motor

Drive Gear

Chain Guide

Inner Rail

Return Wheel

How **SNOW MAKERS** Work

The ski-resort industry is utterly dependent on snow makers, commonly called snow guns. Without these machines, many resorts could only operate a couple months a year, and they would have to leave it up to nature to determine day-to-day skiing conditions. Snow makers let resort owners make good skiing snow whenever they want to, as long as the conditions are right.

One common notion is that machine-made snow is artificial. This is not really the case—it's actually the same stuff that falls out of the sky, it's just created by a machine rather than by weather conditions.

Replicating Nature

The traditional type of snow gun produces water droplets by combining cooled water and compressed air. The snow gun is hooked up to two different lines, leading to two different pumping systems. One pumps in water from a lake, pond, or reservoir and the other pumps in high-pressure air from an air compressor.

The compressed air does three things:

1) It atomizes the water—that is, disrupts the stream so that the water splits into many tiny droplets.
2) It blows the water droplets into the air.
3) It helps cool the water droplets as they fly into the air.

Airless snow guns atomize water into a fine mist by forcing it through nozzles. A high-powered fan blows the mist into the air. The main advantage of this design is that you don't have to hook the snow gun up to a compressed-air supply—you only have to provide water and a power source. Some designs atomize the water with high-speed fans.

Resort owners usually mount snow guns on towers so the water has more time to freeze before it reaches the ground. Elevated snow guns are also less disruptive to skiers.

In most resorts, workers will accumulate a big pile of man-made snow and then spread it along the trail with snow-grooming equipment. Snow groomers are just tractors with very wide tracks that spread the snow around and compact it.

Working with the Weather

The density, or wetness, of snow depends on the temperature and humidity outside, as well as the size of the water droplets launched by the gun. Snow makers have to adjust the proportions of water and air in their snow guns to get the perfect snow consistency for the outdoor weather conditions. Many ski slopes now control their snow guns with a central computer system that is hooked up to weather-reading stations all over the slope.

Snow makers have to take many variables into account to cover a slope with ideal skiing snow. The idea behind man-made snow is extremely simple; but actually getting it to work effectively is quite a feat. Many snow makers describe the job as a challenging marriage of science and art—the basic elements are precise weather measurements and expensive machinery, but you need instinct, improvisation, and creativity to get it exactly right.

HSW Web Links

www.howstuffworks.com

How Ice Rinks Work
How Air Conditioners Work
How Water Blasters Work

How Much Water?

To cover several ski trails with man-made snow, you need a lot of water. It takes about 75,000 gallons (285,000 liters) of water to create a 6-inch blanket of snow covering a 200-square-foot area (61 square meters). The system in a good-sized ski slope can convert 5,000 to 10,000 gallons (18,927 to 37,854 liters) of water to snow every minute!

Resorts generally gather water from a lake, which isn't particularly expensive or harmful to the environment. Most of the water runs down the hill again to fill up the reservoir. Powering the pumps for the snow-maker can be extremely expensive, however.

How **OIL DRILLING** Works

The United States produces and imports hundreds of millions of barrels of crude oil each year. This oil gets refined into gasoline, kerosene, heating oil, and other products. To keep up with our consumption, oil companies must constantly look for new sources of petroleum, as well as improve the production of existing wells.

HSW Web Links

www.howstuffworks.com

How Oil Refining Works
How Gas Prices Work
How Earthquakes Work
How Volcanoes Work

How does a company go about finding oil and pumping it from the ground? You may have seen images of black crude oil gushing out of the ground, or seen an oil well in movies and television shows like *Giant, Oklahoma Crude, Armageddon,* and *The Beverly Hillbillies.* But modern oil production is quite different from the way it's portrayed in the movies.

Oil is a fossil fuel that can be found in many countries around the world. It comes from the remains of tiny plants and animals (plankton) that died in ancient seas between 10 and 600 million years ago. After the organisms died, they sank into the sand and mud at the bottom of the sea. For us to have oil today, here's what happened through many millenia:

1) Over the years, the organisms decayed in the sedimentary layers. In these layers, there was little or no oxygen present, so microorganisms broke the remains into carbon-rich compounds that formed organic layers.
2) The organic material mixed with the sediments, forming fine-grained shale, or *source rock.*
3) As new sedimentary layers were deposited, they exerted intense pressure and heat on the source rock.
4) The heat and pressure distilled the organic material into crude oil and natural gas.
5) The oil flowed from the source rock and accumulated in thicker, more porous limestone or sandstone, called *reservoir rock.*
6) Movements in the earth trapped the oil and natural gas in the reservoir rocks between layers of impermeable rock, or *cap rock,* such as granite or marble.

Finding Oil

The people who search for and find oil today are geologists. Their task is to find the right conditions for an oil trap—the right

New Drilling Technologies

The U.S. Department of Energy and the oil industry are working on new ways to drill oil, including horizontal drilling techniques, which may be able to reach oil under ecologically sensitive areas, and lasers, which may be able to drill deep holes significantly cheaper than current methods.

source rock, reservoir rock, and entrapment. For decades, geologists have done this by interpreting surface features, like surface rock and soil types, and perhaps by taking core samples. Modern oil geologists continue to examine surface rocks and terrain, with the additional help of satellite images. They also use a variety of new methods to find oil. They can use sensitive gravity meters to measure tiny changes in the earth's gravitational field that could indicate flowing oil, as well as sensitive magnetometers to measure tiny changes in the earth's magnetic field caused by flowing oil. They can detect the smell of hydrocarbons using sensitive electronic noses called *sniffers.* Finally, and most commonly, they use seismology, creating shock waves that pass through hidden rock layers and interpreting the waves that are reflected back to the surface.

In seismic surveys, a shock wave is created through one of three different techniques:

- **Compressed-air gun**—This machine shoots pulses of air into the water (for exploration over water).
- **Thumper truck**—This machine slams heavy plates into the ground (for exploration over land).

- **Explosives**—Explosives are drilled into the ground (for exploration over land) or thrown overboard (for exploration over water) and detonated.

The shock waves travel beneath the surface of the earth and are reflected back by the various rock layers. The reflections travel at different speeds depending on the type or density of rock layers they pass through. The reflections of the shock waves are detected by sensitive microphones or vibration detectors—hydrophones over water, seismometers over land. The readings are interpreted by seismologists for signs of oil and gas traps.

Although modern oil-exploration methods are better than previous ones, they still may have only a 10% success rate for finding new oil fields.

Preparing to Drill

After the site has been selected, it must be surveyed to determine its boundaries, and environmental impact studies may be done. Lease agreements, titles, and right-of-way accesses for the land must be obtained and evaluated legally. For offshore sites, legal jurisdiction must be determined.

Once the legal issues have been settled, the crew prepares the land:

- The land is cleared and leveled, and access roads may be built.
- Because water is used in drilling, there must be a source of water nearby. If there is no natural source, a water well is drilled.
- Crews dig a reserve pit, which is used to dispose of rock cuttings and drilling mud during the drilling process, and line it with plastic to protect the environment. If the site is an ecologically sensitive area, such as a marsh or wilderness, then the cuttings and mud must be disposed offsite—trucked away instead of placed in a pit.

Once the land has been prepared, a rectangular pit, called a *cellar*, is dug around the location of the actual drilling hole. The cellar provides a workspace around the hole for the workers and drilling accessories. The crew then begins drilling the main hole, often with a small drill truck rather than the

Oil Rig Systems

The major systems of a land oil rig are:

- **Power system**—Large diesel engines and electrical generators
- **Mechanical system**—Includes the hoisting system and the turntable; driven by electric motors
- **Rotating equipment**—Includes the swivel, a kelly, a turntable, drill string, and drill bits; used for rotary drilling
- **Casing**—A large-diameter concrete pipe that lines the drill hole, prevents the hole from collapsing, and allows drilling mud to circulate

- **Circulation system**—This system pumps *drilling mud* (a mixture of water, clay, weighting material, and chemicals that is used to lift rock cuttings from the drill bit to the surface) under pressure through the kelly, rotary table, drill pipes, and drill collars. This system includes:

  - A pump
  - Pipes and hoses
  - Mud-return line
  - Shale shaker
  - Shale slide
  - Reserve pit
  - Mud pits
  - Mud-mixing hopper

- **Derrick**—A support structure, tall enough to allow new sections of drill pipe to be added to the drilling apparatus as drilling progresses, that holds the drilling apparatus
- **Blowout preventer**—High-pressure valves located under the land rig or on the sea floor that seal the high-pressure drill lines and relieve pressure when necessary to prevent a blowout (an uncontrolled gush of gas or oil to the surface, often accompanied by fire)

main rig. The first part of the hole is larger and shallower than the main portion, and is lined with a large-diameter conductor pipe. Additional holes are dug off to the side to temporarily store equipment. When these holes are finished, the rig equipment can be brought in and set up.

Depending upon the remoteness of the drill site and its access, equipment may be transported to the site by truck, helicopter, or barge. Some rigs are built on ships or barges for work on inland water where there is no foundation to support a rig (as in marshes or lakes). Once the equipment is at the site, the rig is set up.

Drilling the Hole

After the crew sets up the rig, the drilling operations start. First, from the starter hole, a surface hole is drilled down to a preset depth, which is somewhere above where workers think the oil trap is located. Five basic steps are involved in drilling the surface hole. Workers:

1) Place the drill bit, collar, and drill pipe in the hole.
2) Attach the kelly (a piece of pipe) and turntable and begin drilling.

3) Circulate mud through the pipe and out of the bit as drilling progresses to float the rock cuttings out of the hole.
4) Add new sections (joints) of drill pipes as the hole gets deeper.
5) Remove (*trip out*) the drill pipe, collar, and bit when the preset depth (anywhere from a few hundred to a couple thousand feet) is reached.

Once the workers drill down to the preset depth, they must place casing-pipe sections into the hole to prevent it from collapsing in on itself (a process known as "running and cementing" the casing). The casing pipe has spacers around the outside to keep it centered in the hole.

The casing crew puts the casing pipe in the hole. The cement crew pumps cement down the casing pipe using a bottom plug, a cement slurry, a top plug, and drill mud. The pressure from the drill mud causes the cement slurry to move through the casing and fill the space between the outside of the casing and the hole. Finally, the cement is allowed to harden and is then tested for hardness, alignment, and a proper seal.

Drilling continues in stages: Workers drill, then run and cement new casings, then drill again. When the rock cuttings from the mud reveal the oil sand from the reservoir rock, they may have reached the final depth. At this point, they remove the drilling apparatus from the hole and perform several tests to confirm that the final depth has been reached:

- **Well logging**—Lowering electrical and gas sensors into the hole to take measurements of the rock formations there
- **Drill-stem testing**—Lowering a device into the hole to measure the pressures, which will reveal whether reservoir rock has been reached
- **Core samples**—Taking samples of rock to look for characteristics of reservoir rock

Once they have reached the final depth, the crew completes the well to allow oil to flow into the casing in a controlled manner. First, they lower a perforating gun into the well to the production depth. The gun has explosive charges to create holes in the casing through which oil can flow.

After the casing has been perforated, they run tubing into the hole as a conduit for oil and gas to flow up the well. A device called a *packer* is run down the outside of the tubing. When the packer is set at the production level, it is expanded to form a seal around the outside of the tubing. Finally, they connect a multi-valved structure called a *Christmas tree* to the top of the tubing and cement it to the top of the casing. The Christmas tree allows the crew to control the flow of oil from the well.

Once the well is completed, the flow of oil into the well must be started. For limestone reservoir rock, acid is pumped down the well and out the perforations. The acid dissolves channels in the limestone that lead oil into the well. For sandstone reservoir rock, a specially blended fluid containing *proppants* (combinations of sand, walnut shells, aluminum pellets) is pumped down the well and out the perforations. The pressure from this fluid makes small fractures in the sandstone that allow oil to flow into the well, while the proppants hold these fractures open. Once the oil is flowing, the oil rig is removed from the site and production equipment is set up to extract the oil from the well.

Extracting the Oil

After the rig is removed, a pump is placed on the well head. In the pump system, an electric motor drives a gear box that moves a lever. The lever pushes and pulls a polishing rod up and down. The polishing rod is attached to a sucker rod, which is attached to a pump. This system forces the pump up and down, creating a suction that draws oil up through the well.

In some cases, the oil may be too heavy to flow. A second hole is then drilled into the reservoir and steam is injected under pressure. The heat from the steam thins the oil in the reservoir, and the pressure helps push it up the well. This process is called *enhanced oil recovery*.

With all of this oil-drilling technology in use—and new methods in development—the question remains: Will we have enough oil to meet our needs? Current estimates suggest that we have enough oil for about 63 to 95 years to come, based on current and future finds and present demands.

chapter four

NATURE

How **HURRICANES** Work

Every year between June 1 and November 30 (the hurricane season), hurricanes threaten the eastern and gulf coasts of the United States, Mexico, Central America, and the Caribbean. Also known as typhoons or tropical cyclones, these huge storms wreak havoc when they make landfall. They can kill thousands of people and cause billions of dollars of property damage when they hit heavily populated areas.

HSW Web Links

www.howstuffworks.com

How Tornadoes Work
How Floods Work
How Lightning Works
How Radar Works

Did You Know?

The extent of damage a hurricane can do depends on a few things:

- The category of the hurricane
- Whether the storm comes ashore head-on or just grazes the coastline
- Whether the right or left side of the hurricane strikes a given area

It turns out that the right side of a hurricane packs more punch because the wind speed and the hurricane's speed of motion are complementary there. On the left side, the hurricane's speed of motion subtracts from the wind speed.

According to the National Hurricane Center, *hurricane* is a name for a tropical cyclone that occurs in the Atlantic Ocean. *Tropical cyclone* is the generic term used for low-pressure systems that develop in the tropics.

A Hurricane Is Born

Hurricanes form in tropical regions where there is warm water (at least 80°F, 27°C), moist air, and converging equatorial winds. Most Atlantic hurricanes begin off the west coast of Africa, starting as thunderstorms that move out over the warm, tropical ocean waters. A thunderstorm reaches hurricane status in three stages:

- **Tropical depression**—Swirling clouds and rain with wind speeds of less than 38 mph (61 kph/33 knots)
- **Tropical storm**—Wind speeds of 39 to 73 mph (63 to 118 kph/34 to 63 knots)
- **Hurricane**—Wind speeds greater than 74 mph (119 kph/64 knots)

It can take anywhere from hours to several days for a thunderstorm to develop into a hurricane. Although the whole process of hurricane formation is not entirely understood, three things must be present for hurricanes to form:

- A continuing evaporation-condensation cycle of warm, humid ocean air
- Patterns of wind characterized by converging winds at the surface and strong, uniform-speed winds at higher altitudes

Sattelite view of a hurricane.

- A difference in air pressure (a *pressure gradient*) between the surface and high altitudes

Warm, Humid Ocean Air

Warm, moist air from the ocean surface begins to rise rapidly. As this warm air rises, its water vapor condenses to form storm clouds and droplets of rain. The condensation releases heat called *latent heat of condensation*. This latent heat warms the cool air aloft, causing it to rise. This rising air is replaced by more warm, humid air from the ocean below. This cycle continues, drawing more warm, moist air into the developing storm and continuously moving heat from the surface to the atmosphere. A pattern of wind develops, with the wind circulating around a center. This circulation is similar to that of water going down a drain.

Patterns of Wind

Converging winds are winds moving in different directions that run into each other. Converging winds at the earth's surface collide and push warm, moist air upward. This rising air reinforces the air that is already rising from the surface, so the circulation and wind speeds of the storm increase. In the meantime, strong winds blowing at uniform speeds at higher altitudes (up to 30,000 ft/9,000 m) help to remove the rising hot air from the storm's center, maintaining a continual movement of warm air from the surface and keeping the storm organized. If the high-altitude winds do not blow at the same speed at all levels—if *wind shears* are present—the storm loses organization and weakens.

Pressure Gradient

High-pressure air in the upper atmosphere (above 30,000 ft/9,000 m) over the storm's center also removes heat from the rising air, further driving the air cycle and the hurricane's growth. As high-pressure air is sucked into the low-pressure center of the storm, wind speeds increase.

Tall, Dark, and Hurricane

Once a hurricane forms, it has three main parts:

- The *eye* is the low-pressure, calm center of circulation.
- The *eye wall* is the area around the eye with the fastest, most violent winds.
- The *rain bands* are the bands of thunderstorms circulating outward from the eye. The rain bands are part of the evaporation/condensation cycle that feeds the storm.

Hurricanes vary widely in physical size. Some storms are very compact and have only a few trailing bands of wind and rain behind them. Other storms are looser, so the bands of wind and rain spread out over hundreds or thousands of miles. Hurricane Floyd, which hit the eastern United States in September 1999, was felt from the Caribbean islands to New England.

The damage caused by a hurricane comes from several different features of the storm. Hurricanes bring with them huge amounts of rain. A big hurricane can dump dozens of inches of rain in just a day or two, much of it inland. That amount of rain can create inland flooding that can totally devastate a large area around the hurricane's center.

High, sustained winds cause structural damage. These winds can also roll cars, blow over trees, and erode beaches (both by blowing sand and by blowing the waves into the beach). The prevailing winds of a

Saffir-Simpson Hurricane Scale

Once a hurricane forms, it is rated on the Saffir-Simpson Hurricane Scale. There are five categories in this rating system:

- **Category 1**—Wind speed is 74 to 95 mph (119 to 153 kph) with a storm surge 4 to 5 feet (1.2 to 1.5 m) above normal and some flooding. Little or no structural damage occurs.
- **Category 2**—Wind speed is 96 to 110 mph (155 to 177 kph) with a storm surge 6 to 8 feet (1.8 to 2.4 m) above normal and downed trees. Roofs are moderately damaged.
- **Category 3**—Wind speed is 111 to 130 mph (179 to 209 kph) with a storm surge 9 to 12 feet (2.7 to 3.7 m) above normal and severe flooding. Houses have structural damage and mobile homes are destroyed.
- **Category 4**—Wind speed is 131 to 154 mph (210 to 248 kph) with a storm surge 13 to 18 feet (4 to 5.5 m) above normal and severe flooding inland. Some roofs are ripped off and major structural damage occurs.
- **Category 5**—Wind speed is greater than 155 mph (249 kph) with a storm surge at least 18 feet (5.5 m) above normal and severe flooding further inland. Serious damage occurs to most wooden structures.

hurricane push a wall of ocean water, called a *storm surge,* in front of the hurricane. If the storm surge happens to synchronize with a high tide, it causes beach erosion and significant inland flooding. Also, hurricane winds often spawn tornadoes, which are smaller, more intense cyclonic storms that cause additional damage.

The combination of winds, rain, and flooding can level a coastal town and cause significant damage to cities far from the coast. In 1996, Hurricane Fran swept 150 miles (241 km) inland to hit Raleigh, North Carolina. Tens of thousands of homes were damaged or destroyed, millions of trees fell, power was out for weeks in some areas, and the total damage was measured in the billions of dollars.

How **TORNADOES** Work

Tornadoes are one of those amazing, awesome acts of nature, and they can simply leave you dumbfounded. It's difficult to imagine a huge, swirling, 200-mile per hour beast of a storm that appears to have a mind of its own. You have to actually see one with your own eyes to believe it. Yet in certain places, tornadoes appear with amazing regularity. That's why we see them in the news all the time.

HSW Web Links

www.howstuffworks.com

How Hurricanes Work
How Floods Work
How Radar Works
How Radio Works

Tornado Ratings

Tornadoes are rated on what is called the Fujita scale, named for the scale's inventor. Based on the resulting damage, within this scale there are six levels, each with a maximum wind speed:

F0—72 mph (115 kph)

F1—112 mph (180 kph)

F2—157 mph (252 kph)

F3—206 mph (331 kph)

F4—260 mph (418 kph)

F5—318 mph (512 kph)

Tornadoes and Your Bathtub

If you've ever seen a whirlpool form in your bathtub, sink, or toilet when the water is draining, you've seen the fundamentals of a tornado at work. A drain's whirlpool, also known as a *vortex,* forms because of the downdraft that the drain creates in the body of water. The downward flow of the water into the drain makes the water begin to rotate, and as the rotation speeds up the vortex forms.

Why does the water start rotating? There are lots of explanations, but here is one way to think about it. Imagine you are a particle in the water, and you are being pulled toward the suction that the drain creates. You are accelerating toward the point of suction. However, because of your previous momentum, the number of other particles getting sucked toward the point, and so on, chances are that you are going to be off to one side of the point of suction when you arrive. That deflection sets you up on a spiraling path into the point of suction.

Once the spiral has started in one direction, it tends to influence all of the other particles as they arrive. A very strong spiraling tendency is created. Eventually, there's enough spiraling energy to create a vortex.

Given that you see vortexes all the time in tubs and sinks, they're obviously a fairly common phenomenon. In a tornado, the same sort of thing happens—but with air instead of water.

Tornadoes and Thunderstorms

With a tornado, there is no drain to get the process started. Instead, there's a thunderstorm cloud. A typical thunderstorm cloud can accumulate a huge amount of energy. If the conditions are right, this energy creates a massive updraft of air into the cloud. In *supercells* (large, long-lasting thunderstorms) the updrafts are particularly strong. If they are strong enough, a vortex of air can form just like a vortex of water forms in a sink. An air vortex under a thunderstorm cloud is a tornado.

The tornado snakes down out of a thundercloud as a huge, swirling rope of air. Wind speeds in the 200 to 300 mph range (320 to 480 kph) are not uncommon. If the vortex touches ground, the speed of the whirling wind (as well as the updraft and the pressure differences) can cause tremendous damage.

The tornado follows a path that is controlled by the path of its parent thundercloud, and it will often appear to hop. The hops occur when the vortex is disturbed (possibly by an outflow of air). You have probably seen that it is easy to disturb a vortex in the tub, but that it will reform. A tornado's vortex can also be disturbed, causing it to collapse and reform along its path.

Tornadoes are an incredible and fascinating force of nature. All sorts of people spend their lives studying them. Meteorologists, other scientists, and storm chasers work every day to help us gain an even better understanding of how tornadoes work.

How **FLOODS** Work

Water is one of the most useful things on earth. We drink it, bathe in it, clean with it, and use it to cook food. Most of the time, it is beneficial or at least benign. But in large enough quantities, the same stuff you use to rinse your toothbrush can overturn cars, demolish houses, and kill scores of people and animals. Flooding has claimed millions of lives in the last hundred years alone—more than any other weather phenomenon.

Water moves around the earth in a cycle, continually changing form. When the sun heats the oceans, liquid water evaporates to form water vapor in the air. The sun heats this air so that it rises through the atmosphere and is carried along by wind currents. As this water vapor rises, it cools down again, condensing into liquid water droplets (or ice crystals), which form clouds. If a cloud cools, more water may condense onto these droplets, causing them to fall through the air as precipitation. Some precipitation collects in underground reservoirs, but most of it forms rivers and streams that flow into the oceans, bringing the water back to its starting point.

Specific regions generally experience the same sort of weather conditions year to year because wind currents are fairly consistent. But on a day-to-day basis, a huge number of factors combine in an infinite variety of ways, producing all sorts of weather.

What Causes Floods

Occasionally, weather patterns line up so that a huge volume of water lands in one area in a short period of time—such as during a hurricane. Since waterways are formed slowly over time, their size is proportionate to the amount of water that *normally* accumulates in that area. When there is suddenly a much greater volume of water than usual, the normal waterways overflow, and the

water spreads out over the surrounding land. At its most basic level, this is what a flood is—an anomalous accumulation of water in an area of land.

Flooding usually comes from a series of storms bringing massive amounts of rain. Some areas flood every year because of recurring seasonal storms (for an example of seasonal flooding—see the Nile sidebar). Another common source of flooding is unusual tidal activity caused by peculiar wind patterns. Floods may also occur when a man-made dam breaks, releasing a massive wall of water. We build dams to modify the flow of rivers to suit our own purposes. Basically, the dam collects the river water in a large reservoir so that we, rather than nature, can decide when to increase or decrease the river's flow. Engineers build dams that will stand up to any amount of water that is likely to accumulate. Occasionally, however, more water accumulates than the engineers have predicted and the dam breaks.

Rivers and Floods

While a river may appear to us to be a stable, unmovable feature of the landscape, it is really a vibrant, dynamic entity. Over time, these waterways expand, shift their path dramatically and may even change the direction of flow. For this reason, the land around the banks of a river is highly susceptible to flooding.

HSW Web Links
www.howstuffworks.com

How Hurricanes Work
How Tornadoes Work
How Lightning Works
How Earthquakes Work

That's Water Damage!

In 1966, a major storm flooded the Arno, an Italian river that runs through the small city of Florence, which is one of the art capitals of the world. Florence was overrun with water, mud, and general slime. In addition to the loss of life and the damage to buildings, there was a great deal of damage to the city's art collection. Mud and slime covered almost everything that had been stored in the city's basements and ground-level rooms. Through many years of work, scientists and art historians have been able to restore most of the damaged artifacts to good condition.

Unfortunately, rivers are also natural draws for civilization. Among other things, they provide a constant supply of water, rich soil, and an easy means for transportation. At some point, it comes time for the water to shift, and the people who have built along the flood plains quickly discover that they are living on unsound ground. If there is extensive construction in these areas, the flood damage can be devastating.

Flood Damage

In a flood, 2 feet (61 cm) of water can move with enough force to wash a car away, and 6 inches (15 cm) of water can knock you off your feet. It may seem surprising that water can pack such a wallop. After all, you can peacefully swim in the ocean without being knocked around.

Flood waters are more dangerous because they are moving. In a flood, a lot of water may collect in an area while there is hardly any water in another area. Water is fairly heavy, so it moves very quickly to find its own level. The bigger the difference between water volumes across an area, the greater the force of movement. At a particular point, the water may not look so deep, and so it doesn't seem particularly dangerous—until it's too late. Nearly half of all flood deaths result from people attempting to drive their cars through rushing water.

A less catastrophic sort of damage is dampness and mud. If the water level is high enough, loads of water seeps into houses, soaking everything. As water flows over the landscape, it picks up a lot of junk. Everything and everyone in a flood is floating along in one big soup. When the flood is over, the water level drops and everything eventually dries out, but the mud and debris stick around.

Another sort of flood damage is the spread of disease. Flood waters often allow swarms of mosquitoes to hatch and spread disease. Also, as floodwaters flow over an area, they can pick up all sorts of chemicals and waste products, leading to unsanitary conditions. If you are in a flooded area, therefore, it is very important that you drink only bottled or boiled water and observe other sanitation guidelines.

Soak It Up

The severity of a flood depends not only on how much water accumulates, but also on the lands ability to deal with this water. When it rains, soil acts as a sort of sponge. When the land is saturated—that is, when it's soaked up all the water it can hold—any more water that accumulates flows immediately as runoff.

of the most significant changes has been to cover the ground in asphalt and concrete. Obviously, these two surfaces are not the best sponges around: Almost all rainwater that accumulates on them becomes runoff. In an industrialized area without a good drainage system, it may not take much rain to cause significant flooding.

Water, Water Everywhere

To address the problems with concrete and asphalt, some cities, such as Los Angeles, have constructed concrete flood-relief channels that take runoff water out of the city. This can cause flooding wherever the channels end, of course. Again, when you cover an area in concrete and asphalt, you are essentially cutting off part of the earth's natural sponge. The rest of the sponge, therefore, has a lot more water to deal with.

A similar problem can arise with levees—large walls built along rivers to prevent flooding. These structures extend the natural banks of the river so that much more water can flow through it. But while they may be effective at keeping water out of one area, they usually make problems worse for areas further down the river, where there are no levees. Those areas receive the flood waters that would have spread out farther up river. Another danger of levees is that, like dams, they can break. When this happens, a large amount of water flows out onto the land in a short period of time. This can cause some of the most dangerous flood conditions.

We'll never be able to stop flooding. It is an unavoidable element in the complex weather system of our atmosphere. We can, however, work to minimize the damage inflicted by flooding by building flood control systems such as flood-relief channels and levees. But the best way to avoid flood damage may be to back out of flood-prone areas altogether— to get out of the way.

Water evaporates from the ocean, becomes vapor and rises, then cools and falls to the earth as precipitation.

Some materials become saturated much more quickly than others. Soil in the middle of the forest is an excellent sponge. Rock is much less absorbent, and hard clay falls somewhere in between. Generally, tilled crop soil is less absorbent than uncultivated land, so farm areas may be more likely to experience flooding than natural areas.

As civilization has expanded, human beings have altered nature's landscape in a number of ways. In the Western world, one

Flash Floods

Flash floods, the most dangerous flood variety, are caused by a sudden intense accumulation of water. Because there is a great deal of water collected in one area, flash-flood waters tend to move with a great deal of force, knocking people, cars, and even houses out of the way.

One of the worst flash floods in U.S. history occurred in 1976 in Big Thompson Canyon, Colorado. In less than five hours, thunderstorms in nearby areas dumped more rain than the region ordinarily experiences in a year. The Big Thompson River, normally a shallow, slow-moving waterway, abruptly transformed into an unstoppable torrent, dumping 233,000 gallons (882,000 L) of water into the canyon every second. Thousands of campers had gathered in the canyon to celebrate the centennial of the state of Colorado. The flood happened so quickly that there was no time to issue a warning, and 139 people were killed.

How **LIGHTNING** Works

Lightning is one of the most beautiful displays in nature. It's also one of the most deadly natural phenomena known to man. With bolt temperatures hotter than the surface of the sun and shockwaves beaming out in all directions, lightning is a humbling lesson in physical science.

HSW Web Links

www.howstuffworks.com

How Earthquakes Work
How Hurricanes Work
How Surge Protectors Work
How Van de Graaff Generators Work
How Floods Work

It's amazing that one stroke of lightning can produce anywhere from 100 million to as much as 1 billion volts of electricity. According to the National Weather Service a typical flash of lightning provides enough electricity to power a 100-watt lightbulb for three months. Beyond its power and beauty, lightning presents science with an incredible mystery: How does it work?

The Great Mystery

In an electrical storm, storm clouds are charged like giant capacitors in the sky. A *capacitor* is an electrical device that consists of two conductive surfaces separated by an insulating gap. When a voltage is applied to the surfaces, energy is stored in the resulting electric field.

In a cloud capacitor, the upper portion of the cloud is positive and the lower portion is negative. We still don't know exactly how the cloud acquires these charges, but the following process is one possible explanation.

As part of the water cycle, moisture accumulates in the atmosphere. This accumulation is what we see as a cloud. Clouds can contain trillions of water droplets and ice particles suspended in the air. As the process of evaporation and condensation continues, these droplets collide with moisture that's in the process of condensing as it rises. Also, the rising moisture may collide with ice or sleet that is in the process of falling to the earth or that's sitting in the lower portion of the cloud. As a result of these collisions, electrons are knocked off the rising moisture, creating a charge separation.

The newly knocked-off electrons gather at the lower portion of the cloud, giving it a negative charge. The rising moisture that has just lost its electrons carries a positive charge to the top of the cloud.

In addition to the collisions, freezing plays an important role. As the rising moisture encounters colder temperatures in the upper cloud regions and begins to freeze, the frozen portion becomes negatively charged and the unfrozen droplets become positively charged. At this point, rising air currents remove the positively charged droplets from the ice and carry them to the top of the cloud. The remaining frozen portion would likely fall to the lower portion of the cloud or continue to the ground.

With collisions and freezing both affecting charge, you can begin to understand how a cloud may acquire the extreme charge separation that is required for lightning.

The Electric Field

When there is a charge separation in a cloud, there's also an electric field that's associated with the separation. Like the cloud, this field is negative in the lower region and positive in the upper region.

The strength or intensity of the electric field is directly related to the amount of charge build-up in the cloud. As the collisions and freezing continue to occur and the charges at the top and bottom of the cloud increase, the electric field becomes more and more intense—so intense that the electrons at the earth's surface are repelled deeper into the earth by the strong negative charge at the lower portion of the cloud. This repulsion of electrons causes the earth's surface to acquire a strong positive charge.

All that is needed now is a conductive path for the negative cloud bottom to contact the positive earth surface. The strong electric field creates this path.

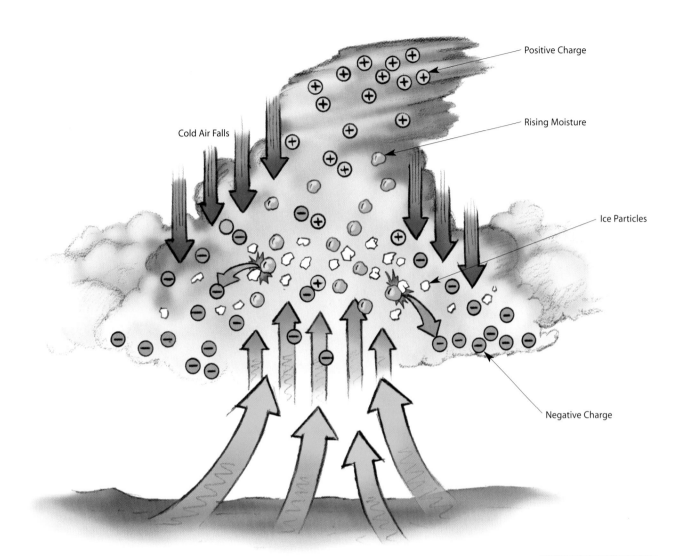

Positive Charge

Cold Air Falls

Rising Moisture

Ice Particles

Negative Charge

Warm Air Rises

Follow the Leader

The strong electric field causes the air around the cloud to "break down," allowing current to flow—in an attempt to neutralize the charge separation. The air breakdown creates a conductive path—as if there were a long metal rod connecting the cloud to the earth.

When the electric field becomes very strong (on the order of tens of thousands of volts per inch), conditions are good for the air to begin breaking down. The electric field causes the surrounding air to separate into positive ions and electrons—in other words, the air is *ionized*. Keep in mind that the ionization does not mean that there's more negative charge (electrons) or more positive charge (positive ions) than before.

This ionization only means that the electrons and positive ions are farther apart than they were in their original molecular or atomic structure. Essentially, the electrons have been stripped from the molecular structure of the non-ionized air.

The importance of this separation/stripping is that the electrons are now free to move much more easily than they could before the separation. So this ionized air, called plasma, is much more conductive than the previous non-ionized air. Plasma is the tool nature uses to neutralize charge separation in an electric field. Think of the ionization process as burning a path through the air for the lightning to follow, much like digging a tunnel through a mountain allows a train to follow.

This path isn't created instantaneously. In fact, there are usually many separate paths of ionized air stemming from the cloud. These paths are typically referred to as *step leaders*.

The step leaders move toward the earth in stages. The air may not ionize equally in all directions. Any object in the air, like dust or impurities, may cause the air to break down more easily in one direction, giving a better chance that the step leader will reach the earth faster in that direction. The step leaders may sprout other leaders in areas where the original leaders bend or turn. Once started, the leader will remain until the current flows, regardless of whether or not it's the leader that reaches the ground first. The leader basically has two possibilities: continue to grow in stages of growing plasma, or wait patiently in its present plasma condition until another leader hits a target.

The leader that reaches the earth first provides a conductive path between the cloud and the earth. This leader is not the lightning strike; it only maps out the course that the strike will follow. The strike is the sudden, massive, electrical current flow from the cloud to the ground.

Lightning with step leaders.

Reaching for the Sky

As the step leaders approach the earth, objects on the surface begin responding to the strong electric field. The objects reach out to the cloud by "growing" positive streamers. Anything on the surface of the earth has the potential to send a streamer—even people. Once produced, the streamers do not continue to grow all the way up to the clouds; bridging the gap is the job of the step leaders as they stage their way down. The streamers wait patiently, stretching upward as the step leaders approach.

After the step leader and the streamer meet, the ionized air (plasma) has completed its journey to the earth, leaving a conductive path from the cloud to the earth. With this path complete, current flows between the earth and the cloud. This discharge of current is nature's way of trying to neutralize the charge separation. The flash we see when this discharge occurs is not the strike—it is the local effects of the strike.

Now you know the mechanics of a lightning strike. It's amazing to realize that all of the activity from the time the ionization begins to the time of the strike occurs in a fraction of a second. High-speed cameras have actually caught the positive streamers on film.

Multiple Strikes

You're sitting in your car and see a flash from a lightning strike. You notice that there are many other branches that flashed at the same time as the main strike. Next you notice that the main strike flickers or dims a few more times. The branches that you saw were actually the step leaders that were connected to the leader that made it to its target.

When the first strike occurs, current flows in an attempt to neutralize the charge separation. The current associated with the energy in the other step leaders must flow to the ground. The electrons in the other step leaders, being free to move, flow through the leader to the strike path. So when the strike occurs, the other step leaders provide current and exhibit the same

heat flash characteristics of the actual strike path. After the original stroke occurs, it's usually followed by a series of secondary strikes. These strikes follow only the path of the main strike; the other step leaders do not participate in this discharge. It's very possible that the main strike can be followed by 30 to 40 secondary strikes.

How **EARTHQUAKES** Work

An earthquake is one of the most terrifying spectacles that nature can dish up. We generally think of the ground we stand on as rock solid and completely stable. An earthquake can shatter that perception instantly.

An earthquake is a vibration that travels through the earth's crust. All kinds of things cause earthquakes: volcanic eruptions, meteor impacts, and underground nuclear tests, for example. Technically, a large truck that rumbles down the street is causing a mini-earthquake, but we tend to think of earthquakes as events that affect a fairly large area, such as an entire city.

The majority of natural earthquakes come from *plate tectonics*—the movement of the huge plates that make up the surface layer of the earth. Think of the plates as tightly packed cracker pieces floating in a bowl of soup. As they move around, they push up against each other, move away from each other, and grind past each other.

Wherever plates meet, you'll find faults—breaks in the earth's crust where the blocks of rock on each side are moving in different directions. Big faults form along boundary lines between different plates. There are four kinds of faults:

- At a *divergent plate boundary*, two plates move away from each other, forming a **normal fault**.
- At a *convergent plate boundary*, plates move against each other, forming a **reverse fault** or a **thrust fault**.

- At a *transform boundary*, plates slide past each other, forming a **strike slip fault**.

In addition, the force of plate movement creates smaller faults within the plates, along the edges. The same thing happens if you twist a sheet of plastic—the force applied to the sheet as a whole causes the plastic to crack along the edges.

Snap Out of It

In all of these types of faults, the different blocks of rock push very tightly together, creating a good deal of friction as they move. If this friction level is high enough, the two blocks become locked—the friction keeps them from sliding against each other. When this happens, the forces in the plates continue to push the rock, increasing the pressure applied at the fault.

If the pressure increases to a high enough level, then it will overcome the force of the friction and the blocks will suddenly snap forward. To put it another way, as the tectonic forces push or pull on the locked blocks, potential energy builds. When the plates finally move, this built-up energy becomes kinetic. Some fault shifts create visible changes at the earth's surface, but other shifts occur in rock well under the surface and do not create a surface rupture.

HSW Web Links

www.howstuffworks.com

How Volcanoes Work
How Skyscrapers Work
How Smart Structures Will
  Work
How Mars Works
How Hurricanes Work
How Nuclear Bombs Work

Liquefaction

Earthquakes can inflict major damage through *liquefaction*. In the right conditions, the violent shaking from an earthquake will make loosely packed sediments and soil behave like a liquid. When a building or house is built on this type of sediment, liquefaction will cause the structure to collapse more easily. Highly developed areas built on loose ground can suffer severe damage from even a relatively mild earthquake. Liquefaction can also cause severe mudslides.

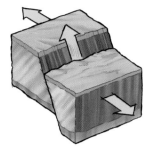

Normal fault line.

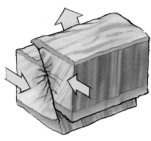

Reverse fault line.

Thrust fault line.

Strike slip fault line.

89

Damage

Earthquakes have caused a great deal of property damage over the years, and they have claimed many lives. In the last hundred years alone, there have been more than 1.5 million earthquake-related fatalities. Usually, it's not the shaking ground itself that claims lives, it's the associated destruction of buildings and the creation of other natural disasters, like tsunamis, avalanches, and landslides.

The initial break that creates a fault, combined with these sudden, intense shifts along already formed faults, are the main sources of earthquakes. Most earthquakes occur around plate boundaries, because this is where the strain from the plate movements is most intense. In a fault zone, the release of kinetic energy at one fault may increase the stress—the potential energy—in a nearby fault, leading to other earthquakes. This is one of the reasons that several earthquakes may occur in an area in a short period of time.

Making Waves

When a sudden break or shift occurs in the earth's crust, the energy radiates out as seismic waves, in the same way that a rock falling in a pond creates water waves.

Earthquakes form several types of seismic waves. *Body waves* move through the inner part of the earth, while surface waves travel over the surface of the earth. *Surface waves,* sometimes known as long waves or L waves, are responsible for most of the damage associated with earthquakes because they cause the most intense vibrations. Surface waves come from body waves that reach the surface.

There are two main types of body waves—*primary waves* and *secondary waves.* Primary waves, also called compressional waves or P waves, travel about 1 to 5 miles per second, depending on the material they're moving through. This speed is greater than the speed of other waves, so P waves arrive first at any surface location. They can travel through solids, liquids, and gas, so they will pass completely through the body of the earth. As they travel through rock, the waves move tiny rock particles back and forth—pushing them apart and then back together—in line with the direction the wave is traveling. These waves typically arrive at the surface as an abrupt thud.

Secondary waves, also called shear waves or S waves, lag a little behind the P waves. As these waves move, they displace rock particles outward, pushing them perpendicular to the path of the waves. S waves create the first period of rolling associated with earthquakes. Unlike P waves, S waves don't move straight through the earth. They only travel through solid material, and so they're stopped at the liquid layer in the earth's core.

Both sorts of body waves do travel around the earth, however, and they can be detected on the opposite side of the planet. At any given moment, there are a number of very faint seismic waves moving all around the globe.

Strong surface waves are something like the waves in a body of water—they move the surface of the earth up and down enough for people to feel them. They rock the foundations of man-made structures. L waves are the slowest moving of all waves, so the most intense shaking usually comes at the end of an earthquake.

Pinpointing the Origin

P waves generally travel 1.7 times faster than S waves. Using this ratio, scientists can calculate the distance between any point on the earth's surface and the earthquake's *focus,* the point where it originated. A seismograph registers the vibration of both waves. The elapsed time between both waves tells you the distance to the earthquake.

If you gather this information from three or more points, you can figure out the location of the focus through the process of *trilateration.* Basically, you draw an imaginary sphere around each seismograph location, with the point of measurement as the center and the measured distance (let's call it X) from that point to the focus as the radius. The surface of the sphere describes all the points that are X miles away from the seismograph. The focus, then, must be somewhere along this sphere. If you come up with two spheres, based on evidence from two different seismographs, you'll get a

Did You Know?

We only hear about earthquakes in the news every once in a while, but they are actually an everyday occurrence on our planet. According to the United States Geological Survey, more than three million earthquakes occur every year. That's about 8,000 a day, or one every 11 seconds! The vast majority of these three million quakes are extremely weak. Also, a good number of the stronger quakes happen in uninhabited places where no one feels them. Only the big quakes in highly populated areas get our attention.

90

two-dimensional circle where they meet. Since the focus must be along the surface of both spheres, all of the possible focus points are located on the circle formed by the intersection of these two spheres. A third sphere will intersect only twice with this circle, giving you two possible focus points. And because the center of each sphere is on the earth's surface, one of these possible points will be in the air, leaving only one logical location for the focus.

Rating Magnitude and Intensity

Scientists rate earthquake magnitude—the waves' energy—on the Richter scale. The Richter scale is logarithmic: Whole-number jumps indicate a tenfold increase in wave amplitude. That is, the wave amplitude of a level 6 earthquake is 10 times greater than that of a level 5 earthquake. The amount of energy released increases 31.7 times between whole number values.

The majority of earthquakes register less than 3 on the Richter scale. These tremors, which aren't usually felt by humans, are called *microquakes*. Generally, you won't see any damage from earthquakes that rate below 4 on the Richter scale. Major earthquakes generally register at 7 or above—the strongest on record was a 9.5. This intense earthquake occurred in Chile in 1960.

An earthquake's destructive power varies depending on the type of soil in an area and the design and placement of buildings. The extent of damage is rated on the Mercalli scale. Mercalli ratings, given as Roman numerals, are subjective interpretations. A low intensity earthquake, one in which only some people feel the vibration and there is no significant property damage, is rated as a II. The highest rating, XII, is applied only to earthquakes in which structures are destroyed, the ground is cracked, and other natural disasters, such as landslides or tsunamis (gigantic seismic sea waves), are initiated.

Predicting Earthquakes

Based on the movement of the plates in the earth and the location of fault zones, scientists can say where major earthquakes are likely to occur. They can also guess when they might occur in a certain area by looking at the history of earthquakes in the region and detecting where pressure is building along fault lines. These predictions are extremely vague, however—typically on the order of decades. We can't yet predict precisely when earthquakes will hit.

Scientists have had some success predicting *aftershocks*—the quakes that follow an initial earthquake. These predictions are based on extensive research of aftershock patterns. Seismologists can make a good guess of how an earthquake originating along one fault will cause additional earthquakes in connected faults.

Another area of study is the relationship between magnetic and electrical charges in rock material and earthquakes. Some scientists have hypothesized that these electromagnetic fields change in a certain way just before an earthquake. Seismologists are also studying gas seepage and the tilting of the ground as warning signs of earthquakes.

Dealing with Earthquakes

The major safety advances over the past 50 years have been in preparedness—particularly in the field of construction engineering. In 1973, the Uniform Building Code, an international set of standards for building construction, added specifications to fortify buildings against the force of seismic waves. The changes included strengthening support material and making buildings flexible enough to absorb vibrations without falling or deteriorating.

It will be a long time, if ever, before we'll be ready for every substantial earthquake that might occur. Just like severe weather and disease, earthquakes are an unavoidable force generated by the powerful natural processes that shape our planet. All we can do is increase our understanding of the phenomenon and develop better ways to deal with it.

Blow to the Midsection

Every now and then, earthquakes do occur in the middle of plates instead of at the boundaries. In 1811 and 1812, an extremely powerful series of quakes shook the middle of the North American continental plate. The earthquakes originated in Missouri, but shook several surrounding states as well. The vibrations of one earthquake were so powerful that they actually rang church bells as far away as Boston!

In the 1970s, scientists found the likely source of the earthquake series: a 600-million-year-old fault zone buried under many layers of rock.

How VOLCANOES Work

*Whenever there is a major volcanic eruption in the world, you'll see a slew of newspaper articles and nightly news stories covering the catastrophe, all stressing a familiar set of words—*violent, raging, *and* breathtaking. *We're in awe of the destructive power of nature. It's certainly unsettling to realize that a peaceful mountain can suddenly become an unstoppable destructive force!*

HSW Web Links
www.howstuffworks.com

How Mars Works
How Fireworks Work
How Hurricanes Work
How Tornadoes Work
How the Sun Works

When people think of volcanoes, the first image that comes to mind is probably a tall, conical mountain with orange lava spewing out the top. Although there are a lot of volcanoes like this, the term *volcano* actually describes a much wider range of geological phenomena. Generally speaking, two things must happen for a volcano to form:

- Huge chunks of the earth move.
- Molten rock explodes from beneath the earth's surface.

Volcano cloud.

Laying the Groundwork

Molten rock, called *magma*, is a partially liquid, partially solid and partially gaseous material that comes from the earth's mantle layer.

The mantle is the largest of the three mega-layers that make up the earth. The second largest layer is the solid core at the center of the planet. We all live on the smallest mega-layer, the rigid outer crust above the mantle. The crust is 3 to 6 miles thick under the oceans and 20 to 44 miles thick under the land. This may seem huge to us, but compared to the rest of the planet, it's very thin—like the outer skin on an apple.

The mantle is extremely hot, but for the most part it stays in solid form because the pressure deep inside the planet is so great that the material can't melt. In certain circumstances, however, the mantle material does melt, forming magma that makes its way through the outer crust.

Moving Plates

The theory of plate tectonics holds that the lithosphere, a layer of rigid material composed of the outer crust and the very top of the mantle, is divided into seven large plates and several smaller plates. These plates drift very slowly over the mantle below. The activity at the boundary between the plates is the primary catalyst for magma production.

If the two plates are moving away from each other, an ocean ridge or continental ridge forms, depending on whether the plates meet under the ocean or on land. As the two plates separate, the mantle rock flows up into the empty space between the separated plates. Because the pressure is not as great at this level, the mantle rock melts, forming magma. As the magma flows out, it cools, hardening to form new crust in the expanding gap. This sort of magma production is called *spreading center volcanism.*

At the point where two plates collide, one plate may push under the other plate, so that it sinks into the mantle. This process, called *subduction,* typically forms a very deep ditch, usually in the ocean floor. As the rigid lithosphere pushes down into the hot, high-pressure mantle, it heats up.

The heat and pressure at this depth forces water out of the plate and into the mantle layer above. The increased water content lowers the melting point of the mantle rock in this wedge, causing it to melt, forming magma. This is called *subduction zone volcanism.*

Magma can also push up under the middle of a lithosphere plate, though this is

Did You Know?

Volcanoes vary a great deal in their destructive power. Some volcanoes explode violently, destroying everything nearby within seconds, while other volcanoes seep out lava so slowly that you can safely walk all around them. The severity of the eruption depends mostly on the composition of the magma.

much less common. This *interplate volcanic activity* is caused by unusually hot mantle material forming in the lower mantle and pushing up into the upper mantle. The mantle material wells up to create a hot spot under a particular point on the earth. Because of the unusual heat of this mantle material, it melts, forming magma just under the earth's crust. The hot spot itself is stationary, but as a continental plate moves over the spot, the magma will create a string of volcanoes, which die out once they move past the hot spot. The Hawaii volcanoes were created by such a hot spot, dating back 70 million years.

land volcanoes are produced by subduction zone volcanism and hot spot volcanism.

In both cases, when the solid rock changes form to a more liquid rock material, it becomes less dense than the surrounding solid rock. Because of this difference in density, the magma pushes upward with great force (for the

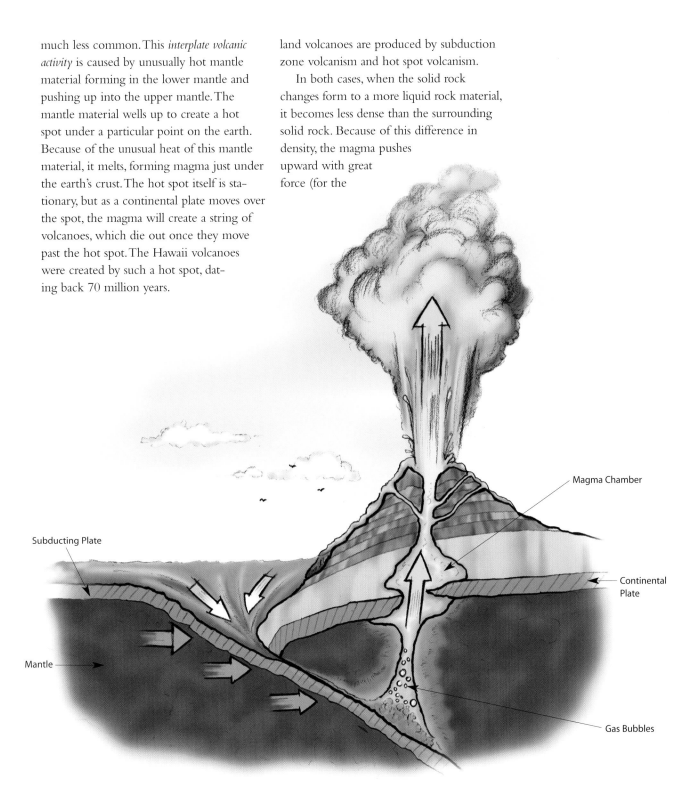

Magma Chamber

Continental Plate

Subducting Plate

Mantle

Gas Bubbles

Hot Rock

What happens to the magma formed by these processes? The magma produced at ocean ridges just hardens to form new crust material, and so it doesn't produce spewing land volcanoes. There are a few continental ridge areas around the world where the magma does spew out onto land. But most

same reason the helium in a balloon pushes up through the denser surrounding air). As the magma pushes up, its intense heat melts some more rock, adding to the molten mixture.

Unless the downward pressure of the surrounding solid rock exceeds the magma's upward pressure, the magma keeps moving

Lava burst.

Flowing lava.

through the crust. The magma collects in chambers below the surface of the earth. If the magma pressure rises to a high enough level or a crack opens up in the crust, the molten rock will spew out at the earth's surface. When this happens, the flowing magma (called *lava* when it reaches the surface) forms a volcano. The structure of the volcano and the intensity of the volcanic eruption are dependent on a number of factors, primarily the composition of the magma.

The material that forms magma contains dissolved gases—gases that have been suspended in the magma solution.

The magma is filled with tiny gas bubbles, which have a much lower density than the surrounding magma and so push out to escape. This is the same thing that happens when you open a bottle of soda, particularly after shaking it up. When you decompress the soda (by opening the bottle), the tiny gas bubbles rush out. If you shake the bottle up first, the bubbles are all mixed up in the soda so they push a lot of the soda out with them. This is true for volcanoes as well. As the bubbles escape, they push the magma out, causing a spewing eruption.

Flow or Fury

The nature of this eruption depends mainly on the gas content and the viscosity of the magma material. Viscosity is the ability to resist flow—it is the opposite of fluidity. If the magma has a high viscosity, meaning it resists flow very well, the gas bubbles will have a hard time escaping from the magma and so will push more material up, causing a bigger eruption. If the magma has a lower viscosity, the gas bubbles will be able to escape from the magma more easily, so the lava won't erupt as violently.

If the magma contains more gas bubbles, it will erupt more violently, and if it contains less gas, it will erupt more calmly. Both gas content and viscosity are determined by the composition of the magma.

If the viscosity and the gas pressure are low enough, lava will flow slowly onto the earth's surface with minimal explosion. While these effusive lava flows can cause considerable damage to wildlife and man-made structures, they are not particularly dangerous to people. They move so slowly that everybody has plenty of time to get out of the way.

If there is a good deal of pressure, however, a volcano will begin its eruption with an explosive launch of material into the air. Typically, this eruption column is composed of hot gas, ash, and pyroclastic rocks—volcanic material in solid form. There are many sorts of explosive eruptions, varying significantly in intensity and duration.

As devastating as it is, volcanic activity is one of the most important, constructive geological processes on earth. After all, volcanoes are constantly rebuilding the ocean floor. As with most natural forces, volcanoes have a dual nature. They can wreak horrible destruction, but they are also a crucial element of the earth's ongoing regeneration.

All Shapes and Sizes

Most land volcanoes have the same basic structure, but volcano shape and size vary considerably. Volcanoes typically have:

- **A summit crater**—The mouth of the volcano, where the lava exists

- **A magma chamber**—Where the lava wells up underground

- **A central vent**—A channel that leads from the magma chamber to the summit crater

The biggest variation in volcano structure is the *edifice*, the structure surrounding the central vent. Volcanic material builds up the edifice as it spews out of the volcano. The nature of the eruption determines the edifice's composition, shape, and structure.

There are three major types of volcano structures, based on the volcano's edifice:

- **Stratovolcanoes**—Created by relatively violent eruptions, stratovolcanoes have steep, symmetrical mountain edifices around a relatively small summit crater.

- **Scoria cone volcanoes**—These volcanoes are created by medium-size eruptions that form steep edifices leading up to a very wide summit crater.

- **Shield volcanoes**—The shield volcano comes from effusive lava flows that form relatively short edifices spreading out for many miles.

How **WILDFIRES** Work

In just seconds, a spark or even the sun's heat alone can set off an inferno in a dry forest. The wildfire quickly spreads, consuming the thick, dried-out vegetation and almost everything else in its path. What was once a forest becomes a virtual powder keg of untapped fuel. In a seemingly instantaneous burst, the wildfire overtakes thousands of acres of surrounding land, threatening the homes and lives of many in the vicinity.

An average of 5 million acres burn every year in the United States, causing millions of dollars in damage. Once a fire begins, it can spread at a rate of up to 14 miles per hour (23 kph), consuming everything in its path. As a fire spreads over brush and trees, it may take on a life of its own—finding ways to keep itself alive, even spawning smaller fires by throwing embers miles away.

Prescribed Fires

Three things are needed for ignition and combustion to occur. A fire requires *fuel* to burn, *air* to supply oxygen, and a *heat* source to bring the fuel up to ignition temperature. Heat, oxygen, and fuel form the fire triangle. Firefighters often talk about the fire triangle when they are trying to put out a blaze. The idea is that if they can take away any one of the pillars of the triangle, they can control and ultimately extinguish the fire.

Many materials have a temperature at which they will burst into flames. This temperature is called a material's *flash point*. Wood's flash point is 572°F (300°C). When wood is heated to this temperature, it releases hydrocarbon gases that mix with oxygen in the air. The gases ignite and create fire.

After combustion occurs and a fire starts to burn, there are three factors that determine how the fire spreads: fuel, weather, and topography. Depending on these factors, a fire can quickly fizzle or turn into a raging blaze that scorches thousands of acres.

Fueling the Flames

Wildfires spread based on the type and quantity of fuel that surrounds them. Fuel can include everything from trees, underbrush, and dry grassy fields to homes. The amount of flammable material that surrounds a fire is referred to as the *fuel load*.

Fuel load is measured by the amount of available fuel per unit area, usually tons per acre. A small fuel load will cause a fire to burn and spread slowly, with a low intensity. If a lot of fuel is available, the fire will burn more intensely, causing it to spread faster. The faster it heats up the material around it, the faster those materials can ignite. The dryness of the fuel can also affect the behavior of the fire. When the fuel is very dry, it is consumed much faster and creates a fire that is much more difficult to contain.

Size, shape, arrangement, and moisture content are the basic fuel characteristics that decide how the fuel will affect a fire.

Flashy fuels, small fuel materials such as dry grass, pine needles, dry leaves, twigs, and other dead brush, burn faster than large logs or stumps (this is why you start a fire with kindling rather than logs). On a chemical level, different fuel materials take longer to ignite than others. But in a wildfire, where most of the fuel is made of the same sort of material, the main variable in ignition time is the ratio of the fuel's total surface area to its volume. Since a twig's surface area is not much larger than its volume, it ignites quickly. By comparison, a tree's surface area is much smaller than its volume, so it needs more time to heat up before it ignites.

As the fire progresses, it dries out the material just beyond it—the heat and smoke that approaches the potential fuel cause the fuel's moisture to evaporate. This makes the fuel easier to ignite when the fire finally reaches it. Fuels that are somewhat spaced out will dry faster than fuels that are packed tightly

HSW Web Links

www.howstuffworks.com

How Volcanoes Work
How Lightning Works
How the Sun Works
How Smoke Detectors
 Work

Did You Know?

While we often look at wildfires as being destructive, many wildfires are actually beneficial. Some wildfires burn the underbrush of a forest, which can prevent a larger fire that might result if the brush were allowed to accumulate for a long time. Wildfires can also benefit plant growth by reducing the spread of disease, releasing nutrients from burned plants into the ground, and encouraging new growth.

together, because more oxygen is available to the thinned-out fuel. More tightly-packed fuels, however, may retain more moisture, so they will be slower to ignite.

Temperature, Wind, and Rain

Weather plays a major role in the birth, growth, and death of a wildfire. Three elements of the weather affect wildfires: temperature, wind, and moisture. Drought and heat lead to favorable conditions for wildfires, and winds aid a wildfire's progress. Weather can spur the fire to move faster and engulf more land. It can also make the job of fighting the fire even more difficult.

Temperature

Temperature has a direct effect on the sparking of wildfires, because heat is one of the three pillars of the fire triangle. The sticks, trees, and underbrush on the ground receive radiant heat from the sun, which heats and dries potential fuels. Warmer temperatures allow for fuels to ignite and burn faster, adding to the rate at which a wildfire spreads. For this reason, wildfires tend to rage in the afternoon, when temperatures are at their hottest.

Wind

Wind probably has the biggest impact on a wildfire's behavior. It is also the most unpredictable factor. Winds supply the fire with additional oxygen, increase the drying of potential fuel, and push the fire across the land at a faster rate.

Dr. Terry Clark, senior scientist at the National Center for Atmospheric Research, has found that not only does wind affect how the fire develops, but that fires themselves can develop wind patterns. When the fire creates its own wind patterns, these patterns can feed back into the fire and control how it spreads. Large, violent wildfires can generate winds called *fire whirls*. Fire whirls, which are like tornadoes, result from the vortices created by the fire's heat. When these vortices are tilted from horizontal to vertical, you get fire whirls. Fire whirls have been known to hurl flaming logs and burning debris over considerable distances.

The stronger the wind blows, the faster the fire spreads. The fire generates winds of its own that are as much as 10 times faster than the ambient wind. It can even throw embers into the air and create additional fires, an occurrence called *spotting*. Wind can also change the direction of the fire, and gusts can raise a ground fire into the tops of trees, creating a crown fire.

Moisture

While wind can help the fire to spread, moisture works against the fire. Moisture in the form of humidity and precipitation can slow the fire down and reduce its intensity. Potential fuel can be hard to ignite if it has high levels of moisture, because the moisture absorbs the fire's heat. The higher the initial humidity, the less likely the fuel is to dry and ignite.

Because moisture can lower the chances of a wildfire igniting, precipitation has a direct impact on fire prevention. Rain and other precipitation raise the amount of moisture in fuels, which keeps potential wildfires from breaking out.

Fire on the Mountain

The third big influence on wildfire behavior, after the available fuel and the weather, is the lay of the land, or *topography*. Although it remains virtually unchanged over time, unlike fuel and weather, topography can either aid or hinder wildfire progression. The most important factor in topography is slope.

Unlike humans, fires usually travel uphill much faster than downhill. The steeper the slope, the faster the fire travels. Fires travel in the direction of the ambient wind, which usually flows uphill. The fire is also able to preheat the fuel farther up the hill because the smoke and heat are rising in that direction. Conversely, once the fire has reached the top of a hill, it must struggle to come back down because it's not able to preheat the downhill fuel as well as it can the uphill.

In addition to the damage that fires cause as they burn, they can also leave behind disastrous problems, the effects of which might not be felt until months after the fire burns out. When fire destroys all the vegetation on a hill or mountain, it can also destroy the organic material in the soil and prevent water from penetrating. One problem that results from this is intense erosion.

Debris Flow

A July 1994 wildfire that burned about 2,000 acres of forest and underbrush on the steep slopes of Storm King Mountain, near Glenwood Springs, Colorado, caused a big problem later on. Two months after the fire, heavy rains caused erosion that poured tons of mud, rock, and other debris onto a 3-mile stretch of Interstate 70, according to the United States Geological Survey. The debris flow engulfed 30 cars and swept two into the Colorado River.

How **RIP CURRENTS** Work

Rip currents are responsible for about 150 deaths every year in the United States. In Florida, they kill more people annually than thunderstorms, hurricanes, and tornadoes combined. They are the number-one concern for beach lifeguards: About 80% of all beach rescues are related to rip currents.

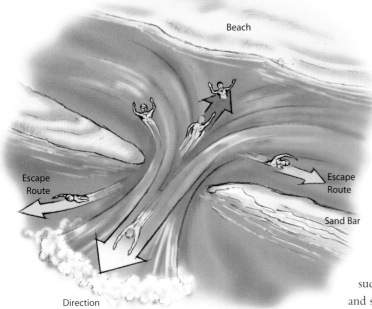

Beach

Escape Route

Escape Route

Sand Bar

Direction of Rip

HSW Web Links

www.howstuffworks.com

How Floods Work
How Barrier Islands Work
How SCUBA Works
How Sharks Work

Despite these startling statistics, many swimmers don't know anything about rip currents, and they have no idea how to survive when caught in one.

In the Flow

A rip current is a narrow, powerful current of water running perpendicular to the beach, out into the ocean. These currents may extend 200 to 2,500 feet (61 to 762 meters) lengthwise, but they are typically less than 30 feet (9 meters) wide. Currents can move 5 miles per hour (8 kph) or faster, carrying swimmers out to sea in a matter of seconds.

Rip currents occur when the ocean's receding flow becomes concentrated in a particular area at a particular time. The most common cause is a break in a sandbar, a long hill of accumulated sand in the ocean.

A sandbar can produce a basin along the shore. Waves move up against the sandbar with enough force to push water into the basin, but the receding water has a hard time making it back over the sandbar to return to sea.

The backward pressure of the receding water may be strong enough to bust a hole in the sandbar. Alternatively, the water might flow along the shore until it reaches a low point on the sandbar. In either case, the water in the basin rushes out to sea once it finds an opening, just as the water in your bathtub rushes out when you unplug the drain.

The resulting current sucks in water from the basin and spits it out on the other side of the sandbar.

Surviving a Rip Current

Rip currents are often invisible, and the sand bar that creates them can be completely submerged. Depending on the severity, though, you may be able to see a rip current from the beach. Strong rip currents disrupt incoming waves and stir up sand from the ocean floor. When you're at the beach, keep an eye out for narrow, muddy streaks in the ocean where there aren't any waves breaking.

People drown when they thrash about in the water or expend all of their energy swimming. To survive a rip current, or any crisis in the water, you have to keep calm, and you have to conserve your energy. If you get caught up in a rip current, you should swim sideways, parallel to the beach. This will get you out of the narrow outward current, so you can swim back in with waves helping you along. If it's too hard to swim sideways while you're being dragged through the water, wait until the current carries you past the sandbar. The water will be much calmer there, and you can get clear of the rip current before heading back in.

Beware of the Undertoad!

Rip currents are often called *riptides,* but this is a misnomer. Tides are the rising and falling of water levels in the ocean. They are primarily caused by the moon's gravitational pull, and they change gradually and predictably every day. Rip currents are caused by the shape of the shoreline itself, and they may be sudden and unexpected.

Rip currents may also be referred to as *undertow,* which is just as inaccurate. Undertow describes a current of water that pulls you down to the ocean bottom. Rip currents move along the surface of the water, pulling you straight out into the ocean, but not underneath the water's surface. A rip current may knock you off your feet in shallow water, however, and if you thrash around and get disoriented, you may end up being pulled along the ocean bottom. But if you relax your body, the current should keep you near the surface.

How **RAINFORESTS** Work

Tropical rainforests are the most diverse ecosystems on earth, and are also the oldest. Today, tropical rainforests cover only 6% of the earth's ground surface, but they are home to over half of the planet's plant and animal species. In this completely unique world, there are thousands of species we haven't even discovered yet.

HSW Web Links

www.howstuffworks.com

How the Nature
 Conservancy Works
How Sharks Work
How Composting Works
How Safaris Work
How Venus Flytraps Work
How Hurricanes Work

A rainforest is an ecosystem that receives high amounts of rainfall and is dominated by tall trees. When people talk about rainforests, they usually mean the tropical rainforests located near the equator. These forests receive nearly the same amount of sunlight, and therefore heat, all year. Consequently, the weather in these regions remains fairly constant.

Rainforests receive between 160 and 400 inches (406 to 1016 centimeters) of rain per year, and receive roughly twelve hours of sunlight every day. With abundant sun and rain, trees can grow 60 to 150 feet (18 to 46 meters) in the air and can live for thousands of years.

The trees' top branches spread out to capture maximum sunlight, creating a thick canopy level at the top of the forest. Some large trees, called *emergents,* tower as much as a hundred feet above the canopy layer.

Under the canopy, there is less sunlight, and therefore less greenery. The spongy forest floor is made up of moss, fungi, and decaying plant matter.

Climbing to the Top

The vast canopy of the rainforest makes for a fascinating ecosystem. New tree seedlings rarely survive to make it to the top unless an older tree dies, creating a "hole" in the canopy. When this happens, all of the seedlings compete to reach the sunlight.

Other plants climb the trees up to the canopy. It's much easier to ascend this way, because the plant doesn't have to form its own supporting structure. At the top of the forest, these climbers may spread from tree to tree, thickening the canopy ceiling.

Some plant species, called *epiphytes,* grow directly on the surface of the giant trees. These plants make up much of the rainforest understory, the layer right below the canopy. Epiphytes are close enough to the top to receive adequate light, and the runoff from the canopy layer provides all the water and nutrients they need.

Strangler epiphytes grow long, thick roots that extend down the tree trunk into the ground. As they continue to grow, the roots form a web all around the tree. At the same time, the strangler's branches extend upward into the canopy. Eventually, the strangler blocks so much light from above and absorbs such a high percentage of nutrients from below that the host tree dies. The sturdy root structure keeps the strangler up as the host decomposes.

Competition over nutrients is almost as intense as competition for light. To gather more nutrients, tree roots grow outward, rather than downward to lower levels. Some trees compensate for this instability by growing natural buttresses—extra tree trunks that extend from the side of the tree to the ground.

All Creatures Great and Small

Most animal species in the world live in the rainforest, and many species in other ecosystems came from the rainforests originally. Researchers estimate that in a large rainforest area, there may be more than 10 million different animal species.

Most of these species have adapted for life in the upper levels of the rainforest, where food is most plentiful. Insects, which can easily climb or fly from tree to tree,

Did You Know?

One of the most remarkable things about rainforest plant life is its diversity. The temperate rainforests of the Pacific Northwest boast a dozen or so tree species. A tropical rainforest, on the other hand, might have 300 distinct tree species. This plant life is spread out over wide areas— in a square acre, an entire species might be represented by just a few individual plants.

make up the largest group. As they travel, insects pick up plant seeds, dropping them some distance away. In this way, insects help to disperse the population of the plant species over a larger area.

The numerous birds of the rainforest also play a major part in seed dispersal. When they eat fruit from a plant, the seeds pass through their digestive system. By the time they excrete the seeds, the birds may have flown many miles away from the fruit-bearing tree. Rainforest bird species come in all shapes and sizes, from tiny hummingbirds to large toucans. They make up over 25% of all bird species in the world today.

Many reptiles and mammals in the rainforests have special adaptations for life in the trees. Some animals have thin webs of arm skin that let them glide from branch to branch. There are gliding lizards, gliding frogs, flying and gliding squirrels, and even gliding snakes. Many mammals, including a wide variety of monkeys, have prehensile tails to grasp hold of tree branches.

Rainforests also have a lot of life on the forest floor. You can find gorillas, wild pigs, big cats, and even elephants in rainforests. People live in the rainforests as well, but their numbers are dwindling. Deforestation is pushing out indigenous tribes at an alarming rate.

Deforestation

Today, people destroy roughly 1.5 acres of rainforest every second to make room for crops and livestock and to gather lumber.

In the current economy, people have a need for all of these resources. But most ecological scientists agree that, over time, the destruction of the rainforests will harm us.

Preservationists note that rainforest land is not particularly suited for crops and livestock. After people clear the forest, there isn't any decomposing plant life to provide nutrients, and the soil becomes too infertile to grow anything. Typically, farmers can only use the land for a year or two.

Cutting large sections of rainforest may be a good source of lumber right now, but it actually diminishes the world's lumber supply. Experts say we could maintain a self-replenishing supply of lumber if we harvested rainforests only on a small scale.

Deforestation might also damage human health. Rainforests are often called "the world's pharmacy," because their diverse plant and animal populations make up a vast collection of potential medicines. More than 25% of the medicines we use today come from plants originating in rainforests, and these plants make up only a tiny fraction of the total collection of rainforest species. Scientists have examined fewer than 1% of rainforest plants for medicinal properties. Our best shot at curing cancer, AIDS, and many other diseases could very well lie in the diminishing rainforests.

The world's rainforests are the main cradle of life on earth, and they hold millions of unique life forms that we have yet to discover. Destroying the rainforests is comparable to destroying an unknown planet—we have no idea what we're losing. If deforestation continues at its current rate, the world's tropical rainforests will be wiped out within 40 years.

Forest Food

Roughly 80% of the food we eat originally came from tropical rainforests. Without rainforests, we wouldn't have the seeds that produce coffee and chocolate. Other rainforest foods include tomatoes, potatoes, rice, bananas, cinnamon, sugar cane, coconuts, oranges, figs, avocados, grapefruit, black pepper, pineapples, and corn.

There are over 3,000 fruits found in rainforests. People in the Western world make use of only about 200 of them, but the indigenous tribes of the rainforest make use of over 2,000.

Emergent Layer

Canopy Layer

Understory Layer

Immature Layer

How **ANIMAL CAMOUFLAGE** Works

In nature, every advantage increases an animal's chances of survival, and therefore its chances of reproducing. This simple fact causes animal species to evolve special adaptations that help them find food and keep them from becoming food. One of the most widespread and varied adaptations is natural camouflage—an animal's ability to hide itself from predator and prey.

HSW Web Links

www.howstuffworks.com

How Safaris Work
How Venus Flytraps Work
How Sharks Work
How Rainforests Work
How Light Works

And Another Thing...

Disruptive camouflage patterns, such as stripes, are particularly effective when animals in a species are grouped together. To a lion, a herd of zebras doesn't look like a whole bunch of individual animals, but more like a big, striped mass. The vertical stripes all seem to run together, making it hard for a lion to stalk and attack one specific zebra. The stripes may also help a single zebra hide in areas of tall grass. Because lions are colorblind, it doesn't matter that the zebra and surrounding environment are completely different colors.

Many fish species are similarly camouflaged. Their vertical stripes may be brightly colored, but when they swim in large schools, their stripes all meld together. This confusing spectacle gives predators the impression of one big, swimming blob.

The simplest camouflage technique is for an animal to match its surroundings. In this case, the elements of the natural habitat are the model for the camouflage. For example, a huge number of land mammals are a brownish, earth-tone color that matches the brown of the trees and soil at the forest ground level. Sharks, dolphins, and many other sea creatures have a grayish-blue coloring, which helps them blend in with the soft light underwater.

Coloration comes from natural chemical pigments in the body. These pigments absorb some colors of light and reflect others. The apparent color of a pigment is a combination of all the visible wavelengths of light that are reflected by that pigment.

Animals may also produce colors via microscopic physical structures. These structures act like prisms, refracting and scattering light to create a certain combination of colors. Polar bears, for example, actually have black skin but appear white because they have translucent hairs. When light shines on the hairs, each hair bends the light a little bit. Collectively, the hairs deflect a lot of the light back out, producing white coloration.

In some animals, the two types of coloration are combined. For example, reptiles with green coloration typically have a layer of skin with yellow pigment and a layer of skin that scatters light to reflect a blue color. Combined, these layers of skin produce green.

Color Change

In the spring and summer, an animal's habitat might be full of greens and browns, while in the fall and winter, everything can be covered with snow. While brown coloration is perfect for a summer environment, it makes an animal an easy target against a white background. In many birds and mammals, changes in the pattern of daylight or shifts in temperature trigger a hormonal reaction that produces different pigments in new hair or feathers. Over time, the animal changes color entirely.

Some animals, such as various cuttlefish species, can change their overall skin color immediately. These animals have pigment-containing cells called *chromatophores.* Each chromatophore is surrounded by a circular muscle. When the cuttlefish constricts the muscle, the chromatophore flattens out. When the muscle relaxes, the chromatophore forms a relatively small blob, which is harder to see than the flattened chromataphore. By constricting all the chromatophores with a certain pigment and relaxing all the ones with other pigments, the animal can change the overall color of its body.

By perceiving the color of a backdrop and constricting the right combination of chromatophores, the animal can blend in with all sorts of surroundings. The most famous color changer, the chameleon, alters its skin color using a similar mechanism, but not usually for camouflaging purposes. Chameleons tend to

Chameleons change their skin color to match their moods.

change their skin color when their mood changes, not when they move into different surroundings.

Nudibranches (small sea creatures) change their coloration by altering their diet. When a nudibranch feeds from a particular sort of coral, its body deposits the pigments from that coral in the skin and outer extensions of the intestines. The pigments show through, and the animal becomes the same color as the coral. Since the coral is not only the creature's food, but also its habitat, the coloration is perfect camouflage. Similarly, some parasite species, such as the fluke, take on the color of their host, which is also their home.

The Element of Disguise

Many animals have patterns and designs on their bodies. These spots, stripes, and patches usually match the animal's surroundings in some important way. For example, animals that inhabit areas with tall, vertical grass often have long, vertical stripes on their bodies. The patterns may also serve as visual disruptions. Usually, the patterns are positioned out of line with the body's contours. That is, the pattern seems to be a separate design superimposed on top of the animal. The conflicting patterns make it hard for the predator to get a clear sense of where the animal begins and ends—the body seems to run off in every direction.

Another camouflage tactic is to mimic an object that other animals wouldn't be interested in. The walking stick insect, for example, looks like an ordinary twig. A predator can easily distinguish a walking stick from its surroundings, but is completely uninterested because it doesn't realize that "twig" could be lunch.

Other animals use more aggressive mimicry. Several moth species have developed

Zebras have long vertical stripes that mimic tall grass.

The young zebra is camouflaged by the surrounding herd.

striking designs on their wings that resemble the eyes of a larger animal. The back of the hawk moth caterpillar actually looks like a snake's head, a frightening visage for most of the predators that the moth would come across.

A simpler variation on this adaptation is simple color mimicry. In many ecosystems, smaller poisonous animals develop a bright coloration and predators learn to steer clear of these colors, lest they get a mouthful of venom. Over time, other, non-poisonous species may develop the same coloration, cashing in on the nasty reputation of their poisonous neighbors.

Often, camouflage can be a more effective survival tool than aggressive weapons such as teeth, claws, and beaks. After all, being entirely overlooked by a predator certainly beats fighting for your life.

How **VENUS FLYTRAPS** Work

The predator waits patiently while its prey wanders about, unaware that danger lurks just inches away. Settling down to taste some sweet-smelling sap, the unsuspecting prey has made a fatal mistake. Swinging swiftly shut, the jaws of the predator close around its body. The struggle is brief, and soon the plant settles down to digest its tasty meal. Plants that eat other creatures? It sounds like a genetic experiment gone awry. But there's actually nothing unnatural about it; carnivorous plants have existed on this planet for thousands of years.

HSW Web Links

www.howstuffworks.com

How the Eden Project
 Works
How Exercise Works
How Food Works
How Cells Work
How Animal Camouflage
 Works

There are more than 500 different kinds of carnivorous plants, with diets ranging from insects and spiders to one- or two-cell aquatic organisms. To be considered carnivorous, a plant must attract, capture, kill, and digest insects or other animal life.

One carnivorous plant in particular has captured the public's imagination: The Venus flytrap (*Dionaea muscipula*). Many people first see this amazing plant in action during their elementary school years, and are fascinated by its strange dietary habits and unique appearance.

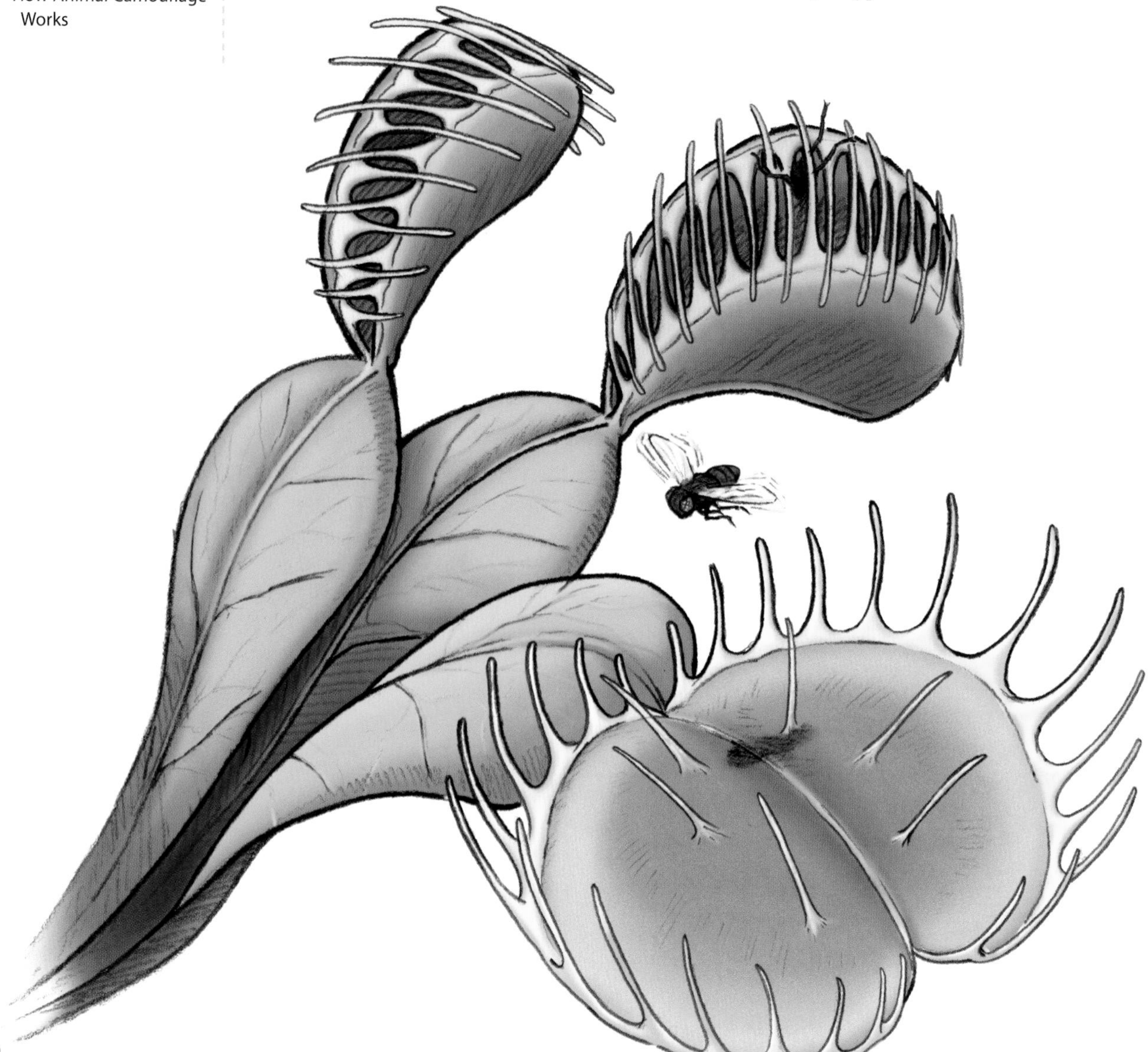

Although the Venus flytrap has captivated people across the world, the plants actually grow in an incredibly small geographic area. In the wild, they are found in a 700-mile region along the coast of North and South Carolina. Within this area, the plants are further limited to living in humid, wet and sunny areas such as bogs and wetlands. Because Venus flytraps are so scarce, some early botanists doubted their existence, despite all the stories that were spread about a flesh-eating plant.

Feed Me, Seymour

Movies such as *Jumanji* and *Little Shop of Horrors* give the Venus flytrap a bad reputation. While Audrey, the mutant flytrap in the movie *Little Shop of Horrors,* developed a taste for humans, real plants prefer insects and arachnids like:

- Spiders
- Flies
- Caterpillars
- Crickets
- Slugs

Ordinary (noncarnivorous) plants depend on the soil around them for survival. They must make amino acids, vitamins, and other cellular components to live. To accomplish this, these plants need nutrients like:

- Nitrogen
- Phosphorus
- Magnesium
- Sulfur
- Calcium
- Potassium

If other plants can thrive on gases in the air and water and minerals from the soil, why do Venus flytraps eat insects? Flytraps actually get a good deal of their sustenance like other plants do, through the process of photosynthesis. However, the soil around them doesn't prove as fortifying.

In the bogs favored by Venus flytraps, the soil is acidic, and minerals and other nutrients are scarce. Most noncarnivorous plants can't survive in this environment because they cannot make enough of the building blocks necessary for growth. The Venus flytrap has evolved the ability to

thrive in this unique ecological niche: Insects provide a good source of the nutrients missing from the soil, and they also contain energy-laden carbohydrates.

Say "Ahhh"

Carnivorous plants have to be able to:

- Attract insects.
- Capture bugs.
- Discriminate between food and nonfood.
- Digest their prey.

All of these steps are accomplished through simple mechanical and chemical processes. Unlike us, plants don't have a brain or nervous system to coordinate their physiological functions and tell them that they are hungry. Plants also don't have complex muscles and tendons to grab food, chew it, swallow it, and process it. The Venus flytrap completes the entire process by way of a specialized set of leaves—leaves that are both mouth and stomach in one.

Most plants have some mechanism to attract animals and insects, regardless of whether or not they plan to feast on their guests. In the case of the Venus flytrap, the leaves forming the trap secrete a sweet nectar that draws in insects searching for food.

When an insect lands or crawls on the trap, it is likely to run into one of six short, stiff hairs on the trap's surface. These are called *trigger hairs,* and they serve as a primitive motion detector for the plant. If two of these hairs are brushed in close succession, or one hair is touched twice, the leaves close down upon the offending insect within half a second. If not, the trap may close part-way until these conditions are met.

What causes the leaves to squeeze shut? Nobody knows exactly how the sequential mechanical stimulation of the trigger hairs translates into closing the trap. The prevailing hypothesis is that:

- Cells in an inner layer of the leaf are very compressed. This creates tension in the plant tissue that holds the trap open.
- Mechanical movement of the trigger hairs puts into motion ATP-driven (for more information on ATP, please see

Did You Know?

There is an upper limit to the size of insect that a Venus flytrap's trap can accommodate. At most, traps are about 1 inch long, and an insect ideally should be about one-third of this size. If an insect is too large, the trap will not be able to form a seal tight enough to keep bacteria and molds out. Once bacteria and mold have a way in, they can proliferate as they feast on the decomposing insect, and the leaves of the trap will succumb to the assault as well. The trap will turn black as the leaves rot, and the whole thing will eventually just drop off the plant.

Venus flytraps can tolerate losing a trap here and there, because the plant can eventually sprout new ones. Nature has actually engineered the traps with planned obsolescence. After about 10 to 12 closures (partial or complete), the traps lose the ability to capture anything. The leaves remain spread wide open, and instead of attracting insects and eating them, the former trap devotes its energy for the remainder of its life span, usually around 2 to 3 months, to the process of photosynthesis. This way, if a trap is repeatedly stimulated by nonedible objects, the plant can recoup some of the energy and ATP lost to opening and closing the trap by focusing solely on photosynthesis.

"How Exercise Works" in the Health Help chapter) changes in water pressure within these cells.

- The cells are driven to expand by the increasing water pressure, and the trap closes as the plant tissue relaxes.

Dinner Time

Even without a brain to analyze what it's eating, the Venus flytrap still manages to differentiate between insects and nonedible debris that might fall into its trap. An insect caught inside the partially closed trap will continue to thrash about in an attempt to escape. It's guaranteed that at least one (if not all) of the trigger hairs will be tweaked by the insect's movement. This serves as the signal to close the trap entirely.

Inanimate objects like stones, twigs, and leaves that fall into the trap, or objects that are placed there (what child can resist sticking the tip of a pencil into the trap to watch it close?), will not move around and fire the trigger hairs. If there is no further stimulation of the hair, the trap stays in its partially shut state until tension can be reestablished in the leaves of the trap. This process takes about 12 hours, and the leaves spread apart again. The unwanted object either falls out as the leaves reopen or is blown out by the wind.

Once the trap fully closes, the leaves form an airtight seal so that:

- Digestive fluids and insect parts are kept inside the trap.
- Bacteria and molds can't get in.

To make sure that the insects are contained within the trap, the edges of the leaves have finger-like growths, called *cilia,* that lace together when the leaves press shut. These long projections make the plant look like it has spiny teeth, but the cilia are really only used to latch the trap shut.

Once the insect is firmly ensconced in the trap, the process of digestion can begin. The trap now serves as a miniature stomach. Just like our stomachs, the trap secretes acidic digestive juices that:

- Dissolve the soft tissues and cell membranes of the food.
- Serve as an antiseptic to kill small amounts of bacteria inadvertently eaten or sealed in with the food.
- Enzymatically digest DNA, amino acids, and other cellular molecules into small pieces that can be taken up by the plant.

These digestive juices are secreted from glands on the inside surface of the trap, right onto the trapped prey. The insect is bathed in these juices over a period of 5 to 12 days, during which the insect is digested and nutrients are extracted. The time it takes depends on:

- **The size of the insect**—The larger it is, the longer it takes to break down.
- **The age of the trap**—The digestive fluid is recycled after each digestion, and an older trap may secrete a somewhat weaker mix of acid and enzymes.
- **The temperature**—The ambient temperature can affect the rate of decomposition; up to a point, increasing temperature makes enzymatic processes much snappier.

The process continues until all that's left of the insect is its hard exoskeleton. After the nutrients are depleted from the acidic bath, the plant reabsorbs the digestive fluid. This serves as a signal to reopen the trap, and the remains of the insect are usually either washed away in the rain or blown away by the wind.

chapter five

SPACE

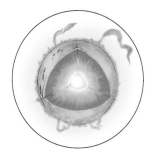

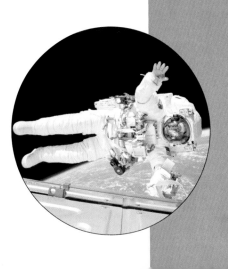

How **STARS** Work

It's a dark, clear, moonless night. You look up into the sky. You see thousands of stars arranged in patterns or constellations. The light from these stars has traveled great distances to reach earth. But what are stars? How far away are they? Are they all the same? Are there other planets around them?

HSW Web Links

www.howstuffworks.com

How Telescopes Work
How Solar Eclipses Work
How Planet Hunting
 Works
How the Sun Works
How SETI Works

Stars are massive, glowing balls of hot gases—mostly hydrogen and helium. Some stars are relatively close and others are incredibly far away. Some stars are alone in the sky, others have companions (binary stars), and some are part of large clusters containing thousands to millions of stars. Not all stars are the same. Stars come in all sizes, temperatures, colors, and degrees of brightness.

Star Features

Stars have many features that can be measured by studying the light that they emit. These include the following:

- Temperature and spectrum
- Brightness, luminosity, and radiance
- Mass and movement

Temperature and Spectrum

Some stars are extremely hot, while others are cool. You can tell a star's temperature by the color of the light that it gives off. If you look at the coals in a charcoal grill, you know that the coals glowing red are cooler than the white-hot ones. The same is true for stars. A blue or white star is hotter than a yellow star, which is hotter than a red star. So, if you look at the strongest color (or wavelength) of light emitted by the star, you can calculate its temperature. A star's spectrum can also tell you the chemical elements that are in that star, because different elements (for example, hydrogen, helium, carbon, and calcium) absorb light at different wavelengths.

Brightness, Luminosity, and Radius

When you look at the night sky, you can see that some stars are brighter than others. Two things control the brightness of a star:

- **Luminosity**—How much energy it puts out in a given time
- **Distance**—How far it is from us

A searchlight puts out more light than a penlight. That is, the searchlight is more luminous. If that searchlight is 5 miles away from you, however, it will not be as bright because light intensity decreases with distance squared. A searchlight 5 miles from you may look as bright as a penlight 6 inches away from you. It's the same with stars.

Astronomers (professional or amateur) can measure a star's brightness by using a

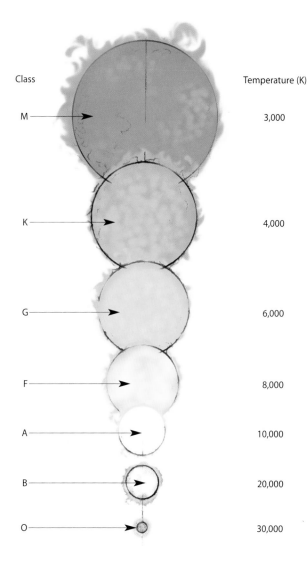

Class	Temperature (K)
M	3,000
K	4,000
G	6,000
F	8,000
A	10,000
B	20,000
O	30,000

photometer or charge-coupled device (CCD) on the end of a telescope. If they know the star's brightness and the distance to the star, they can calculate the star's luminosity [luminosity = brightness × 12.57 × (distance)^2].

Luminosity is also related to a star's size. The larger a star is, the more energy it puts out and the more luminous it is. You can see this on the charcoal grill, too. Three glowing red charcoal briquettes put out more energy than one glowing red charcoal briquette at the same temperature. Likewise, if two stars are the same temperature but different sizes, then the large star will be more luminous than the small one.

Mass and Movement

In 1924, the astronomer A. S. Eddington showed that the luminosity and mass of a star were related. The more massive a star is, the more luminous it is.

Stars around us are moving with respect to our solar system. Some are moving away from us and some are moving toward us. The movement of stars affects the wavelengths of light that we receive from them, much like the high-pitched sound from a fire truck siren gets lower as the truck moves past you. This phenomenon is the Doppler effect. Measuring the star's spectrum and comparing it to the spectrum of a standard lamp, a scientist can measure the amount of the Doppler shift. The amount of the Doppler shift tells us how fast the star is moving relative to us. In addition, the direction of the Doppler shift can tell us the direction of the star's movement. If the spectrum of a star is shifted to the blue end, then the star is moving toward us; if the spectrum is shifted to the red end, then the star is moving away from us. Likewise, if a star is spinning on its axis, the Doppler shift of its spectrum can be used to measure its rate of rotation.

The Life of a Star

As we mentioned before, stars are large balls of gases. New stars form from large, cold (10° Kelvin) clouds of dust and gas (mostly hydrogen) that lie between existing stars in a galaxy.

Usually, some type of gravity disturbance—for example, the passage of a nearby star or the shock wave from an exploding supernova—affects the cloud and causes it to change. The following things happen:

1) The disturbance causes clumps to form inside the cloud.
2) The clumps collapse inward, drawing gas inward by gravity.
3) The collapsing clump compresses and heats up.
4) The collapsing clump begins to rotate and flatten out into a disc.
5) The disc continues to rotate faster, draw more gas and dust inward, and heat up.
6) After about a million years or so, a small, hot (1500° Kelvin), dense core forms in the disc's center. This core is the *protostar.*
7) As gas and dust continue to fall inward in the disc, they give up energy to the protostar, which heats up more.
8) When the temperature of the protostar reaches about 7 million° Kelvin, hydrogen begins to fuse to make helium and release energy.
9) Material continues to fall into the young star for millions of years because the collapse due to gravity is greater than the outward pressure exerted by nuclear fusion. Therefore, the protostar's internal temperature increases.
10) If sufficient mass (0.1 solar mass or greater) collapses into the protostar and the temperature gets hot enough for sustained fusion, then the protostar has a massive release of gas in the form of a jet called a *bipolar flow.* If the mass is not sufficient, the star will not form, but instead the mass becomes a brown dwarf.
11) The bipolar flow clears away gas and dust from the young star. Some of this gas and dust may later collect to form planets.

X-Ray image of Cassiopeia A, a 320-year-old supernova remnant.

Parallax

Measuring the distance to a star is an interesting problem. Triangulation, also known as *parallax*, is one of two methods astronomers use to estimate how far away a star is.

The Earth's orbit around the sun has a diameter of about 186 million miles (300 million kilometers). By looking at a star one day and then looking at it again 6 months later, an astronomer can see a difference in the viewing angle for the star. With a little trigonometry, the different angles yield a distance. This technique works for stars within about 400 light years of the earth.

The young star is now stable in that the outward pressure from hydrogen fusion balances the inward pull of gravity.

Now that the star is stable, it has the same parts as our sun:

- **A core**—Where the nuclear fusion reactions occur
- **A radiative zone**—Where photons carry energy away from the core
- **A convective zone**—Where convection currents carry energy toward the surface

Stars like the sun and those less massive than the sun have the layers in the order that they're listed above. Stars that are several times more massive than the Sun have convective layers deep in their cores and radiative outer layers. Stars that are intermediate-sized, between the sun and the most massive stars, may only have a radiative layer.

Life on the Main Sequence

Stars in the prime of their lives (in their *main sequence*) burn by fusing hydrogen into helium. Large stars tend to have higher core temperatures than smaller stars. Therefore, large stars burn the hydrogen fuel in the core quickly, while small stars burn it more slowly. The length of time that stars spend on the main sequence depends upon how quickly the hydrogen gets used up—massive stars have shorter lifetimes. (The sun, which is an average-sized star, will burn for approximately 10 billion years.) What happens once the hydrogen in the core is gone depends upon the mass of the star.

The Death of a Star

Several billion years after its life starts, a star will die. How the star dies, however, depends on what type of star it is.

When the core of a star like our sun runs out of hydrogen fuel, it will contract under the weight of gravity. However, some hydrogen fusion will occur in the upper layers. As the core contracts, it heats up. This heats the upper layers, causing them to expand. As the outer layers expand, the radius of the star will increase and it will become a red giant. At some point after this, the core will become hot enough to cause the helium to fuse into carbon. When the helium fuel runs out, the core will expand and cool. The upper layers will expand and eject material that will collect around the dying star to form a planetary nebula. Finally, the core will cool into a white dwarf and then eventually into a black dwarf. This entire process takes a few billion years. For stars that are more massive than our sun, the process is different.

When its core runs out of hydrogen, a star bigger than our sun will start to fuse helium into carbon. However, after the helium is gone, the star's mass is enough to fuse carbon into heavier elements such as oxygen, neon, silicon, magnesium, sulfur, and iron. Once the core has turned to iron, it can burn no longer. The star collapses because of its own gravity, and the iron core heats up. The core becomes so tightly packed that protons and electrons merge to form neutrons. In less than a second, the iron core, which is about the size of the Earth, shrinks to a neutron core with a radius of about 6 miles (10 kilometers). The outer layers of the star fall inward on the neutron core, crushing it further. The core heats to billions of degrees and explodes in a supernova, releasing large amounts of energy and material into space. The shock wave from the supernova can initiate star formation in other interstellar clouds. The remains of the core can form a neutron star or a black hole, depending on the mass of the original star.

How the **SUN** Works

It warms our planet every day, provides the light by which we see, and is absolutely necessary for life on Earth. The sun is one of those things we take for granted, but life would be impossible without it!

If you think about the sun, you come up with lots of questions, such as:

- If the sun is in the vacuum of space, how does it burn?
- What keeps all that gas from leaking into space?
- How big is the sun?
- Why does it send out solar flares?
- When will it stop burning?
- Is the sun like other stars?

The answers to these questions are what make the sun so interesting!

The sun is a star, just like the other stars we see at night. The difference is distance. The other stars we see are light years away, while our sun is only about 8 light minutes away—it is many thousands of times closer.

Officially, the sun is classified as a G2 type star because of its temperature and the color of light that it emits. Stars range in color from red through orange, yellow, and white, to blue. The color is directly related

HSW Web Links

www.howstuffworks.com

How Solar Eclipses Work
How Solar Sails Will Work
How Barrier Islands Work
How Solar Cells Work
How Solar Yard Lights
 Work

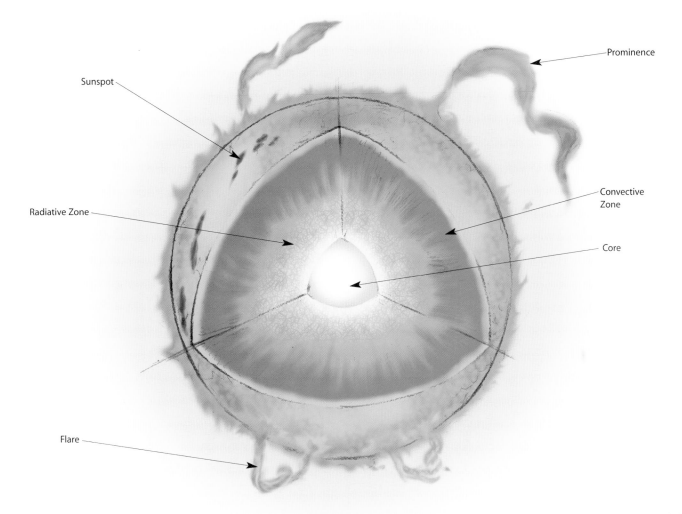

Sunspot

Prominence

Radiative Zone

Convective Zone

Core

Flare

to the temperature. For example, a blue or white star is hotter than a yellow star, which is hotter than a red star. Type G stars are yellow to white with a surface temperature of 5,000 to 6,000° Kelvin. The sun, therefore, is an "average" star, merely one of billions of stars that orbit the center of our galaxy.

The sun has "burned" for more than 4.5 billion years and will continue to light our planet for several billion more. It is a massive collection of gas, mostly hydrogen and helium, about 875,000 miles in diameter. Over 500,000 planet earths could fit inside the sun! Because it is so massive, it has immense gravity—enough gravitational force to hold all of that hydrogen and helium together and to hold all of the planets in their orbits around the sun.

The sun does not "burn" like wood burns. Instead the sun is a gigantic nuclear reactor, as you will soon learn.

Parts of the Sun

The sun is made entirely of gas and has no solid surface like earth does. However, it still has a defined structure. The three major surface areas are:

- The core
- The radiative zone
- The convective zone

The Core

The core starts from the center and extends to 25% of the sun's radius. Here, gravity pulls all of the mass inward and creates an intense pressure. The pressure is high enough to force atoms of hydrogen to come together in nuclear fusion reactions. Two atoms of hydrogen combine in several steps to create helium-4 and energy:

1) Two protons combine to form a *deuterium* (a hydrogen atom with one neutron), *a positron* (similar to an electron, but with a positive charge), and a neutrino (an almost massless uncharged particle).

2) A proton and a deuterium atom combine to form a helium-3 atom (two protons with one neutron) and a gamma ray.

3) Two helium-3 atoms combine to form a helium-4 (two protons and two neutrons) and two protons.

These reactions create 85% of the sun's energy. The remaining 15% comes from other fusion reactions. When hydrogen fuses to form helium-4 atoms, for instance, the helium atom has less mass than the two hydrogen atoms that started the process. The difference in mass converts to energy as described by Einstein's theory of relativity ($E=mc^2$). The energy leaves the sun as various forms of light (ultraviolet light, X-rays, visible light, infrared, microwaves, and radio waves). The sun also sends out energized particles—neutrinos and protons—that make up the solar wind. This energy strikes earth, where it warms the planet, drives our weather, and provides energy for life. We are not harmed by most of the radiation or solar wind because the earth's atmosphere and magnetic field protects us.

Radiative Zone

The sun's radiative zone extends from the core to about 55% of the sun's radius. In this zone, the energy from the core moves outward as photons. As one photon is made, it travels about 1 micron (1 millionth of a meter) before being absorbed by a gas molecule. The gas molecule is heated and re-emits another photon of the same wavelength. The new photon travels another micron before being absorbed by another gas molecule, and the cycle repeats itself. Each interaction between photon and gas molecule takes time. Approximately 1,025 absorptions and re-emissions take place in this zone before a photon reaches the surface, so there is a significant time delay between a photon made in the core and one that reaches the surface.

Convective Zone

The convective zone, which is the final 30% of the sun's radius, uses convection currents to carry the energy outward to the surface. These convection currents are rising movements of hot gas next to falling movements of cool gas, much like what you can see if you placed glitter in a simmering pot of water. The convection currents carry photons outward to the surface faster than the radiative transfer that occurs in the core and radiative zone. With so many interactions occurring between photons and gas molecules in the radiative and convection zones, it takes a photon approximately 100,000 to 200,000 years to reach the surface!

Spots, Prominences, and Flares

Through telescope images, we can see several interesting features on the sun—sunspots, solar prominences, and solar flares. Both solar prominences and solar flares can have significant effects here on earth. Let's look at sunspots, solar prominences, and solar flares.

Sunspots

Dark, cool areas called sunspots appear on the sun. Sunspots always appear in pairs and are intense magnetic fields (about 5,000 times greater than the earth's magnetic field) that break through the surface. Field lines leave through one sunspot and reenter through the other one. The magnetic field is caused by movements of gases in the sun's interior. Sunspot activity occurs as part of an 11-year solar cycle where there are periods of maximum and minimum activity; in our current cycle, 2002 is a high point of sunspot activity and 2007 is a low point.

It is not known what causes this 11-year cycle, but two hypotheses have been proposed:

- Uneven rotation of the sun distorts and twists magnetic field lines in the interior. The twisted field lines break through the surface, forming sunspot pairs. Eventually, the field lines break apart and sunspot activity decreases. The cycle starts again.
- Huge tubes of gas circle the sun's interior at high latitudes and begin to move toward the equator. When they roll against each other, they form spots. When they reach the equator, they break up and sunspots decline.

Solar Prominences

Occasionally, clouds of gases will rise and orient themselves along the magnetic lines from sunspot pairs. These arches of gas are called *prominences*. Prominences can last two to three months and can extend 30,000 miles (50,000 km) or more above the sun's surface. Upon reaching this height above the surface, they can erupt for a few minutes or even for hours and send large amounts of

material racing through the corona and outward into space at 600 miles per second (1,000 km/s); these eruptions are called *coronal mass ejections.*

Erupting prominences, when directed toward Earth, can affect communications, navigation systems, and even power grids, while producing auroras visible in the night sky.

Solar Flares

Sometimes in complex sunspot groups, abrupt, violent explosions from the sun occur. These are called solar flares. Scientists believe that solar flares are caused by sudden magnetic field changes in areas where the sun's magnetic field is concentrated. Solar flares are accompanied by the release of gas, electrons, visible light, ultraviolet light, and X-rays. When this material reaches the earth's magnetic field, it interacts with the earth's magnetic fields at the poles to produce the auroras borealis and australis. Solar flares can disrupt communications, satellites, navigation systems, and even power grids. The radiation and particles ionize the atmosphere and prevent the movement of radio waves between satellites and the ground or between the ground and the ground. The ionized particles in the atmosphere can induce electric currents in power lines and cause power surges. These power surges can overload a power grid and cause blackouts.

Our sun is both complex and interesting. It produces the light and heat that all life on earth depends on. Many people worry about the fate of the sun. There's really no need for immediate concern, though. The sun has enough hydrogen fuel to "burn" for about 10 billion years. It's been shining for about 4.5 billion years, so that leaves another 5.5 billion to go.

How **SOLAR ECLIPSES** Work

Solar eclipses do not occur very often, but they're incredibly interesting to watch when they do. On those rare occasions when you're in the right place at the right time for a full solar eclipse, it is amazing!

HSW Web Links

www.howstuffworks.com

How the Sun Works
How Telescopes Work
How the Hubble Space
 Telescope Works
How Stars Work
How Pinhole Cameras
 Work

A solar eclipse occurs when the moon passes in a direct line between the earth and the sun. The moon's shadow travels over the earth's surface and blocks out the sun's light.

Because the moon orbits the earth at an angle of approximately 5 degrees relative to the earth-sun plane, the moon crosses the earth's orbital plane only twice a year. These times are called *eclipse seasons,* and they are the only times when eclipses can occur. For an eclipse to take place, the moon must be in the correct phase during an eclipse season; for a solar eclipse, it must be a new moon. It's not often that these conditions all line up, so solar eclipses are rare events.

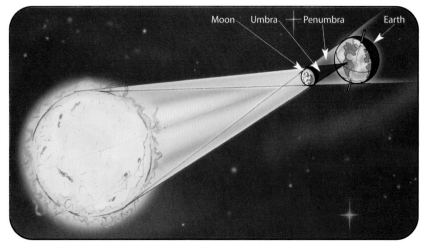

Moon Umbra Penumbra Earth

Types of Eclipses

The moon's shadow has two parts: a central region (the *umbra*) and an outer region (the *penumbra*). Depending upon which part of the shadow passes over you, you will see one of three types of solar eclipses:

- **Total**—The entire central portion of the sun is blocked out.
- **Partial**—Only part of the sun's surface is blocked out.
- **Annular**—Only a small, ring-like sliver of light is seen from the sun's disc.

If the umbra passes over you, the entire central portion of the sun will be blocked out.

You will see a total solar eclipse, and the sky will darken as if it were night. During a total solar eclipse, you can see the sun's outer atmosphere, called the corona. In fact, this is the only time that you can see the corona, which is why astronomers get so excited when a total eclipse is about to occur. Many astronomers travel the world chasing eclipses.

If the penumbra passes over you, only part of the sun's surface will be blocked out. You will see a partial solar eclipse, and the sky may dim slightly depending on how much of the sun's disc is covered.

In some cases, the moon is far enough away in its orbit that the umbra never reaches the earth at all. In this case, there is no region of totality, and what you see is an annular solar eclipse. In an annular eclipse, only a small, ring-like sliver of light is seen from the sun's disc.

Observing Safely

Never look at the sun directly—doing this will damage your eyes. The best way to observe the sun is by projecting its image. Here is one way to project the sun's image:

1) Get two pieces of cardboard (the flaps from a box or the backs of paper tablets, for example).
2) With a pin or pencil point, poke a small hole (no bigger than the pin or pencil point) in the center of one piece.
3) Take both pieces in your hands.
4) Stand with your back to the sun.
5) In one hand, hold the piece of cardboard with the pinhole; with your other hand, place the second piece of cardboard (the screen) behind the first piece.
6) The sunlight will pass through the pinhole and form an image on the screen.
7) Adjust the distance between the two pieces to focus and change the size of the image.

Enjoy observing!

How **BLACK HOLES** Work

You may have heard someone say, "My desk has become a black hole!" Or you may have seen an astronomy program on television or read a magazine article on black holes. These exotic objects have captured our imagination ever since Einstein's theory of general relativity predicted them in 1915.

Stars are huge, amazing fusion reactors. Because stars are so large and are made out of gas, an intense gravitational field is always trying to collapse the star. The fusion reactions happening in the core are like a giant fusion bomb that is trying to explode the star. The balance between the gravitational forces and the explosive forces is what defines the size of the star.

A black hole is what remains when a massive star dies. A massive star usually has a core that is at least three times the mass of our sun.

As the star dies, the nuclear fusion reactions stop because the fuel for these reactions gets used up. At the same time, the star's gravity pulls material inward and compresses the core. As the core compresses, it heats up and eventually creates a supernova explosion in which material and radiation blasts out into space. What remains is the highly compressed, and extremely massive, core.

Because the core's gravity is so strong, the core sinks through the fabric of space-time and literally disappears from view, creating a hole in space-time—this is why the object is called a *black hole*. What was simply the core of the original star now becomes the central part of the black hole—it's called the *singularity*. The opening of the hole is called the *event horizon*.

You can think of the event horizon as the mouth of the black hole. Once something passes the event horizon, it is gone for good. Once inside the event horizon, all *events* (points in space-time) stop, and nothing—not even light—can escape. There are two types of black holes.

The Schwarzschild black hole is the simplest black hole, in which the core does not rotate. This type of black hole only has a singularity and an event horizon.

The Kerr black hole, which is probably the more common form, rotates because the star from which it was formed was rotating. When the rotating star collapses, the core continues to rotate, and this carries over to the black hole. The Kerr black hole has the following parts:

- **Singularity**—The collapsed core
- **Event horizon**—The opening of the hole
- **Ergosphere**—An egg-shaped region of distorted space around the event horizon (the distortion is caused by the spinning of the black hole, which "drags" the space around it)
- **Static limit**—The boundary between the ergosphere and normal space

HSW Web Links

www.howstuffworks.com

How Stars Work
How the Sun Works
How Hubble Space
 Telescope Works
How Solar Eclipses Work
How Comets Work

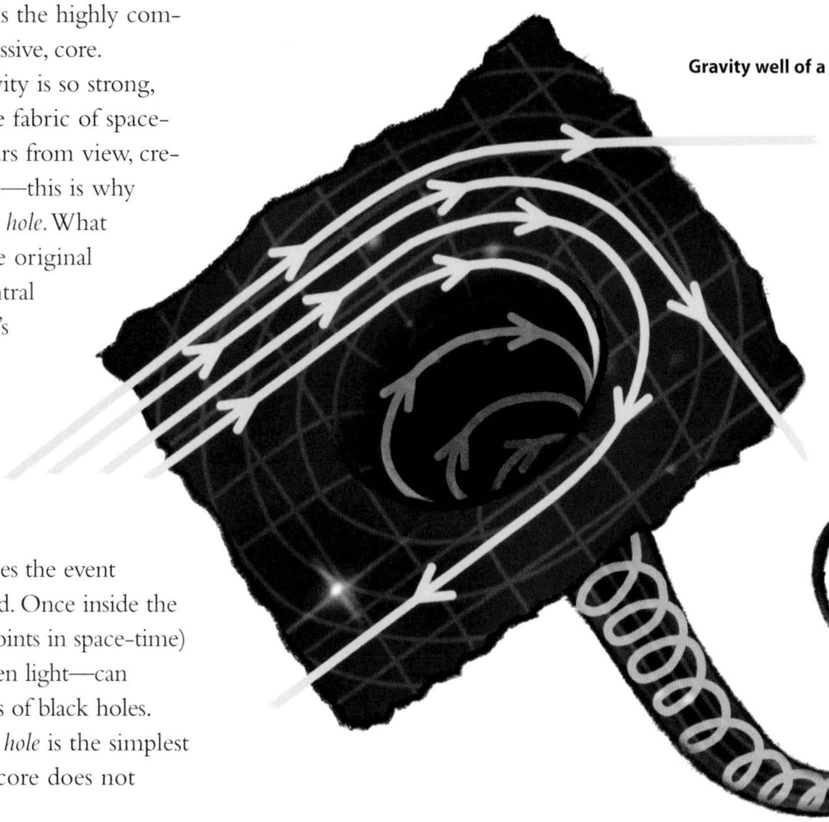

Gravity well of a black hole.

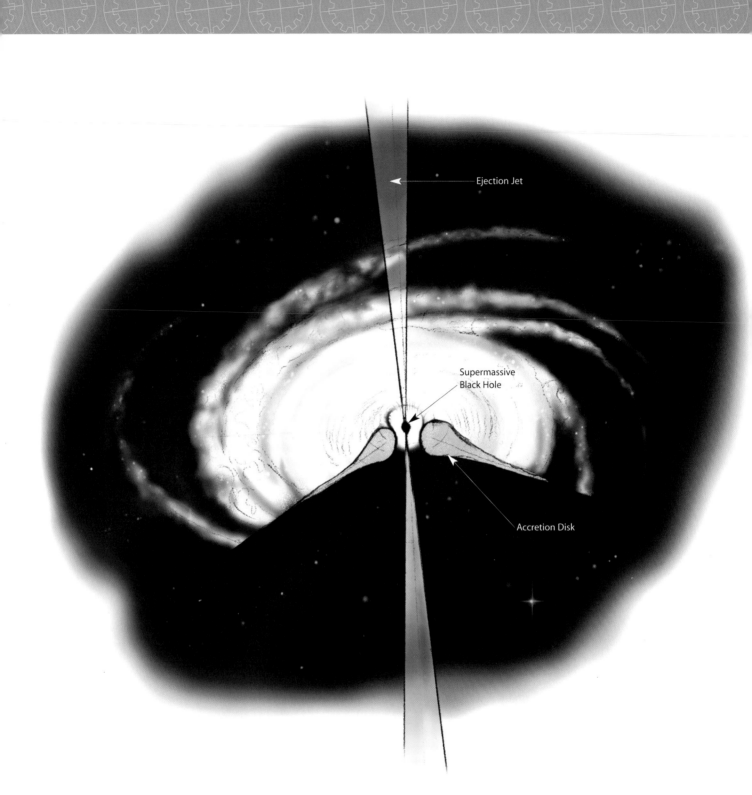

Ejection Jet

Supermassive
Black Hole

Accretion Disk

If an object passes into the ergosphere, it can still be ejected from the black hole by gaining energy from the hole's rotation. However, if an object crosses the event horizon, it will be sucked into the black hole and will never escape. What happens inside the black hole is unknown; even our most current theories of physics do not apply in the vicinity of a singularity.

Even though we cannot see a black hole, it does have three properties that can or could be measured:

- Mass
- Electric charge
- Rate of rotation (angular momentum)

As of now, we can only measure the mass of the black hole reliably by the movement of other objects around it. If a black hole has a companion (another star or disk of material), it's possible to measure the radius of rotation or speed of orbit of the material around the unseen black hole. The mass of the black hole can be calculated using rotational motion or Kepler's modified third law of planetary motion ($P^2 = Ka^3$).

How We Detect Black Holes

Although we cannot see black holes, we can detect or guess the presence of one by measuring its effects on objects around it. The following effects may be measured:

- Mass estimates from objects orbiting a black hole or spiraling into the core
- Gravitational lens effects
- Emitted radiation

Mass

Many black holes have objects around them, and by looking at the behavior of the objects you can detect the presence of a black hole. You then use measurements of the movement of objects around a suspected black hole to calculate the black hole's mass.

What you look for is a star or a disk of gas that is behaving as though there were a large mass nearby. For example, if a visible star or disk of gas is wobbling or spinning, and there is not a visible reason for such a motion, and the invisible reason appears to be caused by an object with a mass greater than three solar masses (in other words, an object that is too big to be a neutron star), then it is possible that a black hole is causing the motion. You then estimate the mass of the black hole by looking at the effect it has on the visible object.

For example, in the core of galaxy NGC 4261, a brown, spiral-shaped disk is rotating. The disk is about the size of our solar system, but weighs 1.2 billion times as much as the sun. Such a huge mass for a disk might indicate that a black hole is present within the disk.

Gravity Lens

Einstein's general theory of relativity predicted that gravity could bend space. This was later confirmed during a solar eclipse when a star's position was measured before, during, and after the eclipse. The star's position shifted because the light from the star was bent by the sun's gravity. Therefore, an object with immense gravity (like a galaxy or black hole) between the Earth and a distant object could bend the light from the distant object into a focus, much like a lens can.

Emitted Radiation

When material falls into a black hole from a companion star, it gets heated to millions of degrees Kelvin and accelerated. The superheated materials emit X-rays, which can be detected by X-ray telescopes, such as the orbiting Chandra X-ray Observatory.

The star Cygnus X-1 is a strong X-ray source and is considered to be a good candidate for a black hole. It is believed that stellar winds from the companion star, HDE 226868, could blow material onto the accretion disk (a disk-shaped area of matter) surrounding the black hole area. As this material falls into the black hole, it emits X-rays.

In addition to X-rays, black holes can also eject materials at high speeds to form jets. Many galaxies have been observed with such jets. Currently, it is thought that these galaxies have supermassive black holes (billions of solar masses) at their centers that produce the jets, and that also produce strong radio emissions.

It is important to remember that black holes aren't cosmic vacuum cleaners—they will not consume everything. Although we can't see black holes, there is indirect evidence that they exist. They have been associated with time travel and wormholes, and remain fascinating objects in the universe.

How **ASTEROIDS** Work

You've probably heard about the possibility of an asteroid smashing into Earth in the future, as in the movie Armageddon. *From the title alone, it's easy to see the sense of catastrophe that surrounds the scenario. Why so much concern?*

HSW Web Links

www.howstuffworks.com

How the Sun Works
How Stars Work
How Black Holes Work
How Comets Work
How Mars Works
How Comets Work

There are tens of thousands of asteroids circling the sun. Most are grouped inside the asteroid belt, which is between the orbits of Mars and Jupiter. Some asteroids that stray from this orbit, though, fly close to earth on occasion—hence the worry about one colliding with our planet. It turns out there are obvious craters on the earth (and the moon) that show us a long history of large objects hitting our planet.

The most famous asteroid ever is the one that crashed into earth 65 million years ago. It is thought that this asteroid threw so much moisture and dust into the atmosphere that it cut off sunlight, lowering temperatures worldwide and causing the extinction of the dinosaurs. So there's a reason for the worry. But what exactly is an asteroid?

What Is an Asteroid?

Asteroids are small, rocky bodies that orbit the sun, typically between the orbits of Mars and Jupiter. There are more than 20,000 known asteroids. They are irregularly shaped and vary in size from a radius of about 0.6 miles (1 km) to over 200 miles. Ceres is the largest asteroid, with a radius of 284 miles (457 km). By measuring fluctuations in their brightness, we know that many asteroids rotate in periods of 3 to 30 days. Beyond their size, shape, and rotation, we know relatively little about asteroids. Estimating their mass is difficult because they are not large

Asteroid Gaspra, as viewed from the Galileo spacecraft.

enough to disturb the gravity of Mars or Jupiter, but Ceres is thought to be about 2.6 billion trillion pounds (1.2×10^{21} kg). By using telescopic spectroscopy to examine the spectra of light reflected from these objects, we can classify asteroids as follows:

C—Dark, probably carbon-containing (carbonaceous)

S—Twice as bright as C, these contain deposits of nickel, iron, and magnesium

M—Similar to iron meteorites, they contain nickel and iron

P and D—Low brightness, reddish

In addition to iron, nickel, and magnesium, scientists think water, oxygen, gold, and platinum also exist on some asteroids.

Asteroids appear to be of two different origins:

- Primitive, essentially unchanged pieces of the early solar system (the C asteroids)
- Smashed remnants of differentiated pieces of the solar system

Scientists think that asteroids are the remainders of *planetismals* (early pieces of the solar system) that formed between Mars and Jupiter. Some of the planetismals began to form into planets, but were smashed apart by Jupiter's immense gravity. Others did not begin to form planets (for unknown reasons). Many questions remain about asteroids, because we have never been able to study them closely, although this is changing.

Project NEAR

Project NEAR was the first spacecraft to orbit a small body of the solar system. Launched in February 1996, NEAR flew by the asteroid Mathilde in June 1997, coming to within 753 miles (1,212 km) of the surface. It continued on its journey to eventually orbit Eros (asteroid 433) in February 2000.

Eros is one of the largest known asteroids, discovered by Gustav Witt and August Charlois in 1898. Eros is potato shaped and is 21 miles (33 km) long, 8 miles (13 km) wide, and 8 miles thick. It rotates every five hours and orbits the sun at about 1.5 AU (about 140 million miles, or 225 million km). Eros is an S-type asteroid.

NEAR orbited Eros for almost a year, passing as close as 4 miles (6 km) and as far as 300 miles (500 km) from the surface. During this time, it measured the asteroid's gravity, photographed the asteroid, and mapped and made chemical measurements of the surface. After a year in orbit, NEAR landed on the surface of Eros.

NEAR Lands on Eros

Credit NASA with another milestone in space exploration: On February 12, 2001, a spacecraft landed on the surface of an asteroid for the first time in history. After a year orbiting Eros, NEAR was almost out of fuel. It was designed to only orbit the asteroid and eventually crash onto the surface (because NEAR was never designed to land, it was not equipped with landing legs). However, NEAR's scientists decided to try to land the spacecraft rather than let it crash. The landing procedure allowed scientists to test complex maneuvers for a spacecraft and to get close-up pictures of the surface. These pictures enabled scientists to see objects on Eros as small as 4 inches (10 cm) in diameter.

Scientists ordered NEAR to slow from its circular orbit and execute a series of braking turns as it approached the surface. The landing site was in the saddle-shaped middle of the asteroid. As NEAR approached the surface, it sent back pictures of Eros taken from ranges of 1,650 feet (500 m) down to 396 feet (120 m).

Eros's gravity is weak, with an escape velocity of a mere 22 mph (Earth's escape velocity is 25,000 mph), but it could hold NEAR, which survived the landing and could still radio information back to Earth.

NEAR spacecraft during preflight preparation.

Project NEAR has provided an amazing opportunity for scientists to closely study an asteroid. As we move forward in the human exploration of space, asteroids could prove to be an incredible resource. If you enjoy science fiction, then you know that the thought of colonizing the moon makes for some incredibly imaginative stories. Right now, one of the biggest problems with the idea of a moon colony is the question of building supplies. There is no Home Depot on the moon, so the building supplies have to come from somewhere. Asteroids may be a much better place to get the supplies than earth.

Early evidence suggests that there are trillions of dollars' worth of minerals and metals buried in asteroids that come close to the earth. Asteroids are so close that many scientists think an asteroid mining mission is easily feasible. Several international organizations are developing plans for going up to get these natural space resources.

How **COMETS** Work

Comets have fascinated humankind since we first noticed the distinctive tail streaking across the night sky. Astronomers find comets especially fascinating. Comets are remarkable pieces of our universe's past, and they tell us a great deal about how the universe was formed.

HSW Web Links

www.howstuffworks.com

How the Sun Works
How Light Works
How Solar Sails Will Work
How Asteroids Work
How Stars Work

Comets are small members of the solar system, usually a few miles or kilometers in diameter. They have been described as "dirty snowballs" by astronomer Fred Whipple and are thought to be made of:

- Dust
- Ice (formed of water, ammonia, methane, carbon dioxide)
- Some carbon-containing (organic) materials (for example, tar)
- A rocky center (in the case of some comets)

Scientists believe comets are made from the earliest materials of the solar system. When the sun first formed, it blew lighter material like gases and dust out into space. Some of this material (mainly gas) condensed to form the outer planets (Jupiter, Saturn, Uranus, and Neptune) and some remains in orbit in two areas far from the sun:

- **The Oort cloud**—A sphere about 50,000 AU from the sun
- **The Kuiper belt**—An area within the plane of the solar system outside the orbit of Pluto

The Path of a Comet

Comets are thought to orbit the sun in either the Oort cloud or Kuiper belt. When another star passes by the solar system, its gravity pushes the Oort cloud and/or Kuiper belt and causes comets to descend toward the sun in a highly elliptical orbit

Halley's comet.

with the sun at one focus of the ellipse. Comets can have *short-period orbits* (less than 200 years, such as Halley's comet) or *long-period orbits* (greater than 200 years, such as comet Hale-Bopp).

As the comet passes within 6 AU of the sun, the ice begins to go directly from the solid to the gas state. (*AU* stands for *astronomical unit*; 1 AU is the mean distance from the earth to the sun, or about 93 million miles, 150 million kilometers.) When the ice sublimes like this, the gas and dust particles flow away from the sun to form the comet's tail.

Parts of a Comet

As a comet approaches the sun, it warms up. During this warming, you can observe several distinct parts:

- The nucleus
- The coma
- The hydrogen envelope
- The dust tail
- The ion tail

The *nucleus* is the main, solid part of the comet. It is usually 0.6 to 6.0 miles (1 to 10 kilometers) in diameter, but it can be as big as 60 miles (100 kilometers).

The *coma* is a halo of evaporated gas (water vapor, ammonia, carbon dioxide) and dust that surrounds the nucleus. The coma is made as the comet warms up and is often 1,000 times larger than the nucleus. It can

Did You Know?

As comets pass through the inner solar system, they can be broken into pieces by the gravity of large objects that they pass. Comet Shoemaker-Levy 9 was broken into 20 pieces by Jupiter's gravity, each of which collided with Jupiter in one of the most spectacular examples of interplanetary impacts in recorded history.

Recently, comet LINEAR was broken into fragments by the sun's gravity as it passed the sun.

even become as big as Jupiter or Saturn (60,000 miles in diameter). The coma and nucleus together form the head of the comet.

Surrounding the coma is an invisible layer of hydrogen called the *hydrogen envelope;* the hydrogen may come from water molecules. The hydrogen envelope usually has an irregular shape because it is distorted by the solar wind. The envelope gets bigger as the comet approaches the sun.

The comet's *dust tail* always faces away from the sun. The tail is made of small (one micron) dust particles that have evaporated from the nucleus and are pushed away from the comet by the pressure of sunlight. The dust tail is the easiest part of the comet to see because it reflects sunlight and because it is long—several million kilometers (several degrees of the sky). The dust tail is often curved because the comet is moving in its orbit at the same speed that the dust is moving away, much as water curves away from the nozzle of a moving hose.

Comets often have a second tail called an *ion tail.* The ion tail is made of electrically charged gas molecules that are pushed away from the nucleus by solar wind. Sometimes the gas tail disappears and later reappears when the comet crosses a boundary where direction of the sun's magnetic field is reversed.

Comet Folklore

Not only has humankind been fascinated by comets; they've been fearful of them as well. It turns out that comets have more than one tail—tale, that is. There's an abundance of folklore and mythology surrounding these extraterrestrial events. Long considered precursors to calamity by many cultures throughout history, comets are believed to be responsible for or at least related to destruction, disaster, and even death.

A portion of the Bayeaux Tapestry represents one of the most famous examples of a comet as a harbinger of great change. Reportedly, Halley's Comet was sighted just prior to the Battle of Hastings in 1066 A.D. At that time, many people believed this sighting was directly related to the Norman Conquest. In fact, the Bayeaux Tapestry,

which commemorates the Norman Conquest, includes a scene depicting a number of men staring up at a comet.

The birth and death of royalty has also been attributed to these celestial sightings. A number of notable figures have died during the year of a visible comet, including Julius Caesar and Diana, former Princess of Wales. And, incredibly, Halley's Comet heralded both the birth and death of famed author Mark Twain.

Observing Comets

Many comets are actually discovered by amateur astronomers. To look for comets, here are things to keep in mind:

- Go to a place where there are few lights.
- Learn what a comet looks like (observe as many comets as you can) and what a comet does not look like (observe other deep sky objects, because they also appear as small fuzzy objects).
- Use binoculars or a telescope (low magnification, 20 to 40x).
- Look toward the east about 30 minutes before sunrise or to the west about 20 minutes after sunset, because comets are often spotted by their dust tails reflecting in the fading light.
- Sweep the sky slowly near the horizon.

This type of observing takes discipline, long hours, and patience. On average, comet hunters spend several hundred hours of observing time in order to find a new comet. Comets are named after their discoverers, so many people think it's worth the effort. For a discussion of comet hunting, consult *The Sky: A User's Guide* by David H. Levy, who has discovered several comets, including comet Shoemaker-Levy 9, which hit Jupiter.

How **METEORS** Work

If you've spent much time looking up at the night sky, you've probably seen at least one or two shooting stars. Most shooting stars come from dust floating in space. The earth is orbiting the sun at a very high speed—about 66,000 miles per hour (over 100,000 kph). When the earth runs into the dust in space, the dust particles streak through the atmosphere and heat to the point where they incandesce—in other words, they give off bright light as they burn up in the atmosphere. We see this light as a shooting star. It turns out that this flash of light is actually a meteor.

⚙ HSW Web Links

www.howstuffworks.com

How Stars Work
How Asteroids Work
How Comets Work
How Hubble Space
  Telescope Works

The Leonid Shower

Every year in November we are treated to an unusual and fascinating sky show called the Leonid meteor shower. If you live in the right place and stand outside at the right time, you can see hundreds or thousands of shooting stars in just an hour. This amazing event happens because the earth is passing through an especially dusty area of space.

The extra dust is there because the orbit of the Temple-Tuttle comet passes near earth's orbit every 33 years and leaves behind a dusty trail. The shower in 2001 was especially intense because the earth actually passed through two dust trails—one left by the comet's 1699 orbit and another left in 1866.

Once you find out the most optimal viewing days, it's pretty easy to witness this incredible spectacle of shooting star activity. Simply go outside at the appropriate hour, look toward the constellation Leo (hence the name *Leonid*) and you should be able to see a lot of shooting stars!

It Came from Outer Space

Discussing meteor activity can be tricky, because the terminology is confusing. The term *meteor* actually refers to the streak of light caused by a piece of space debris burning up in the atmosphere. The pieces of debris are called *meteoroids,* and remnants of the debris that reach the earth's surface (or another planet's) are called *meteorites.*

Meteoroids have a pretty big size range. They include any space debris bigger than a molecule and smaller than about 330 feet (100 meters); any space debris bigger than this is considered an asteroid. Most of the debris the earth comes in contact with is dust shed by comets traveling through the solar system. This dust tends to be made up of very small particles.

One of the most amazing things about a meteor or meteor shower is that the majority of the space debris that causes visible meteors is tiny—between the size of a grain of sand and a small pebble!

The Fast Track

So how can we see a meteor caused by such a small bit of matter? It turns out that what these meteoroids lack in mass they make up for in speed. Their extremely high speed is what causes the flash of light in the sky.

Meteoroids enter the atmosphere at 7 to 45 miles per second (11 to 72 kilometers per second). They can travel at this rate very easily in space because space is a vacuum—there's nothing to stop them.

Barringer Meteor Crater located in Northern Arizona.

The earth's atmosphere, on the other hand, is full of matter, which creates a great deal of friction on a traveling object. This friction generates enough heat (up to 3,000°F, or 1,649°C) to raise the meteoroid's surface to its boiling point, so the meteoroid is vaporized, layer-by-layer, as it moves through earth's atmosphere.

The friction breaks the molecules of both the meteoroid material and the surrounding atmosphere into glowing ionized particles, which then recombine, releasing light energy to form a bright "tail." A meteor tail caused by a grain-sized meteoroid is only a few feet wide but, because of the high speed of the debris, it may be many miles long.

Surprisingly, most of the meteoroids that reach the ground are especially small—from microscopic debris to pieces the size of dust particles. They don't get vaporized because they are light enough that they slow down very easily. Moving about 1 inch per second through the atmosphere, they don't experience the intense friction that larger meteoroids do. In this sense, almost all the meteoroids that enter the atmosphere make it to the ground—they just do so in the form of microscopic dust.

The meteorites a person is likely to find on the ground probably came from significantly larger meteoroids—pieces of debris at least the size of a basketball, typically, since larger meteoroids usually break up into smaller chunks as they travel through the atmosphere.

How **HUBBLE SPACE TELESCOPE** Works

Imagine having a telescopic eye on the universe, and being able to look out at a distant star or nebula with amazing clarity. With such a telescope, you could peer billions of light years away and see things that happened billions of years ago. Astronomers are doing just that with the Hubble Space Telescope!

The major problem with observing the light from distant stars using ground-based telescopes is that the light must pass through the earth's atmosphere. Besides clouds and the weather, the earth's atmosphere is a messy place—dust, currents of warm air rising, currents of cold air falling, and water vapor can impede the view. All of these factors can produce fuzzy images of the stars and limit the usefulness of ground-based telescopes. A telescope in space can help solve many of these problems.

Inside the Hubble

Like any telescope, the Hubble Space Telescope (HST) has a long tube that is open at one end. It has mirrors to gather and bring the light to a focus where its "eyes" are located. The HST has several types of eyes in the form of various instruments. In fact, it is these various scientific instruments that make HST such an amazing astronomy tool. In addition, the HST is also a spacecraft. So it has to have power and be able to move in orbit. Before we look at the spacecraft systems, let's start with the telescope functions.

Optics

Light enters the HST through the opening and bounces off a primary mirror to a secondary mirror. The secondary mirror reflects the light through a hole in the center of the primary mirror to a focal point behind the primary mirror. At the focal point, smaller, half-reflective, half-transparent mirrors distribute the light to the various scientific instruments.

HST's mirrors (like most large telescope mirrors) are made of special low-expansion glass that does not expand or contract very much with changes in temperature. The glass is coated with layers of pure aluminum (three-millionths of an inch thick) and magnesium fluoride (one-millionth of an inch

thick) to make it reflect visible, infrared, and ultraviolet light. The primary mirror weighs 1,825 pounds (828 kg). The secondary mirror weighs 27.4 pounds (12.3 kg).

Scientific Instruments

By looking at the different wavelengths, or the spectrum of light, from a celestial object, you can tell many of its features or properties. HST is equipped with various scientific instruments that enable you to look at all the different wavelengths. Each instrument uses charge-coupled devices (CCD), rather than photographic film, to capture the light. The light detected by the CCDs becomes a digital signal, which is then stored in on-board computers and relayed to earth. The digital data gets transformed into the amazing pictures that we see on the news and in magazines. Here is a look at each instrument:

- **The Wide Field Planetary Camera 2 (WFPC2)**—This is the main eye, or camera, of the HST. WFPC2 has four CCD chips to catch the light: three low-resolution, wide-field CCD chips arranged in an L shape and one high-resolution planetary camera CCD chip inside that L. All four CCD chips are exposed to the target at the same time, and the target image is centered on the desired CCD chip, either at high or low resolution. This camera sees visible and ultraviolet light. WFPC2 can take images through various filters (red, green, blue) to make natural color pictures.
- **The Near Infrared Camera and Multi-Object Spectrometer (NICMOS)**—Interstellar gas and dust can block visible light. However, it is

HSW Web Links

www.howstuffworks.com

How Telescopes Work
How Stars Work
How Satellites Work
How Space Stations Work
How Mars Works
How Black Holes Work

Cool Facts

The two solar panels (also known as arrays) on the Hubble are designed so that they can be easily replaced. They're a lot like a giant portable projection screen. There's a support structure or fixed frame. The solar panels attach to the fixed frame. If necessary, these panels can be rolled up for easy transport, much like the screen on a portable projection system.

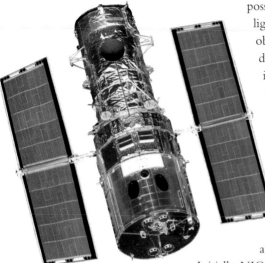

possible to see the infrared light, or heat, from the objects hidden in the dust and gas. To see this infrared light, HST has three sensitive cameras that make-up the (NICMOS).

Because they are so sensitive to heat, the NICMOS sensors must be kept in a large thermos-like bottle at −321°F (77° Kelvin). Initially, NICMOS was cooled with a 230-pound (104-kg) block of frozen nitrogen, but now, NICMOS is actively cooled with a machine that acts like a refrigerator.

• **The Space Telescope Imaging Spectrograph (STIS)**—It's one thing to look at the light from a celestial object—but how can you tell what the object is made of? The colors, or spectrum of light, coming from a star or other celestial object are a chemical fingerprint of that object. The specific colors tell us what elements are present in the object, and the intensity of each color tells us how much of that element is present. So to identify the colors—the specific wavelengths of light—the STIS separates the incoming colors of light much like a prism makes a rainbow.

In addition to the chemical composition, the spectrum can tell us about the temperature and motion of a celestial object.

• **The Faint Object Camera (FOC)**—How does the HST zoom in on an object? To do this, HST uses the Faint Object Camera (FOC), which is a high-resolution camera. The FOC magnifies or intensifies the image through three stages. At each stage

within the camera, sensitivity and resolution are increased severalfold. The FOC is sensitive to both visible and ultraviolet light.

For example, when aimed at Betelgeuse (a red star in the shoulder of the constellation Orion), the FOC can actually image the surface of the star. This is the first time that the surface of a star other than our sun has been imaged. From the image, scientists determined that Betelgeuse has a mysterious hot spot on its surface that is 2000° Kelvin hotter than the rest of the star's surface.

• **Fine Guidance Sensors (FGS)**—These sensors are used to point the telescope and to make detailed, precise measurements of the positions of stars, the separation of binary stars, and the diameter of stars. There are three FGS in the Hubble. Two of the FGS are used to point the telescope and keep it fixed on its target, looking for guide stars in the HST field near the target; when each FGS finds a guide star, it locks on to it and feeds information back to the HST steering system to keep that guide star in its field. While two FGS are steering the telescope, one is free to make astrometric measurements (star positions). Astrometric measurements are important for detecting planets, because orbiting planets cause their parent stars to wobble in their motion across the sky.

Now let's look at the spacecraft systems.

Spacecraft Systems

As mentioned earlier, the HST is also a spacecraft. It must have power and be able to communicate with the ground and change its *attitude* (its orientation).

All of the instruments and computers on board the HST need electrical power. This electrical power comes from two large solar panels, each panel measuring 24.79 ft (7.6 m) by 8.05 ft

This image was created from 3 distinct pointings of the Hubble.

(2.45 m). The solar panels provide 3,000 watts of electricity, which is equal to the electricity used by 75 40-watt light bulbs. When the HST is in the earth's shadow, electrical power is provided by 6 nickel-hydrogen batteries, which provide the same storage as 20 car batteries. The batteries are recharged by the solar panels when the HST comes around to sunlight again.

Communications

The HST must be able to talk with controllers on the ground to relay data from its observations and receive commands for its next targets. To communicate, the HST uses a series of relay satellites called the Tracking and Data Relay Satellite System (TDRSS), which is the same system used by the International Space Station.

Incoming light from an object gets received by the HST and converted to digital data. The data is then sent to the TDRSS in orbit, which then transmits it to the ground receiving station in White Sands, New Mexico. The White Sands facility transmits the data to NASA's Goddard Spaceflight Control Center, where HST operations are centered. The data are then analyzed by scientists at the nearby Space Telescope Science Institute in Baltimore, Maryland. Most of the time, commands are relayed to the HST in advance of a planned observing run; however, real-time commands are possible when necessary.

Steering

The HST must remain fixed on a target while it takes an image, and it normally takes several hours (or even several days) to collect enough light. Bear in mind that the HST is moving around the earth every 97 minutes, so focusing on a target is like keeping sight of a small object on the shore from the deck of a boat that is rapidly moving along the coast, bobbing up and down in the waves. To remain fixed on an object, the HST has three on-board systems:

- **Gyroscopes**—To sense small to large motions
- **Reaction wheels**—To move the telescope
- **FGS**—To sense fine motion

The gyroscopes keep track of the gross movements of the HST. Like a compass, they sense the motion of the HST, telling the flight computer that the HST has moved away from the target. The flight computer then calculates how much and in what direction the HST must move to remain on target. The flight computer then directs the reaction wheels to move the telescope.

The HST cannot use rocket engines or gas thrusters to steer like most satellites do, because the exhaust gases would hover near the telescope and cloud the surrounding field of view. Instead, the HST has reaction wheels oriented in the three directions of motion (x/y/z or pitch/roll/yaw). The reaction wheels are flywheels. When the HST needs to move, the flight computer tells one or more flywheels which direction to spin in and how fast, which provides the action force. In accordance with Newton's third law of motion—for every action there is an equal and opposite reaction—the HST spins in the opposite direction of the flywheels until it reaches its target.

As mentioned above, the FGS help keep the telescope fixed on its target by sighting on guide stars. Two of the three FGS find guide stars around the target within their respective fields of view. Once found, they lock onto the guide stars and send information to the flight computer to keep the guide stars within their field of view. The FGS are more sensitive than the gyroscopes; the combination of gyroscopes and FGS can keep the HST fixed on a target for hours, despite the telescope's orbital motion.

Despite its flawed early history, the HST has performed well, yielding much scientific data and beautiful images. However, the HST will not last forever. Plans are underway for a new space telescope, called the Next Generation Space Telescope (NGST). NGST will be even more sensitive than HST and provide better images of even more distant objects. The age of optical space telescopes started by HST promises to revolutionize astronomy, as much or more than Galileo's first use of the telescope did long ago.

Hubble instrument upgrade.

Astronauts install a new Power Control Unit on the Hubble Space Telescope.

Did You Know?

The HST has two main computers that fit around the telescope's tube above the scientific instrument bays. One computer talks to the ground to transmit data and receive commands. The other computer is responsible for steering the HST, as well as various housekeeping functions. There are also backup computers in the event of an emergency.

Each instrument on board the HST also has microprocessors built in to move filter wheels, control the shutters, collect data, and talk to the main computers.

How the **INTERNATIONAL SPACE STATION** Works

From the early days of science fiction and space exploration, we have dreamed of space stations. Visionaries have proposed space stations as outposts in orbit, much like the U.S. western frontier's forts and outposts of the eighteenth and nineteenth centuries. The International Space Station is the world's latest and largest attempt to maintain a permanent presence in space.

HSW Web Links

www.howstuffworks.com

How Hubble Space
  Telescope Works
How Weightlessness
  Works
How Planet Hunting
  Works
How Comets Work
How Black Holes Work

For some, space stations are a place to do cutting edge scientific research in an environment that cannot be duplicated on earth. For others, space stations are a place for business, where unique materials (crystals, semiconductors, pharmaceuticals) can be manufactured in better forms than on earth. Still others dream of space stations as staging points for expeditions to the planets and stars, as tourist attractions, or even as new cities and colonies that could relieve an overpopulated planet.

The assembly of the International Space Station (ISS) in orbit began in 1998. The ISS will ultimately consist of more than 100 modules and will require 44 spaceflights to deliver the components into orbit. One-hundred sixty spacewalks, totaling 1,920 man-hours, will be required to assemble and maintain the ISS, which is scheduled for completion in 2006 and will have an anticipated life of 10 years and a projected total cost of $35 to $37 billion. When completed, the ISS will be able to house up to seven astronauts. It will have the following major components:

International Space Station. This photo was taken by a crew member from the Discovery in 1999.

- **Zarya control module**—The control module contains propulsion (two rocket engines), command, and control systems.
- **Nodes**—Three nodes connect major portions of the ISS.
- **Zvezda service module**—The service module contains living quarters and life support for early parts of the ISS, docking ports for Progress resupply ships, and rocket engines for attitude control and re-boost.
- **Scientific laboratories**—The station will have six labs containing scientific equipment and a robotic arm to move payload on an outside platform.
- **Laboratory module**—Shirtsleeve environment facility for research on microgravity, life sciences, earth sciences, and space sciences.
- **Truss**—The truss is a long, tower-like spine for attaching modules, payloads, and systems equipment.
- **Mobile servicing system**—This is a robotic system that will move along the truss and be equipped with a remote arm for assembly and maintenance activities.
- **Transfer vehicles**—For emergency evacuation, the station will have a Soyuz capsule and a crew return vehicle (X-38).
- **Electrical power system**—The station has solar panels and equipment for generating, storing, managing, and distributing electrical power.

How the ISS Works

To sustain a permanent environment where people can live and work, the ISS must be able to provide the following things: life support, propulsion, communications and tracking, navigation, electrical power, computers, re-supply, and an emergency escape route.

Uses of the ISS

The ISS will be used mostly for scientific research in the unique environment of microgravity. Four times larger than Mir, the ISS is capable of staying in orbit much longer than the space shuttle, which orbits for three weeks. Researchers from governments, industry, and educational institutions will be able to use the facilities on the ISS. The types of research that will be done include:

- Microgravity science
- Life science
- Earth science
- Space science
- Engineering research and development
- Commercial product development

Microgravity Science

Gravity influences many physical processes on earth. For example, gravity alters the way that atoms come together to form crystals. In microgravity, near-perfect crystals can be formed. Such crystals can yield more efficient drugs to combat diseases or better semiconductors for faster computers.

Another effect of gravity is that it causes convection currents to form in flames, which leads to unsteady flames. This makes the study of combustion very difficult. However, microgravity makes flames simple, steady, and slow-moving; these types of flames make studying the combustion process easier. The resulting information could yield a better understanding of the combustion process, and lead to better designs of furnaces or the reduction of air pollution.

The ISS will be equipped with a state-of-the-art laboratory for studying the effects of microgravity on these processes.

Life Science

Life as we know it has evolved in a world of gravity. Our body shape and plan have been influenced by gravity. We have skeletons to help support us against the force of gravity. Our senses can tell us which direction is up or down, because we can sense gravity. But how exactly does gravity influence living things? The ISS gives us the opportunity to study plants and animals in the absence of gravity. For example, when a plant seed sprouts, the roots grow down and the shoots or leaves grow up (called *gravitropism*); somehow, the young plant must sense gravity. So what would happen if seeds were to grow in microgravity? This type of experiment will be done on the ISS.

Long-term exposure to weightlessness causes our bodies to lose calcium from bones, tissue from muscles, and fluids from all over. These effects of weightlessness are similar to the effects of aging (decreased muscle strength, osteoporosis), so exposure to microgravity may give us new insights into the aging process. If we can develop countermeasures to prevent the degrading effects of microgravity, perhaps we can prevent some of the physical effects of aging. The ISS will provide long-term exposure to microgravity that could not be obtained by using spacecraft.

The ISS will allow us to test ecological life-support systems that are similar to the ecology on earth. We can grow plants in large quantities in space to make oxygen, remove carbon dioxide, and provide food. The information gathered on the ISS will be important when scientists are planning long interplanetary space voyages, such as a trip to Mars or Jupiter.

International Space Station with solar array panels deployed. (Photo taken in 2000.)

Earth Science

The ISS's orbit will cover 75% of earth's surface for observation. With on-board instruments, the astronauts will be able to:

- Study climate and weather.
- Study geology.
- Gather information on atmospheric quality.
- Map vegetation, land use, and mineral resources.
- Monitor the health of rivers, lakes, and oceans.

The data gathered from these studies will help us understand how the earth's biosphere works, and how to minimize humankind's influences on it.

Space Science

The ISS will be an orbiting platform above the earth's atmosphere. Like the Hubble space telescope, the telescopes on board the ISS will have clear views of the sun, stars, and planets, without the interference of the

And Another Thing...

The United States placed its first and only space station, called Skylab 1, in orbit in 1973. During the launch, the station was damaged. A critical meteoroid shield and one of the station's two main solar panels were ripped off and the other solar panel was not fully stretched out. That meant Skylab had little electrical power, and the internal temperature rose to 126°F (52°C). The first crew was launched 10 days later to fix the ailing station. The astronauts stretched out the remaining solar panel and set up an umbrella-like sunshade to cool the station. With the station repaired, that crew and two subsequent crews spent a total of 112 days in space, conducting scientific and biomedical research.

earth's atmosphere. Instruments on board the ISS will look for planets around other stars, and search in distant galaxies for clues to the origin of the universe. Instruments on the ISS will be able to be repaired and interchanged more easily than those on the Hubble space telescope.

Engineering Research and Development

Much of the engineering research and development on the ISS will go toward studying the effects of the space environment on materials and developing new technologies for space exploration. Some of the things to be studied include:

- New construction techniques for building things in space
- New space technologies, including solar cells and storage
- New satellite and spacecraft communications systems
- Advanced life-support systems for future spacecraft

Before the ISS, NASA launched a satellite called the Long Duration Exposure Facility (LDEF) to study the effects of the space environment (atomic oxygen in the upper atmosphere, cosmic rays, micrometeoroids). Materials were mounted on the outside of the satellite. After several years in orbit, the satellite was retrieved by the space shuttle, brought back to earth, and analyzed.

This type of experiment is incredibly easier with the ISS. Materials can be placed on the ISS in open platforms, to be exposed to the space environment for years. These materials can be interchanged for analysis more easily than if they were on satellites. The information retrieved will help scientists design better materials for making satellites last longer in the space environment.

Commercial Product Development

As mentioned above, more-perfect crystals can be grown aboard the space station, and these crystals will help to develop better drugs, catalysts for extracting oil, and semiconductors. Again, the ISS will have dedicated laboratories for manufacturing these products, and will provide a much longer time in orbit than could be achieved by the space shuttle.

The Future of Space Stations

We are just beginning the development of space stations. The ISS will be a vast improvement over Salyut, Skylab, and Mir, but we are still a long way from the realization of the large space stations or colonies envisioned by science fiction writers. None of our space stations thus far have had any gravity, for two reasons:

1) We want a place without gravity so that we can study the effects of gravity.
2) We lack the technology to practically rotate a large structure, like a space station, to produce artificial gravity.

In the future, artificial gravity will be a requirement for space colonies with large populations.

The ISS will need periodic reboosting because of its position in low earth orbit. However, future space stations may be able to take advantage of Lagrange Points L-4 and L-5, two places between the earth and the moon. At these points, the earth's gravity and the moon's gravity are counter-balanced so that an object placed there would not be pulled toward the earth or moon. The orbit would be stable and require no boosting. The L5 Society was formed more than 20 years ago to push the idea of placing space stations in orbit at these points. As we learn more from our experiences on the ISS, we may build larger and better space stations that would enable us to live and work in space, and the dreams may become reality.

Digital rendering of what the International Space Station will look like once all additions are completed in 2003.

The Beginnings of the ISS

In 1984, President Ronald Reagan proposed that the United States, in cooperation with other countries, build a permanently inhabited space station. Reagan envisioned a station that would have government and industry support. The U.S. forged a cooperative effort with 14 other countries (Canada, Japan, Brazil, and the countries of the European Space Agency—the United Kingdom, France, Germany, Belgium, Italy, the Netherlands, Denmark, Norway, Spain, Switzerland, and Sweden). During the planning of the ISS and after the fall of the Soviet Union, the United States invited Russia to cooperate in the ISS in 1993; this brought the number of participating countries to 16. NASA is taking the lead in coordinating the ISS's construction.

How **WEIGHTLESSNESS** Works

You've probably seen pictures of astronauts floating around inside the space shuttle, the International Space Station, or Mir. While weightlessness looks like fun, it can also be a pain. Initially, you feel nauseated, dizzy, and disoriented. Your head and sinuses swell and your legs shrink. In the long term, your muscles weaken and your bones become brittle. These effects on your body could do severe damage if you were on a long space voyage, such as a trip to Mars.

Imagine that you're dressed in your spacesuit and lying on your back in the flight deck of the space shuttle. You have been on your back in the chair for several hours as the pilots and mission control have gone through the preflight launch preparations. Normally, when you're standing upright, gravity pulls blood downward so it pools in the veins of your legs. However, because you've been lying on your back, the blood is distributed differently through your body, shifting slightly toward your head because your feet are elevated. Your head may feel a little stuffy, much like it does at night when you sleep.

The rocket engines fire and you feel the acceleration. You get pushed back into your seat as the shuttle ascends. You feel heavy as the G-forces of the shuttle's acceleration increase to up to three times normal gravity (some roller coaster rides can briefly achieve this level of acceleration). Your chest feels compressed and you may have some difficulty breathing. In about eight and a half minutes, you are in outer space, experiencing an entirely different sensation: weightlessness.

Encountering Microgravity

Weightlessness is more correctly termed *microgravity*. You aren't actually weightless, because the earth's gravity is holding you and everything in the shuttle in orbit. You are actually in a state of free fall, much like if you jumped from an airplane, except that you are moving so fast horizontally (5 miles per second, or 8 kilometers per second) that, as you fall, you never touch the ground because the earth curves away from you. It's like this: When you stand on a bathroom scale, it measures your weight because gravity pulls down on you and the scale. Because the scale is resting on the ground, it pushes up on you with an equal force—this equal force is your weight. However, if you were to jump off a cliff while standing on a bathroom scale, both you and the scale would be pulled down equally by gravity. You would not push on the scale and it would not push back against you. Therefore, your weight would read 0.

Because the shuttle and all of the objects in it are falling around the world at the same rate, everything in the shuttle that is not secured floats. If you have long hair, it floats around your face. If you pour a glass of water out, it creates a large, spherical drop that you can break up into separate, smaller drops. Food and candy gently float to your mouth if you push them in that direction. While sitting in your seat, you have no sense

HSW Web Links

www.howstuffworks.com

How Spacesuits Work
How Space Shuttles Work
How Space Stations Work
How Mars Works
How Your Kidneys Work
How Your Heart Works

Fluid Loss

One countermeasure to deal with fluid loss is a device called lower body negative pressure (LBNP), which applies a vacuum-cleaner-like suction below your waist to keep fluids down in your legs. This device might be attached to an exercise device, such as a treadmill. You might spend 30 minutes per day in the LBNP to keep your circulatory system in near-earth condition.

Also, just prior to your return to earth, you can drink large volumes of water or electrolyte solutions to help replace the fluids you've lost. This can prevent you from fainting when you stand up and step out of the shuttle.

that you are seated because your body does not press against the seat. If you are not secured to something, you float. If you cannot reach a wall, handhold, or foothold, you cannot move from your position because you have nothing to push against. For this reason, NASA has placed many restraints, handholds, and footholds throughout the cabin of the shuttle.

How You Feel in Microgravity

When most people first encounter microgravity, they can suffer from: nausea, disorientation, headache, loss of appetite, and congestion.

In addition, the longer you stay in microgravity, the more your muscles and bones weaken. These sensations are caused by changes in various systems of your body. Let's take a closer look at how your body responds to microgravity.

Spacesickness

The nausea and disorientation that you feel are like that sinking feeling in your stomach when your car hits a dip in the road or you experience a drop on a roller coaster ride, only you have that feeling constantly for

several days. This is the feeling of space sickness, or space motion sickness, which is caused by conflicting information that your brain receives from your eyes and the vestibular organs located in your inner ear. Your eyes can see which way is up and down inside the shuttle. However, because your vestibular system relies on the downward pull of gravity to tell you which way is up versus down and in which direction you are moving, it does not function in microgravity. So your eyes may tell your brain that you are upside-down, but your brain does not receive any interpretable input from your vestibular organs. Your confused brain produces the nausea and disorientation, which in turn may lead to vomiting and loss of appetite. Fortunately, after a few days your brain adapts to the situation by relying solely on the visual inputs, and you begin to feel better. NASA issues medication patches to help astronauts deal with the nausea until their bodies adapt.

Puffy Face and Bird Legs

In microgravity, your face will feel full and your sinuses will feel congested, which may contribute to headaches as well as space motion sickness. You feel the same way on earth when you bend over or do a head or hand stand, because blood rushes to your head. On earth, gravity pulls on your blood, causing significant volumes to pool in the veins of your legs. Once you encounter microgravity, the blood shifts from your legs into your chest and head. Your face tends to get puffy and your sinuses swell. The fluid shift also shrinks the size of your legs.

When the blood shifts to the chest, your heart increases in size and pumps more blood with each beat. Your kidneys respond to this increased blood flow by producing more urine, much like they do after you drink a large glass of water. Also, the increase in blood and fluid decreases anti-diuretic hormone (ADH) secretion by the pituitary gland, which makes you less thirsty. Therefore, you do not drink as much water as you might on earth. Overall, these two factors combine to help rid your chest and head of the excess fluid, and in a few days, your body's fluid levels are less than what they were on earth. Although you still have a slightly puffy head and stuffy sinuses, the symptoms aren't as bad after the first couple of days. Upon your return to earth, gravity will pull those fluids back down to your legs and away from your head, which will cause you to feel faint when you stand up. But you will also begin to drink more, and your fluid levels will return to normal in a couple of days.

Space Anemia

As your kidneys eliminate the excess fluid, they also decrease their secretion of erythropoietin, a hormone that stimulates red blood-cell production by bone marrow cells. The decrease in red blood-cell production matches the decrease in plasma volume so that the *hematocrit* (percentage of blood volume occupied by red blood cells) is the same as on earth. Upon your return to earth, your erythropoietin levels will increase, as will your red blood-cell count.

Weak Muscles

When you are in microgravity, your body adopts a *fetal posture*—you crouch slightly, with your arms and legs half-bent in front of you. In this position, you do not use many of your muscles, particularly those muscles that help you stand and maintain posture (the anti-gravity muscles).

As your stay in space lengthens, your muscles change. The mass of your muscles decreases, which contributes to the bird-leg appearance. The muscle fiber types change from slow-twitch to fast-twitch. Your body no longer needs slow-twitch endurance fibers, such as those used in standing. Instead, more fast-twitch fibers are needed as you push yourself quickly off of space station surfaces. The longer you stay in space, the less muscle mass you will have. This loss of muscle mass makes you weaker,

presenting problems for long-duration space flights and upon returning home to earth's gravity.

Brittle Bones

On earth, your bones support the weight of your body. The size and mass of your bones are balanced by the rates at which certain bone cells lay down new mineral layers and other cells chew up those mineral layers. In microgravity, your bones do not need to support your body, so all of your bones, especially the weight-bearing bones in your hips, thighs, and lower back, are used much less than they are on earth. In these bones, the rate at which your bone cells deposit new bone layers is reduced while the rate at which your other bone cells chew up bone stays the same. The result is that the size and mass of these bones continue to decrease as long as you remain in microgravity, at a rate of approximately 1% per month. These changes in bone mass make your bones weak and more likely to break upon your return to earth's gravity. It is not known how much of the bone loss is recoverable after return to earth, although it is probably not 100%.

You can see that, long-term, the human body responds to weightlessness in very interesting ways, and some of these ways are not always good. A huge amount of research and thought is going into different ways to avoid these problems on long-term space missions.

Deterioration of Muscles and Bones

NASA and the Russian space agency have found that the best way to minimize loss of muscle and bone mass in space is to exercise frequently. Doing so trains your muscles, prevents them from deteriorating, and places stress on your bones to produce a sensation similar to weight. In a weightless state, you should exercise as much as 2 hours every day on various machines (treadmill, rowing machine, bicycle). You have to be restrained during your exercise, usually by tension-producing straps, such as bungee cords, that hold you to the machine.

Much more research needs to be done to develop countermeasures to the body's changes in microgravity. This research must be conducted both on the ground and in outer space—aboard the international space station—using both humans and animals. The results of such research will help to improve the health of astronauts and pave the way for long-term space exploration, such as a trip to Mars.

How **PLANET HUNTING** Works

Until 1991, our sun was the only star known to have planets around it. This changed when astronomer Alex Wolszczan discovered two planets orbiting a pulsar in the constellation Virgo. Since his discovery, there have been over 50 planets found in orbit around other stars. These orbiting bodies are called extrasolar planets. *How is it that scientists are able to search for and ultimately discover these planets?*

HSW Web Links

www.howstuffworks.com

How SETI Works
How Stars Work
How the Sun Works
How Telescopes Work
How Mars Works
How Satellites Work
How Hubble Space
 Telescope Works

Habitable Zone

If life exists outside of our solar system, it will be on these extrasolar planets. Light from a star warms the orbiting planet and supplies the energy necessary for life. In addition to energy, life seems to need a liquid chemical solvent of some type in which to develop. On earth, this solvent is water, but it is conceivable that other solvents (such as ammonia, methane, or hydrogen fluoride) might also work. With this in mind, it seems that the planet must lie within a certain range of distances from the star so that the solvent can remain in liquid form—if the planet is too close to the star, the solvent will boil away, and if it is too far from the star, the solvent will freeze. For our sun, the habitable zone appears to be between the orbits of Venus and Mars.

It's amazing what human ingenuity can come up with. There are hundreds, if not thousands of astronomers peering at the heavens every night using a relatively small set of tools—sometimes just a telescope. So, they spend a great deal of time thinking of different ways to use these tools. To be able to sense objects as small as a planet at a distance of trillions of miles away using a modest collection of tools is a truly a major accomplishment.

What Is a Planet?

There are eight other planets in our solar system besides earth. But what exactly is a planet? By definition, a planet is a large body that orbits a star and shines by reflecting starlight from its surface. Planets vary in their mass, composition, and distance from the star. In our solar system, the planets are divided into three major categories:

- **Inner-terrestrial planets**—Mercury, Venus, earth, and Mars. These planets are made of rock and orbit close to the sun.
- **Outer gas giants (or Jovian planets)**—Jupiter, Saturn, Uranus, and Neptune. These planets are massive (with hundreds of times the mass of earth). They have dense, gaseous, hydrogen-rich atmospheres that also contain helium, ammonia, and methane. These atmospheres probably surround inner cores made of rock.
- **Other bodies**—Pluto, comets, asteroids, and Kuiper Belt objects. These bodies

are made of rock and ice mixtures. Despite controversy, Pluto is still categorized as a planet, though its composition is more similar to asteroids and comets than to other planets.

The planets in our solar system were made from the disc of swirling gas and dust that formed our sun. As the hydrogen gas and dust of the early solar system fell into the center of this disc, forming the protosun, the gas and dust heated up to a temperature that could sustain nuclear fusion. At the same time, smaller clumps of dust and gas, called *planetismals,* formed in the outer parts of the disc. When the protosun "ignited," it blew the dust and gas away from its immediate vicinity. The planetismals coalesced to form the planets (see "How Stars Work" for more details). Scientists believe that other solar systems were and are being formed in the same way.

Searching for Extrasolar Planets

It's difficult to find planets around other stars because the light from the star is so bright that the glare drowns out the light reflected from the planet. It's like trying to see a lighted birthday candle placed in front of a searchlight. The only way to detect extrasolar planets right now is to measure their effects on their parent stars. There are two ways in which planets affect their parent stars: They tug on the star as they orbit it, and they can dim the light from the star if they pass directly between the star and our field of view (eclipsing part of the star's light). The effects of these planetary motions on the star can be detected from earth by three methods:

- **Astrometry**—Measuring the star's precise position in the heavens
- **Doppler spectroscopy**—Measuring the wavelength spread of light emitted from the star
- **Photometry**—Measuring the intensity or brightness of the light emitted from the star

As the star moves toward the earth, the light waves coming from it are compressed and shifted toward the blue (shorter wavelength) end of the spectrum. As the star moves away from us, the light waves are stretched out toward the red (longer wavelength) end of the spectrum. These shifts in the spectrum of light coming from the

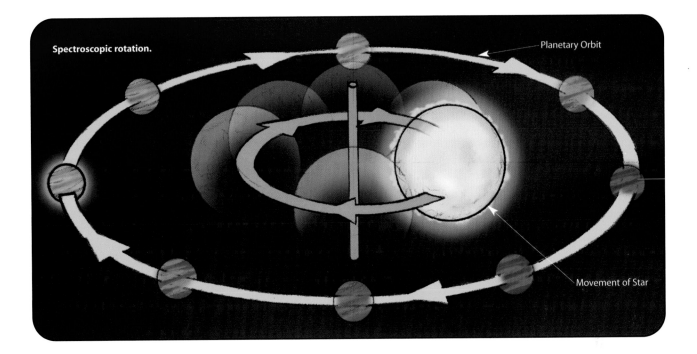

Spectroscopic rotation.

Planetary Orbit

Movement of Star

Astrometry

As a planet tugs on a star with its gravitational pull, it causes the star to wobble in its path across the sky. By making careful, precise measurements of the star's position in the sky, we can detect this extremely slight wobble. When we know the *period* (peak-to-peak time or trough-to-trough time) of the wobble, we can calculate the period of the planet's orbit, the distance or radius of the planet's orbit, and the mass of the planet.

Doppler Spectroscopy

As a planet orbits a star, it periodically pulls the star closer to and farther away from earth (our observation point). This motion has an effect on the spectrum of light coming from the star.

star are called *Doppler shifts*. By making measurements of the star's spectrum over time, we can detect shifts that would indicate the presence of a planet. We can also use Doppler shifts to measure the *radial velocity* of the star's movement, which is how fast the star moves toward us and away from us.

Conceptually, we can deduce the size of the planet from the radial velocity. A massive planet will tug on the star with more gravitational force than a small one, causing the star to have a greater radial velocity. If we graph the radial velocity versus time, we get a sine curve. From the period (peak-to-peak time or trough-to-trough time) and the star's mass, we can get the distance of the planet from the star—the planet's orbital radius. From the amplitude of the curve, we can calculate the planet's mass.

131

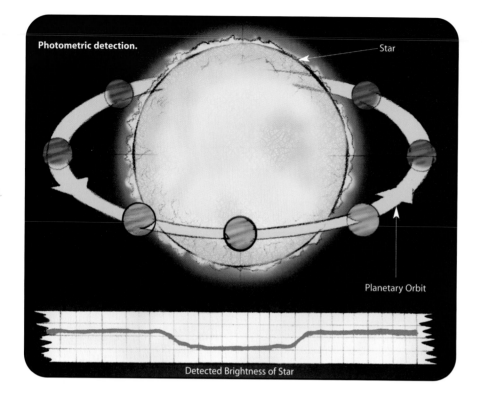

Photometric detection.

Star

Planetary Orbit

Detected Brightness of Star

dimmer (by about 2% to 5%). The planet eclipses the star. As the planet passes behind the star, the star's normal brightness returns. By constantly measuring the star's light intensity over time, we can detect changes in its brightness that might indicate the presence of a planet or planets.

Future Planet Hunting

NASA's chief administrator, Daniel Goldin, has set a major goal for NASA to find earth-like planets orbiting other stars. NASA plans to launch a set of telescopes called the Terrestrial Planet Finder (TPF) to aid in achieving this goal. The TPF will be an array of four optical telescopes and a combiner instrument. Each telescope in the array will detect light from the target star. The light will be combined in such a way as to cancel out the bright glare from the star, a technique known as *nulling interferometry*. Precision flying methods, which are currently being developed at NASA, will keep the array in formation.

After the star's light is cancelled, the infrared spectrum of the planet's light can be examined for the presence of substances in the planet's atmosphere that would indicate an earth-like environment.

The TPF mission is in development stages, and hopefully will be launched within the next decade. Once operational, this space-based telescope system will revolutionize planet hunting and the search for life in the universe.

Photometry

If the orbit of an extrasolar planet is in a straight line of sight with earth, the planet will pass directly between the star it's orbiting and earth. When the planet passes in front of the star, it blocks some portion of the star's light and the star gets slightly

And Another Thing...

In July 1995, two astronomers from the University of Geneva, Didier Queloz and Michael Mayor, found the first planet orbiting a normal star in the constellation of Pegasus. They used the spectroscopy method. The discovery of 51 Pegasus was confirmed by astronomers Geoff Marcy and Paul Butler of San Francisco State University. Marcy and Butler have since found numerous planets around other stars using the spectroscopy method. As of May 2000, scientists have found over 50 extrasolar planets. All of the planet-hunting methods tend to detect large planets—about half the size of Jupiter to several times the size of Jupiter. These planets tend to orbit their parent stars within about 3 astronomical units (AU).

chapter six

SCIENCE

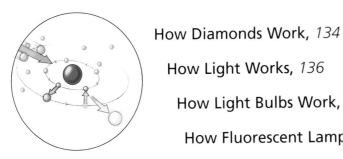

How **DIAMONDS** Work

On your next trip to the local shopping mall, stop by one of the jewelry stores. Notice the diamond jewelry that takes up the majority of the showcase and the number of people hovering over the counters trying to pick out a diamond for their loved one. There will surely be a salesperson explaining the 4 Cs—cut, clarity, carat, and color—to a young shopper, and explaining why one diamond is better than the one right next to it. Why all the fuss over diamonds?

HSW Web Links

www.howstuffworks.com

How Volcanoes Work
How Moissanite Jewels Work
How Carbon-14 Dating Works
How Oil Refining Works

Space Diamonds

Diamonds are not exclusive to earth. There is some scientific evidence that diamonds may be found in abundance on Neptune and Uranus.

Neptune and Uranus contain a great abundance of the hydrocarbon gas methane. Researchers at the University of California, Berkeley, have demonstrated that focusing a laser beam on pressurized liquid methane can produce diamond dust. Neptune and Uranus contain about 10% to 15% methane under an outer atmosphere of hydrogen and helium. Scientists think that this methane could possibly turn to diamond at fairly shallow depths.

Diamonds are just carbon in its most concentrated form. That's it—carbon, the element that makes up 18% of the weight of your body. In many countries, including the United States and Japan, there is no other gemstone as cherished as the diamond, but in truth, diamonds are no more rare than many other precious gems are.

The Origin of Diamonds

Carbon is one of the most common elements in the world, and it's one of the four essentials for the existence of life as we know it. When occurring in nature, carbon exists in three basic forms:

- **Diamond**—An extremely hard, clear crystal.
- **Graphite**—A soft, black mineral made of pure carbon. The molecular structure is not as compact as diamond's, which makes it weaker than diamond.
- **Fullerite**—A mineral made of perfectly spherical molecules consisting of exactly 60 carbon atoms.

Diamonds form about 100 miles (161 km) below the earth's surface, in the molten rock of the earth's mantle. This location provides the right amounts of pressure and heat to transform carbon into a diamond. In order for a diamond to be created, carbon must be placed under at least 435,113 psi (pounds per square inch) of pressure at a temperature of at least 752°F (400 °C). If either of these conditions isn't met, graphite will be created instead of diamond. At depths of 93 miles or more, pressure builds to about 725,189 psi and heat can exceed 2,192°F (1200°C).

Most diamonds that we see today were formed millions (if not billions) of years ago. Powerful magma eruptions brought the diamonds to the surface, creating kimberlite

pipes. Most of these eruptions occurred between 1,100 million and 20 million years ago. Kimberlite is named after Kimberly, South Africa, where these pipes were first found. It is a bluish rock that diamond miners look for when seeking out new diamond deposits.

Kimberlite pipes are created as magma flows through deep fractures in the earth. The magma inside the kimberlite pipes acted like an elevator, pushing the diamonds and other rocks and minerals through the mantle and crust in just a few hours. These eruptions were short, but many times more powerful than volcanic eruptions that happen today.

The magma eventually cooled inside these kimberlite pipes, leaving behind conical veins of kimberlite rock that contain diamonds. The surface area of diamond-bearing kimberlite pipes ranges from 5 to 361 acres.

Diamonds may also be found in riverbeds, which are called *alluvial* diamond sites. These diamonds originate in kimberlite pipes but get moved by erosion. Glaciers and water can also move diamonds thousands of miles from their original location. Today, most diamonds are found in Australia, Borneo, Brazil, Russia, and in several African countries, including South Africa and Zaire.

Diamonds are found as rough stones and must be cut and polished to create a sparkling gem that is ready for purchase.

Diamond Properties

As mentioned earlier, diamonds are the crystallized form of carbon created under extreme heat and pressure. It's this process of intense heat and pressure that makes diamonds the hardest mineral we know of. A diamond ranks a 10 on the Mohs Hardness Scale (explained in "The Mohs Scale" sidebar). In fact, diamonds are so hard that they can be anywhere from 10 to hundreds of times harder than the next hardest mineral on the scale, corundum, a mineral that can form rubies and sapphires.

The molecular structure of diamonds is what makes them so hard. Diamonds are made of carbon atoms linked together in a lattice structure. Each carbon atom shares electrons with four other carbon atoms, forming a tetrahedral unit. This tetrahedral bonding of five carbons forms an incredibly strong molecule. Graphite, another form of carbon, isn't as strong as diamond because the carbon atoms in graphite link together in rings, where each atom is only linked to one other atom.

Square Cut or Pear Shaped . . .

Special techniques are used to cut and shape a diamond before it gets to the jewelry store. Diamond cutting creates the facets that visually bring a diamond to life. Diamond cutters use these four basic techniques:

- **Cleaving**—In order to remove any impurities or irregularities in the diamond, a rough diamond is placed in quick-drying cement. A sharp groove is then carved into the diamond along planes of weakness, using another diamond or a laser. Then a steel blade is placed in the groove and a sharp blow to the blade splits the stone. The diamond is then removed from the cement.
- **Sawing**—Sometimes diamonds have to be cut against a cleavage plane, which cannot be done with cleaving. Using a phosphor-bronze blade rotating at about 15,000 rpm, the saw slowly cuts through the diamond. Lasers are also used to saw diamonds.
- **Bruting**—The diamond is placed in a lathe, and another diamond in the lathe is rubbed against it to create the rough finish of the *girdle,* the outside rim of the diamond at the point of largest diameter.
- **Polishing**—To give the diamond its finished look, it is placed onto the arm above a rotating polishing wheel. The wheel is coated with diamond powder that smoothes the diamond as it is pressed against the wheel.

A Girl's Best Friend

Diamonds are judged on several factors. If you've ever purchased a diamond, you've heard of the 4 Cs:

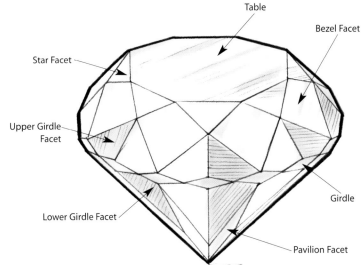

- **Cut**—This refers to how the diamond has been cut and its geometric proportions. When a diamond is cut, facets are created and the diamond's finished shape is determined. Ideal diamonds have specific dimensions to maximize brilliance. Diamonds that are too tall or too flat do not reflect the same way and are not as brilliant.
- **Clarity**—This is the measurement of a diamond's flaws, or inclusions that are seen in the diamond. Clarity levels begin with Flawless and move down to Very Very Slight (VVS), Very Slight (VS), and Slightly Included (SI).
- **Carat**—This is the weight of the diamond. One carat is equal to about 200 milligrams.
- **Color**—The color scale of transparent diamonds runs from D to Z, beginning with icy white—the color of the most expensive diamonds—and ending with a light yellow.

Most diamonds never reach the consumer market because they are too flawed. Often, these diamonds are used for industrial purposes—as an abrasive, for drill bits, or for cutting diamonds and other gems.

Other unique qualities of the diamond include its transparency, luster, and dispersion of light. A diamond that is created from 100% carbon will be completely transparent. Diamonds often contain other elements that can affect the color. Although we think of diamonds as being clear, there are also blue, red, black, pale green, pink, and violet diamonds. These colored diamonds are the truly rare ones.

The Mohs Scale

The Mohs Scale is used to determine the hardness of solids, especially minerals. It is named after the German mineralogist Friedrich Mohs. The scale reads as follows, from softest to hardest:

1) Talc—Easily scratched by the fingernail
2) Gypsum—Just scratched by the fingernail
3) Calcite—Scratches and is scratched by a copper coin
4) Fluorite—Not scratched by a copper coin and does not scratch glass
5) Apatite—Just scratches glass and is easily scratched by a knife
6) Orthoclase—Easily scratches glass and is just scratched by a file
7) Quartz (amethyst, citrine, tiger's-eye, aventurine)—Not scratched by a file
8) Topaz—Scratched only by corundum and diamond
9) Corundum (sapphires and rubies)—Scratched only by diamond
10) Diamond—Scratched only by another diamond

How **LIGHT** Works

We see things every day, from the moment we get up in the morning until we go to sleep at night. We look at everything around us using light. We appreciate illustrations in a book, swirling computer graphics, gorgeous sunsets, a blue sky, shooting stars, and rainbows. We rely on mirrors to make ourselves presentable and sparkling gemstones to show affection. But did you ever stop to think that when we see any of these things, we're not directly connected to it? We are, in fact, seeing light instead—light that somehow left objects far or near and reached our eyes. Light is all our eyes can really see.

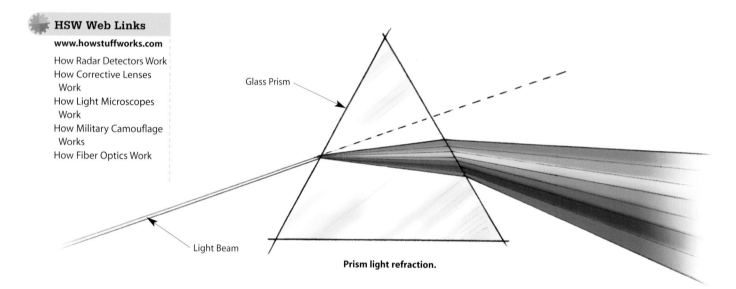

Glass Prism

Light Beam

Prism light refraction.

You have probably heard two different ways of talking about light:

- There is the wave theory, expressed by the term *light wave*.
- There is the particle theory, expressed in part by the word *photon*.

Light is a form of energy. Modern physicists believe that this energy can behave as both a particle and a wave, but they also recognize that either view is a simple explanation for something more complex.

Making Waves

To understand light waves, it helps to start by discussing a more familiar kind of wave—the kind we see in water. The most important point to keep in mind about a water wave is that it's not made up of water: The wave is made up of energy traveling through the water. If a wave moves across a pool from left to right, this does not mean that the water on the left side of the pool is moving to the right side of the pool. The water has actually stayed about where it was. It is the wave that has moved. When you move your hand through a filled bathtub, you make a wave because you're putting your energy into the water. The energy travels through the water in the form of the wave.

All waves are traveling energy, and they are usually moving through some medium, such as water. A water wave consists of water molecules that vibrate up and down at right angles to the direction of motion of the wave.

Light waves are a little more complicated, and they do not need a medium to travel through. They can travel through a vacuum. A light wave consists of energy in the form of electric and magnetic fields. The fields vibrate at right angles to the direction of movement of the wave and at right angles to each other. Because light has both electric and magnetic fields, light is also referred to as *electromagnetic radiation*.

Light waves come in many sizes. The size of a wave is measured as its wavelength, which is the distance between any two corresponding points on successive waves, usually peak-to-peak or trough-to-trough. The wavelengths of the light we can see range from 400 to 700 billionths of a meter. But the full range of wavelengths included in the definition of electromagnetic radiation extends from one billionth of a meter, as in gamma rays, to centimeters and meters, as in radio waves. Visible light is one small part of the spectrum.

Light waves also come in many frequencies. The frequency is the number of waves that pass a point in space during any time interval, usually one second. It is measured in units of cycles (waves) per second, or Hertz (Hz). The frequency of visible light is referred to as color, and ranges from 430 trillion Hz, seen as red, to 750 trillion Hz, seen as violet. Again, the full range of frequencies extends beyond the visible spectrum, from less than one billion Hz, as in radio waves, to greater than 3 billion billion Hz, as in gamma rays.

The amount of energy in a light wave is proportionally related to its frequency: High frequency light has high energy; low frequency light has low energy. Thus gamma rays have the most energy, and radio waves have the least. Of visible light, violet has the most energy, red the least.

Light not only vibrates at different frequencies, it also travels at different speeds. Light waves move through a vacuum at their maximum speed, 186,000 miles per second (300,000 kilometers per second) which makes light the fastest phenomenon in the universe. Light waves slow down when they travel inside substances, such as air, water, glass, or a diamond. The way different substances affect the speed at which light travels is key to understanding the bending of light, or *refraction,* which we will discuss later.

Producing Particles

In particle theory, light is a form of energy that can be released by an atom. It's made up of many small particle-like packets that have energy and momentum but no mass. These particles, called *light photons,* are the most basic units of light.

Atoms release light photons when their electrons become excited. *Electrons* are the negatively charged particles that move around an atom's nucleus (which has a net positive charge). An atom's electrons have different levels of energy, depending on several factors, including their speed and distance from the nucleus. Electrons of different energy levels occupy different orbitals. Generally speaking, electrons with greater energy move in orbitals farther away from the nucleus.

When an atom gains or loses energy, the change is expressed by the movement of electrons. When something—heat, for example—passes energy on to an atom, an electron may be temporarily boosted to a higher orbital (an orbital farther away from the nucleus). The electron only holds this position for a tiny fraction of a second; almost immediately, it is drawn back toward the nucleus, to its original orbital. As it returns to its original orbital, the electron releases

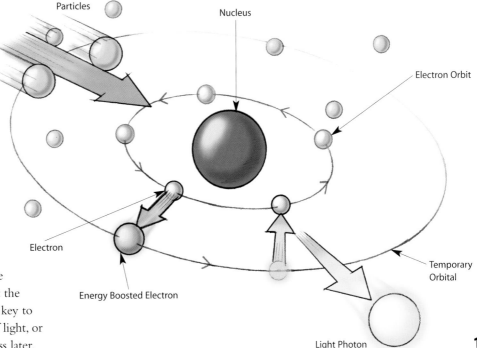

Particles

Nucleus

Electron Orbit

Electron

Energy Boosted Electron

Temporary Orbital

Light Photon

137

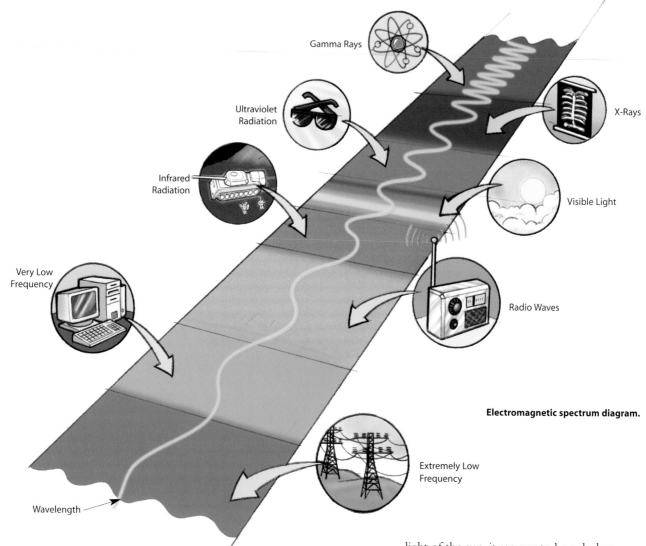

Gamma Rays

Ultraviolet
Radiation

X-Rays

Infrared
Radiation

Visible Light

Very Low
Frequency

Radio Waves

Electromagnetic spectrum diagram.

Extremely Low
Frequency

Wavelength

the extra energy in the form of a photon, in some cases a light photon.

This is the basic mechanism at work in nearly all light sources. The main difference between these sources is the process of exciting the atoms. In an incandescent light source, such as an ordinary light bulb or gas lamp, atoms are excited by heat; in a light stick, atoms are excited by a chemical reaction.

The wavelength of the emitted light depends on how much energy is released, which depends on the particular position of the electron. Consequently, different sorts of atoms will release different sorts of light photons. In other words, the color of the light is determined by what kind of atom is excited.

Creating Color

Visible light is light that can be perceived by the human eye. When you look at the visible light of the sun, it appears to be colorless, which we call white. And although we can see this light, white is not considered to be part of the visible spectrum. This is because white light is not the light of a single color, or frequency. Instead, it is made up of many color frequencies. When sunlight passes through a glass of water to land on a wall, we see a rainbow on the wall. This would not happen unless white light were a mixture of all of the colors of the visible spectrum. Isaac Newton was the first person to demonstrate this aspect of light. Newton passed sunlight through a glass prism to separate the colors into a rainbow spectrum. He then passed sunlight through a second glass prism and combined the two rainbows. The combination produced white light. This proved conclusively that white light is a mixture of colors, or a mixture of light of different frequencies. The combination of every color in the visible spectrum produces a light that is colorless, or white.

Another way to make colors is to absorb some of the frequencies of light, thereby removing them from the white light combination. The absorbed colors are the ones you will not see—you see only the colors that come bouncing back to your eye. This is what happens with paints and dyes. The paint or dye molecules absorb specific frequencies and bounce back, or reflect, other frequencies to your eye. The reflected frequency (or frequencies) is what you see as the color of the object. For example, the leaves of green plants contain a pigment, chlorophyll, that absorbs the blue and red colors of the spectrum and reflects the green.

So there are two basic ways by which we can see colors. Either an object can directly emit light waves in the frequency of the observed color or an object can absorb all other frequencies, reflecting back to your eye only the light wave (or combination of light waves) that appears as the observed color. For example, when you see a yellow object, either the object is directly emitting light waves in the yellow frequency or it is absorbing the blue part of the spectrum and reflecting the red and green parts back to your eye, which perceives the combined frequencies as yellow.

When Light Hits an Object

When a light wave hits an object, what happens to it depends on the energy of the light wave, the natural frequency at which electrons vibrate in the material it strikes, and the strength with which the atoms in the material hold on to their electrons. Based on these three factors, four different things can happen when light hits an object:

- The waves can *pass through* the object with no effect.
- The waves can be *absorbed* by the object.
- The waves can be *reflected* or *scattered* off the object.
- The waves can be *refracted* through the object.

And more than one of these things can happen at once.

- **Transmission**—If the frequency or energy of the incoming light wave is much higher or much lower than the frequency needed to make the electrons in the material vibrate, then the electrons will not capture the energy of the light and the wave will pass through the material unchanged. As a result, the material will be transparent to that frequency of light.

- **Absorption**—In absorption, the frequency of the incoming light wave is at or near the vibration frequency of the electrons in the material. The absorption of light makes an object dark or opaque to the frequency of the incoming wave.

- **Reflection**—The atoms in some materials hold on to their electrons loosely. When the electrons in this type of material absorb energy from an incoming light wave, they do not pass that energy on to other atoms. The energized electrons merely vibrate and then send the energy back out of the object as a light wave with the same frequency as the incoming wave.

- **Scattering**—Scattering is merely reflection off a rough surface. Incoming light waves get reflected at all sorts of angles because the surface is uneven.

- **Refraction**—Refraction occurs when the energy of an incoming light wave matches the natural vibration frequency of the electrons in a material. The light wave penetrates deeply into the material, and causes small vibrations in the electrons. The electrons pass these vibrations on to the atoms in the material, and they send out light waves of the same frequency as the incoming wave. But this all takes time. The part of the wave inside the material slows down while the part of the wave outside the object maintains its original frequency. This has the effect of bending the portion of the wave inside the object toward what is called the *normal line,* an imaginary straight line that runs perpendicular to the surface of the object. The amount of bending, or *angle of refraction,* of the light wave depends on how much the material slows down the light.

Everything we see is a product of, and is affected by, the nature of light. Light is a form of energy that travels in waves. Our eyes are attuned only to those wave frequencies that we call visible light. Intricacies in the wave nature of light explain the origin of color, how light travels, and what happens to light when it encounters different kinds of materials.

Rainbows in Soap Bubbles

Have you ever wondered why soap bubbles are rainbow colored, or why an oil spill on a wet road has rainbow colors in it? The rainbow effect happens when light waves pass through an object with two reflective surfaces. When two incoming light waves of the same frequency strike a thin film of soap, parts of the light waves are reflected from the top surface while other parts of the light pass through the film and are reflected from the bottom surface. Because the parts of the waves that penetrate the film interact with the film longer, they get knocked out of sync with the parts of the waves reflected by the top surface. Physicists refer to this state as being *out of phase.* When the two sets of waves strike the photoreceptors in your eyes, they interfere with each other; interference occurs when waves add together or subtract from each other and so form a new wave of a different frequency, or color. In order for this to happen, the two reflective surfaces have to be a fraction of a light wave apart. That's why you see rainbows from thin films like bubbles and oil on water.

Basically, when white light, which is a mixture of different colors, shines on a film with two reflective surfaces, the various reflected waves interfere with each other to form rainbow fringes. The fringes change colors when you change the angle at which you look at the film, because you are changing the path by which the light must travel to reach your eye. If you decrease the angle at which you look at the film, you increase the amount of film the light must travel through for you to see it. This causes greater interference.

How **LIGHT BULBS** Work

The first practical light bulb, invented in the 1870s, was such an improvement over oil lamps and torches that the world has never looked back. The amazing thing about this giant leap in history is that the light bulb itself could hardly be simpler.

The modern light bulb, which is very similar to the original model, is made up of only a handful of parts.

Light bulbs have two metal contacts at their base that connect to the ends of an electrical circuit. Two stiff wires extend from the contacts up into the bulb, to each end of a thin metal filament. The glass bulb is filled with an inert gas, such as argon.

When you hook the bulb up to a power supply, electric current flows from one contact to the other, through the wires and the filament.

As the electrons zip along, they are constantly bumping into the atoms that make up the filament. The energy of each impact vibrates an atom—in other words, the current heats the atoms up. A thinner conductor heats up more easily than a thicker conductor, because it is more resistant to the movement of electrons.

Bound electrons in the vibrating atoms may be boosted temporarily to higher energy levels. When they fall back to their normal levels, the electrons release the extra energy, in the form of light photons. Metal atoms release mostly infrared light photons, which are invisible to the human eye. But if they are heated to a high enough level—around

4,000°F, in the case of a light bulb—they will emit a good deal of visible light also.

The Right Materials

Most metals melt before they heat up enough to glow. Light bulbs are manufactured with tungsten filaments because tungsten has an abnormally high melting temperature.

But tungsten will catch on fire if the conditions are right. In the first light bulbs, all the air was sucked out of the bulb to create a near vacuum—an area with no matter at all. Since there wasn't any gaseous matter present (or hardly any), the material could not combust.

But vacuum bulbs have a major problem. At extreme temperatures, the occasional tungsten atom will vibrate enough to detach from the atoms around it. In a vacuum bulb, free tungsten atoms fly off the filament in a straight line and collect on the inside of the glass. As more and more atoms evaporate, the filament starts to disintegrate, and the glass starts to get darker.

In a modern light bulb, an inert gas, typically argon, greatly reduces tungsten loss. When a tungsten atom evaporates, chances are it will collide with an argon atom and bounce right back toward the filament, rejoining the solid structure.

Incandescent light bulbs give off most of their energy in the form of heat-carrying infrared light photons. Only about 10% of the light produced is in the visible spectrum. Cool light sources, such as fluorescent lamps and LEDs, don't waste so much energy generating heat—they give off mostly visible light. For this reason, they are slowly edging out the old reliable light bulb.

Cheap, effective, and easy to use, the light bulb has proved a monstrous success. It is still the most popular method of bringing light indoors and extending the day after sundown. But by all indications, it will eventually give way to more advanced, energy-efficient technologies.

Support Wires
Filament
Inert Gas
Thread
Contact Wire
Screw Thread
Contact
Foot Contact
Wire
Foot Contact

How **FLUORESCENT LAMPS** Work

You see fluorescent lamps everywhere—in offices, homes, warehouses, street corners. But even though they're all around us, these devices are a total mystery to most people. Just what is going on inside those white tubes? How do they produce so much light without heating up?

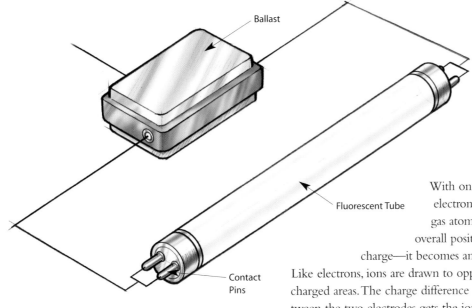

Ballast

Fluorescent Tube

Contact Pins

HSW Web Links

www.howstuffworks.com

How Light Works
How Atoms Work
How Light Bulbs Work
How Light Sticks Work
How Lightning Works
How Gas Lanterns Work

Fluorescent lamps do basically the same thing as conventional incandescent lamps, but they go about it a very different way. An incandescent lamp (a standard light bulb, for example) drives electrical current through a solid filament, exciting atoms with heat so they release light photons. A fluorescent lamp conducts current through a plasma—an ionized gas—to generate photons.

The central element in a fluorescent lamp is a sealed glass tube. The tube contains a small amount of mercury and an inert gas, typically argon, kept under very low pressure. One thing to note is that mercury is funny; it can spontaneously evaporate. The tube also contains a phosphor powder, coated along the inside of the glass. The tube has two electrodes, one at each end. Both are wired to an electrical circuit.

The electrodes in a standard rapid-start fluorescent lamp have a filament design— they are relatively high-resistance wires connected to the electrical circuit. When you switch on the fluorescent lamp, both electrode filaments heat up, boiling off electrons. The free electrons collide with the gas atoms in the tube, knocking loose bound electrons.

With one less electron, each gas atom has an overall positive charge—it becomes an ion. Like electrons, ions are drawn to oppositely charged areas. The charge difference between the two electrodes gets the ions and electrons moving through the tube—the gas becomes a conductive plasma with an electrical arc running through it.

When a charged particle collides with a mercury atom, the collision may boost one of the atom's electrons to a higher level. The electron immediately falls back to its original level, releasing the extra energy in the form of a light photon.

The Phosphor

Energized mercury atoms emit mostly ultraviolet photons, which are invisible to the human eye. In order to be of much use, the lamp needs to convert the ultraviolet light into visible light.

This is where the tube's phosphor powder coating comes in. Phosphors are substances that give off light when they're energized. When a photon hits a phosphor atom, one of the phosphor's electrons jumps to a higher energy level and the atom heats up. When the electron falls back to its normal level, it releases energy in the form of another photon. This photon has less energy than the original photon, because the phosphor atom gives off some of the original energy as heat.

Sources of Light

Fluorescent lamps are just one lighting application of a gas discharge tube. Black lights are essentially fluorescent lamps without a phosphor coating. They mostly emit ultraviolet light, which causes phosphors outside of the lamp to emit visible light. White clothes glow under a black light because most laundry detergents contain phosphorescent material.

Neon lights are gas discharge lamps containing gases, such as neon, that release colored visible light when stimulated by electrons and ions. Many streetlights use a similar system.

141

These lower energy photons are in the visible spectrum—the phosphor gives off white light we can see. Manufacturers can vary the color of the light by using different combinations of phosphors.

Conventional incandescent light bulbs produce their light using heat. They also emit a good bit of ultraviolet light, but they do not convert any of it to visible light. Consequently, a lot of the energy used to power an incandescent lamp is wasted. A fluorescent lamp puts this invisible light to work and produces very little heat, so it's more efficient. Incandescent lamps also lose more energy through heat emission than do fluorescent lamps. Overall, a typical fluorescent lamp is four to six times more efficient than an incandescent lamp.

this way, current will climb on its own in a gas discharge as long as there is adequate voltage.

A fluorescent lamp's *ballast* works to control runaway current increases. The simplest sort of ballast, generally referred to as a *magnetic ballast,* works like a basic inductor. An inductor consists of a coil of wire in a circuit, which may be wound around a piece of metal. When you send electrical current through a wire, it generates a magnetic field. Positioning the wire in concentric loops amplifies this field.

This sort of field affects not only objects around the loop, but also the loop itself. Increasing the current in the loop increases the magnetic field, which applies a voltage opposite the flow of current in the wire. In short, a coiled length of wire in a circuit (an inductor) opposes change in the current flowing through it. The transformer elements in a magnetic ballast use this principle to regulate the current in a fluorescent lamp.

Magnetic ballasts modulate electrical current at a relatively low cycle rate, which can cause a noticeable flicker. Magnetic ballasts may also vibrate at a low frequency. This is the source of the audible humming sound people associate with fluorescent lamps.

Modern ballast designs use advanced electronics to regulate the current more precisely. Since they use a higher cycle rate, you don't generally notice a flicker or humming noise coming from an electronic ballast. Different lamps require specialized ballasts designed to maintain the specific voltage and current levels that are needed for varying tube designs.

Fluorescent lamps come in all shapes and sizes, but they all work on the same basic principle: An electric current stimulates mercury atoms, which causes them to release ultraviolet photons. These photons in turn stimulate phosphor, which emits visible light photons. At the most basic level, that's all there is to it!

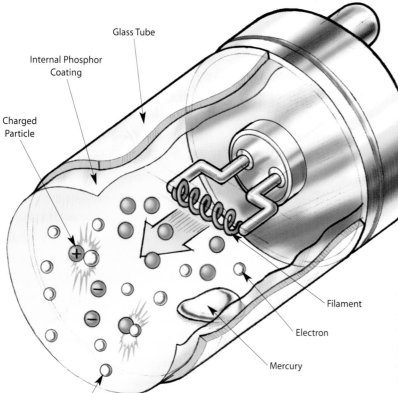

Glass Tube

Internal Phosphor Coating

Charged Particle

Mercury Particle

Filament

Electron

Mercury

The Ballast

In a solid conductor, such as a wire, electrical resistance is a constant at any given temperature. In a gas discharge tube, such as a fluorescent lamp, current causes resistance to decrease. This happens because, as more electrons and ions flow through a particular area, they bump into more atoms, which frees up electrons, creating more charged particles. In

How **LIGHT STICKS** Work

Since their invention in the 1970s, light sticks have become a Halloween staple. They're perfect as safety lights for trick-or-treaters because they're portable, cheap, and they emit a ghostly glow. Light sticks also make an ideal lamp for SCUBA divers, spelunkers, and campers.

Just like any other light source, light sticks glow when individual atoms become excited and release photons. But instead of using heat to excite atoms, as in a light bulb, light sticks use a chemical reaction.

The light stick itself is simply a housing for two liquid chemical compounds. Before you activate the light stick, the two solutions are in separate chambers. The stick itself contains a phenyl oxalate ester and dye solution. A small glass vial in the middle of the stick contains a hydrogen peroxide solution, called the *activator*.

When you bend the plastic stick, the glass vial snaps open, and the two solutions flow together. The atoms in the different compounds are attracted to each other, so they rearrange themselves to form new compounds.

This chemical reaction causes a substantial release of energy. Just as in an incandescent light bulb, atoms in the materials are excited, causing electrons to rise to a higher energy level and then return to their normal levels. When the electrons return to their normal levels, they release energy as light. This process is called *chemilumenesence*.

Here's the exact sequence of events:

- The hydrogen peroxide oxidizes the phenyl oxalate ester, creating a chemical called phenol and an unstable peroxyacid ester.
- The unstable peroxyacid ester decomposes, creating additional phenol and a cyclic peroxy compound.
- The cyclic peroxy compound decomposes to carbon dioxide.
- This decomposition releases energy to the dye.

- The electrons in the dye atoms jump to a higher level then fall back down, releasing energy in the form of light. The color of the light depends on the composition of the dye.

Depending on which compounds are used, the chemical reaction may go on for a few minutes or for many hours. If you heat the solutions, the extra energy will accelerate the reaction, and the stick will glow brighter, but for a shorter amount of time. If you cool the light stick, the reaction will slow down, and the light will dim. If you want to preserve your light stick for the next day, put it in the freezer—it won't stop the process, but it will drag out the reaction considerably.

Light sticks aren't particularly useful as indoor lighting—they don't glow as brightly as a light bulb or fluorescent lamp. But they are handy lightweight alternatives to flashlights, and light sticks make terrific toys.

HSW Web Links

www.howstuffworks.com

How Light Works
How Atoms Work
How Halloween Works
How Television Works

Cool Facts

Light sticks are just one application of an important natural phenomenon—luminescence. Generally speaking, luminescence is any emission of light that is not caused by heating. Among other things, luminescence is used in televisions, neon lights and glow-in-the-dark stickers. It's also the principle that lights up a firefly!

It turns out that fireflies (also known as lightning bugs) make light within their bodies. This process is called bioluminescence and is shared by many other organisms, mostly sea-living or marine organisms. Fireflies light up to attract a mate. To do this, the fireflies contain specialized cells in their abdomen that make light.

The cells contain a chemical called luciferin and make an enzyme called luciferase. To make light, the luciferin combines with oxygen to form an inactive molecule called oxyluciferin. The luciferase speeds up the reaction. The oxygen is supplied to the cells through a tube in the abdomen called the abdominal trachea. It's not known whether the on-off switching of the light is controlled by nerve cells or the oxygen supply.

How **LASERS** Work

Lasers show up in an amazing range of products and technologies. You will find them in everything from CD players to dental drills to high-speed metal cutting machines to measuring systems.

All of these devices use lasers. But what is a laser? And what makes a laser beam different from the beam of a flashlight?

The Basics of an Atom

There are only 92 different kinds of naturally occurring atoms in the universe. Everything we see is made up of these 92 atoms in an unlimited number of combinations. How these atoms are arranged and bonded together determines whether the atoms make up a cup of water, a piece of metal, or the fizz that comes out of your soda can.

A laboratory laser experiment.

Atoms are constantly in motion. They continuously vibrate, move, and rotate. Even the atoms that make up the chairs that we sit in are moving around. Atoms can be in different states of excitation. In other words, they can have different energies. If we apply a lot of energy to an atom, it can leave what is called the *ground-state energy level* and go to an *excited level*. The level of excitation depends on the amount of energy that is applied to the atom via heat, light, or electricity.

An atom, in the simplest model, consists of a nucleus, which contains the protons and neutrons, and a cloud of orbiting electrons. It's helpful to think of the electrons in this cloud circling the nucleus in many different orbits. Although more modern views of the atom do not depict discrete orbits for the electrons, it can be useful to think of these orbits as the different energy levels of the atom. In other words, if we apply some heat to an atom, we might expect that some of the electrons in the lower-energy orbitals would transition to higher-energy orbitals farther away from the nucleus.

Once an electron moves to a higher-energy orbit, it eventually wants to return to the ground state. When it does, it releases its energy as a photon—a particle of light. You see atoms releasing energy as photons all the time. For example, when the heating element in a toaster turns bright red, the red color is caused by atoms, excited by heat, that are releasing red photons. When you see a picture on a TV screen, what you're seeing is phosphor atoms, excited by high-speed electrons, emitting different colors of light. Anything that produces light—fluorescent lights, gas lanterns, incandescent bulbs—does it through the action of electrons changing orbits and releasing photons.

The Laser/Atom Connection

A laser is a device that controls the way that energized atoms release photons. *Laser* is an acronym for *light amplification by stimulated emission of radiation*, which describes very succinctly how a laser works.

Laser light is very different from normal light—for example, the light from an incandescent light bulb or the sun. Laser light has the following properties:

• **The light released is monochromatic.** It contains one specific wavelength of light (one specific color). The wavelength of light is determined by the amount of energy released when the electron drops to a lower orbit.

- **The light released is coherent.** It is "organized"—each photon moves in step with the others. This means that all of the photons have wave fronts that launch in unison.
- **The light is very directional.** Laser light has a very tight beam and is very strong and concentrated. A flashlight, on the other hand, releases light in many directions, and the light is very weak and diffuse when compared to a laser.

The way that a laser produces its light explains these differences.

Although there are many types of lasers, all have certain essential features. In a laser, the lasing medium is "pumped" to get the atoms into an excited state. Typically, very intense flashes of light or electrical discharges pump the lasing medium and create a large collection of excited-state atoms (atoms with higher-energy electrons). It is necessary to have a large collection of atoms in the excited state for the laser to work efficiently. In general, the atoms are excited to a level that is two or three levels above the ground state. This increases the degree of population inversion. *Population inversion* is the number of atoms in the excited state versus the number in ground state.

Once the lasing medium is pumped, it contains a collection of atoms that have some electrons sitting in excited levels. The excited electrons have energies greater than the more relaxed electrons. Just as the electron absorbed some amount of energy to reach this excited level, it will also release this energy. This emitted energy comes in the form of photons (light energy). The photons that are emitted have a very specific wavelength (color) that depends on the state of the electrons' energy when the photon is released. Two identical atoms with electrons in identical states will release photons with identical wavelengths.

The unique quality of laser light comes from something called *stimulated emission.* Stimulated emission doesn't occur in your ordinary flashlight—in a flashlight, all of the atoms release their photons randomly. In stimulated emission, photon emission is organized.

Types of Lasers

There are many different types of lasers. Lasers are commonly designated by the type of lasing material employed. The laser medium can be a solid, gas, liquid, or semiconductor. Here are the most common types of lasers:

- **Solid-state lasers**—These lasers have lasing material that's distributed in a solid matrix.
- **Gas lasers**—Helium and helium-neon (HeNe) are the most common gas lasers. These lasers have a primary output of visible red light. CO_2 lasers emit energy in the far infrared and are used for cutting hard materials.
- **Excimer lasers**—These lasers use reactive gases, such as chlorine and fluorine, mixed with inert gases, such as argon, krypton, or xenon. When electrically stimulated, a pseudo molecule (a *dimer*) is produced. When lased, the dimer produces light in the ultraviolet range. (The name *excimer* is derived from the terms *excited* and *dimer.*)
- **Dye lasers**—These lasers use complex organic dyes, such as rhodamine 6G, in liquid solution or suspension as lasing media. They are tunable over a broad range of wavelengths.
- **Semiconductor lasers**—Sometimes called *diode lasers,* these lasers are not solid-state lasers. These electronic devices are generally very small and use low power. They are commonly found in laser printers and CD players.

The photon that any atom releases has a wavelength that is dependent on the energy difference between the excited state and the ground state. If this photon (possessing a certain energy and phase) should encounter another atom that has an electron in the same excited state, stimulated emission can occur. The first photon can stimulate the second atom to emit a photon that vibrates with exactly the same frequency and direction as the incoming photon.

The other key to a laser is a pair of mirrors, one at each end of the lasing medium. Photons with a very specific wavelength and phase reflect off the mirrors to travel back and forth through the lasing medium. In the process, they stimulate other atoms to emit photons. A cascade effect occurs, and soon there are many photons of the same wavelength and phase. The mirror at one end of the laser is *half-silvered,* meaning it reflects some light and lets some light through. The light that makes it through is the laser light.

Lasers truly are amazing—they can concentrate an incredible amount of energy into a point of light. The next time you listen to your favorite CD, notice your friend's sparkling diamond engagement ring, or visit the dentist, you'll have a new appreciation and understanding of just how useful and incredible this technology is.

How **SEMICONDUCTORS** Work

Semiconductors have had a monumental impact on our society. Modern electronics are built around semiconductors. You find semiconductors at the heart of microprocessor chips and transistors. Anything that's computerized or uses radio waves depends on semiconductors.

HSW Web Links

www.howstuffworks.com

How Restaurant Pagers
Work

How Electronic Gates
Work

How Microprocessors
Work

How Batteries Work

How EUV Chipmaking
Works

Today, most semiconductor chips and transistors are created with silicon. You may have heard expressions like *Silicon Valley* and the *silicon economy,* and that's why—silicon is the heart of any electronic or computer device.

Understanding Silicon

Silicon is a very common element—for example, it is the main element in sand and quartz. If you look silicon up in the periodic table, you will find that it sits next to aluminum, below carbon, and above germanium. Carbon, silicon, and germanium (which, like silicon, is also a semiconductor) have a unique property in their electron structure—each has four electrons in its outer orbital. This allows them to form nice

An inactive semiconductor.

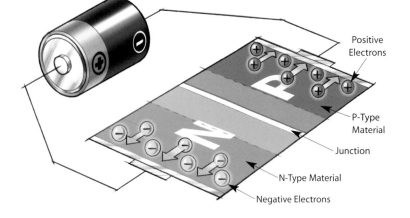

Positive
Electrons

P-Type
Material

Junction

N-Type Material

Negative Electrons

crystals. The four electrons form perfect covalent bonds with four neighboring atoms, creating a lattice. In carbon, we know the crystalline form as diamond. In silicon, the crystalline form is a silvery, metallic-looking substance. Metals tend to be good conductors of electricity because they usually have free electrons that can move easily between atoms, and electricity involves the flow of electrons. While silicon crystals look metallic, they are not, in fact, metals. All of the outer electrons

in a silicon crystal are involved in perfect covalent bonds, so they can't move around.

A pure silicon crystal is nearly an insulator—very little electricity will flow through it.

Doping Silicon

You can change the behavior of silicon and turn it into a conductor by doping it. In *doping,* you mix a small amount of an impurity into the silicon crystal. There are two types of impurities:

- **N-type**—In N-type doping, phosphorus or arsenic is added to the silicon in small quantities. Phosphorus and arsenic each have five outer electrons, so they're out of place when they get into the silicon lattice. The fifth electron has nothing to bond to, so it's free to move around. It takes only a very small quantity of the impurity to create enough free electrons to allow an electric current to flow through the silicon. N-type silicon is a good conductor. Electrons have a negative charge, hence the *N* in the name *N-type.*

- **P-type**—In P-type doping, boron or gallium is the dopant. Boron and gallium each have only three outer electrons. When mixed into the silicon lattice, they form "holes" in the lattice where a silicon electron has nothing to bond to. These holes can conduct current. A hole happily accepts an electron from a neighbor, moving the hole over a space. P-type silicon is a good conductor.

A minute amount of either N-type or P-type doping turns a silicon crystal from a good insulator into a viable (but not great) conductor—hence the name *semiconductor.*

N-type and P-type silicon are not that amazing by themselves, but when you put them together, you get some very interesting behavior at the junction.

Creating a Diode

A diode is the simplest possible semiconductor device. A diode allows current to flow in one direction but not the other. You may have seen turnstiles at a stadium or a subway station that let people go through in only one direction. A diode is a one-way turnstile for electrons.

Diodes can be used in a number of ways. For example, a device that uses batteries often contains a diode that protects the device if you insert the batteries backward. The diode simply blocks any current from leaving the battery if it is reversed—this protects the sensitive electronics in the device.

When reverse-biased, an ideal diode would block all current. A real diode lets perhaps 10 microamps through—not a lot, but still not perfect. And if you apply enough reverse voltage, the junction breaks down and lets current through. Usually, the breakdown voltage is a lot more voltage than the circuit will ever see, so it is irrelevant.

When forward-biased, there is a small amount of voltage necessary to get the diode going. In silicon, this voltage is about 0.7 volts. This voltage is needed to start the hole-electron combination process at the junction.

It is putting N-type and P-type silicon together that gives a diode its unique properties.

Even though N-type silicon by itself is a conductor, and P-type silicon by itself is also a conductor, the combination does not conduct any electricity. The negative electrons in the N-type silicon get attracted to the positive terminal of the battery. The positive holes in the P-type silicon get attracted to the negative terminal of the battery. No current flows across the junction because the holes and the electrons are each moving in the wrong direction.

If you flip the battery around, the diode conducts electricity just fine. The free electrons in the N-type silicon are repelled by the negative terminal of the battery. The holes in the P-type silicon are repelled by the positive terminal. At the junction between the N-type and P-type silicon, holes and free electrons meet. The electrons fill the holes. Those holes and free electrons cease to exist, and new holes and electrons spring up to take their place. The effect is that current flows through the junction.

Transistors and Chips

A transistor is created by using three layers rather than the two layers used in a diode. You can create either an NPN or a PNP sandwich. A transistor can act as a switch or an amplifier.

A transistor looks like two diodes back-to-back. You'd imagine that no current could

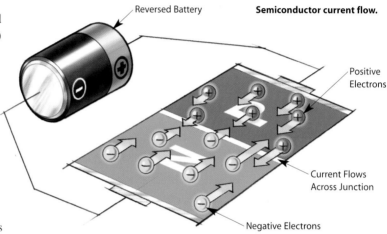

Reversed Battery

Semiconductor current flow.

Positive Electrons

Current Flows Across Junction

Negative Electrons

flow through a transistor because back-to-back diodes would block current both ways. And this is true. However, when you apply a small current to the center layer of the sandwich, a much larger current can flow through the sandwich as a whole. This gives a transistor its switching behavior. A small current can turn a larger current on and off.

A silicon chip is a piece of silicon that can hold thousands of transistors. With transistors acting as switches, you can create Boolean gates, and with Boolean gates you can create microprocessors.

The natural progression from silicon to doped silicon to transistors to chips is what has made microprocessors and other electronic devices so inexpensive and ubiquitous in today's society. The fundamental principles are surprisingly simple. The miracle is the constant refinement of those principles to the point where, today, tens of millions of transistors can be inexpensively formed onto a single chip.

How **LIGHT-EMITTING DIODES** Work

Light-emitting diodes, commonly called LEDs, are real unsung heroes in the electronics world. They do dozens of different jobs and are found in all kinds of devices. Among other things, they form the numbers on digital clocks, transmit information from remote controls, light up watches, and tell you when your appliances are turned on. Collections of LEDs can form images on a jumbo television screen or illuminate a traffic light!

HSW Web Links

www.howstuffworks.com

How Semiconductors
  Work
How Atoms Work
How Light Works
How Digital Clocks Work
How Fluorescent Lamps
  Work

If you've read "How Semiconductors Work" in this chapter, you know that a diode is an N-type semiconductor layer bonded to a P-type semiconductor layer, with electrodes on each end. The N-type layer has an excess of free electrons and the P-type layer has open holes where the electrons can go. The diode conducts current if you hook the negative end of a circuit up to the N-type layer and the positive end up to the P-type layer. The free electrons in the N-type layer move from hole to hole, carrying charge across the diode.

The electrons move in the atoms' conduction bands, which are at a higher energy level than the open holes. To fill an available hole, then, the electron has to lose some energy. Just like the excited electrons in a light bulb or fluorescent lamp, the traveling electrons release this energy in the form of a light photon.

This process happens in any diode, but you can only see the photons when the diode is composed of certain material. The atoms in a standard silicon diode are arranged in such a way that the electron drops a relatively short distance. As a result, the photon's frequency is so low that it is invisible to the human eye—it is in the infrared portion of the light spectrum. This isn't necessarily a bad thing, of course: Infrared LEDs are ideal for remote controls and other machines.

The materials in visible light emitting diodes (VLEDs) have a wider gap between the conduction band and the lower orbitals. The size of the gap determines the frequency of the photon—in other words, it determines the color of the light.

Typically, the diode in an LED is housed in a plastic bulb. The bulb concentrates the light from the diode in a particular direction. Most of the light bounces off the sides of the bulb, traveling through to the rounded end.

LEDs have several advantages over conventional incandescent lamps. For one thing, they don't have a filament that will burn out, so they last much longer. Additionally, their small plastic bulb makes them a lot more durable. They also fit more easily into modern electronic circuits.

But the main advantage is efficiency. In conventional incandescent lamps, the light-production process generates a lot of heat because the filament must be hot to create light. The heat is completely wasted energy, unless you're using the lamp as a heater, because a huge portion of the available electricity isn't going toward producing visible light. LEDs generate very little heat, relatively speaking. A much higher percentage of the electrical power is going directly to generating light, which cuts down on the electricity demands considerably.

Up until recently, LEDs were too expensive to use for most lighting applications because they're built from advanced semiconductor materials. The price of semiconductor devices has plummeted over the past decade, however, making LEDs a cost-effective lighting option. While they may be more expensive than incandescent lights upfront, LEDs are often a better buy in the long run because of their efficiency.

How **SOLAR CELLS** Work

For the past two decades, people have been talking about the solar revolution—the change that will happen when we all use free electricity from the sun. Solar power has seductive promise: On a bright, sunny day, the sun shines approximately 1,000 watts of energy per square meter of the planet's surface, and if we could collect all of that energy we could easily power our homes and offices for free. Solar cells are what we currently use to harness this energy.

Solar cells are everywhere. Lots of things—from satellites to calculators—have solar cells. Solar powered calculators never need batteries, and in some cases they don't even have an off button. As long as there's enough light, they seem to work forever. There are larger solar cells, called *solar panels,* on emergency road signs, call boxes, buoys, and even in parking lots to power lights. But how do these cells convert the sun's energy into electricity?

Converting Photons to Electrons

The solar cells that you see on calculators and satellites are *photovoltaic cells* or *modules* (a module is simply a group of cells electrically connected and packaged in one frame). Photovoltaics convert sunlight directly into electricity (as the name implies: *photo* = light and *voltaic* = electricity). Once used almost exclusively in space, photovoltaics are used more and more in less exotic ways. They could even power your house. So, how do these devices work?

Photovoltaic (PV) cells are made of special materials called *semiconductors.* Currently, the most commonly used semiconductor is silicon. Basically, when light strikes the cell, a certain portion of it is absorbed within the semiconductor material. This means that the energy of the absorbed light is transferred to the semiconductor. The energy knocks electrons loose, allowing them to flow freely. PV cells also all have one or more electric fields that act to force electrons freed by light absorption to flow in a certain direction. This flow of electrons is a current, and when metal contacts are placed on the top and bottom of the PV cell, that current can be drawn off to be used externally. For example, the current can power a calculator. This current, together with the cell's voltage (which is a result of its built-in electric field or fields), defines the power (or *wattage*) that the solar cell can produce.

That's the basic process, but there's really much more to it. Let's take a deeper look into one example of a PV cell: the single crystal silicon cell.

Silicon in Solar Cells

Pure silicon is a poor conductor of electricity because none of its electrons are free to move about. Instead, the electrons are all locked in the crystalline structure. You can change the behavior of silicon and turn it into a conductor through a process known as *doping,* in which impurities are added to the silicon.

HSW Web Links

www.howstuffworks.com

How the Sun Works
How Solar Sails Will Work
How Solar Eclipses Work
How Solar Yard Lights Work

Generic PV solar cell configuration.

Back Contact

N-Type Silicon

Anti-Reflective Coating

P-Type Silicon

Contact Grid

Sunlight

Glass Cover Plate

149

Other Solar Cell Materials

Single crystal silicon isn't the only material used in PV cells. Polycrystalline silicon is also used in an attempt to cut manufacturing costs, although the resulting cells aren't as efficient as single crystal silicon. Amorphous silicon, which has no crystalline structure, is also used, again in an attempt to reduce production costs. Other materials used include gallium arsenide, copper indium diselenide, and cadmium telluride.

Phosphorous and then boron are added to the silicon, forming N-type and then P-type silicon. For more information, see "How Semiconductors Work" in this chapter.

When you put the N-type silicon and the P-type silicon together, an amazing thing happens. Right at the junction where they mix, a barrier forms—resulting in an electric field separating the two sides.

Remember that every PV cell has at least one electric field. Without an electric field, the cell wouldn't work, and this field forms when the N-type and P-type silicon are in contact. Suddenly, the free electrons in the N side, which have been looking for holes to fall into, see all the free holes on the P side, and there's a mad rush to fill them in.

Before the electrons fill the holes, the silicon is all electrically neutral. The extra electrons are balanced out by the extra protons in the phosphorous. The missing electrons (holes) are balanced out by the missing protons in the boron. When the holes and electrons mix at the junction between N-type and P-type silicon, however, that neutrality is disrupted. Do all the free electrons fill all the free holes? No. If they did, then the whole arrangement wouldn't be very useful. Right at the junction, however, they do mix and form a barrier, making it harder and harder for electrons on the N side to cross to the P side.

Eventually, equilibrium is reached, and we have an electric field separating the two sides.

This electric field acts as a diode, allowing (and even pushing) electrons to flow from the P side to the N side, but not the other way around. It's like a hill—electrons can easily go down the hill (to the N side), but can't climb it (to the P side).

So, now what you have is an electric field acting as a diode in which electrons can only move in one direction. But what happens when light hits the cell?

When Light Hits the Cell

When light, in the form of photons, hits the solar cell, its energy frees electron-hole pairs.

Each photon with enough energy will normally free exactly one electron, and

result in a free hole. If this happens close enough to the junction, or if a free electron and a free hole happen to wander into its range of influence, the field will send the electron to the N side and the hole to the P side. This causes further disruption of electrical neutrality, and if provided with an external current path, electrons will flow through the path to their original side (the P side) to unite with holes that the electric field sent there, doing work along the way. The electron flow provides the current, and the cell's electric field causes a voltage. Voilà!—the current and voltage produce power.

Finishing the Cell

Silicon happens to be a very shiny material, which means that it is very reflective. Photons that are reflected can't be used by the cell. For that reason, an antireflective coating is applied to the top of the cell to reduce reflection losses to less than 5%.

The final step is the glass cover plate that protects the cell from the elements. PV modules are made by connecting several cells (usually 36 of them) in series and parallel to achieve useful levels of voltage and current, and putting the cells in a sturdy frame complete with a glass cover and positive and negative terminals on the back.

Photovoltaics are a wonderful source of energy, but it isn't yet feasible for the entire world to run on solar power. While it's true that sunlight is free, the electricity generated by PV systems is not. Although costs are coming down as research progresses, right now PV systems simply can't compete with the utilities. Researchers are confident that PV will one day be cost-effective in both rural and urban areas. Part of the problem is that manufacturing needs to be done on a large scale to reduce costs as much as possible. That kind of demand for PV, however, won't exist until prices fall to competitive levels. Even so, demand and module efficiencies are constantly rising, prices are falling, and the world is becoming increasingly aware of environmental concerns associated with conventional power sources, making photovoltaics a technology with a bright future.

How **LIGHT MICROSCOPES** Work

Ever since their invention in the late 1500s, light microscopes have enhanced our knowledge in basic biology, biomedical research, medical diagnostics, and materials science. Light microscopes can magnify objects up to 1,000 times, revealing details that are impossible to see with the naked eye.

Light-microscopy technology has evolved far beyond the first microscopes of Robert Hooke and Antoni van Leeuwenhoek. Special techniques and optics have been developed to reveal the structures and biochemistry of living cells. Microscopes have even entered the digital age, using charge-coupled devices (CCDs) and digital cameras to capture images. Yet the basic principles of these advanced microscopes are a lot like those of the student microscope you may have used in your first biology class.

The Basics

A microscope must gather light from a tiny area of a thin, well-illuminated specimen that is close-by. To manage this, a microscope has an *objective lens*—a small and spherical lens that brings the image of the object into focus at a short distance within the microscope's tube. The image is then magnified by a second lens, called an *ocular lens* or *eyepiece,* as it is brought to your eye.

In addition to a light source, a microscope has a *condenser.* The condenser is a lens system that focuses the light from the source onto a tiny, bright spot of the specimen, which is the same area that the objective lens examines.

Microscopes typically have interchangeable objective lenses and fixed eyepieces. By changing the objective lenses (going from relatively flat, low-magnification objectives to rounder, high-magnification objectives), a microscope can bring increasingly smaller areas into view.

Image Quality

When you look at a specimen using a microscope, the quality of the image you see depends on:

- **Brightness**—How light or dark is the image? Brightness is controlled by the illumination system and can be changed by changing the voltage to the lamp and adjusting the condenser and diaphragm/pinhole apertures (openings). Brightness is also related to the numerical aperture of the objective lens (the larger the numerical aperture, the brighter the image).
- **Focus**—Is the image blurry or well-defined? Focus is related to focal length and can be controlled with the focus knobs. The thickness of the cover glass on the specimen slide can also affect your ability to focus the image—it can be too thick for the objective lens. The correct cover-glass thickness is written on the side of the objective lens.
- **Resolution**—How close can two points in the image be before they are no

HSW Web Links

www.howstuffworks.com

How Light Works
How Telescopes Work
How Cells Work
How Sunglasses Work
How Corrective Lenses Work

Specimen Preparation

When observing a specimen by transmitted light, light must pass through the specimen in order to form an image. The thicker the specimen, the less light passes through. The less light that passes through, the darker the image. Therefore, the specimens must be thin (0.1 to 0.5 mm). Many living specimens must be cut into thin sections before observation. Specimens of rock or semiconductors are too thick to be sectioned and observed by transmitted light, so they are observed by the light reflected from their surfaces.

longer seen as two separate points? Resolution is related to the numerical aperture of the objective lens (the higher the numerical aperture, the better the resolution) and the wavelength of light passing through the lens (the shorter the wavelength, the better the resolution).

- **Contrast**—What is the difference in lighting between adjacent areas of the specimen? Contrast is related to the illumination system and can be adjusted by changing the intensity of the light and the diaphragm/pinhole aperture. Also, chemical stains applied to the specimen can enhance contrast.

Types of Microscopy

A major problem in observing specimens under a microscope is that their images do not have much contrast. This is especially true of living things (such as cells), although natural pigments, such as the green in leaves, can provide good contrast. One way to improve contrast is to treat the specimen with colored pigments or dyes that bind to specific structures within the specimen.

Different types of microscopy have been developed to improve the contrast in specimens. The specializations are mainly in the illumination systems and the types of light passing through the specimen. For example, a darkfield microscope uses a special condenser to block out most of the bright light and illuminate the specimen with oblique light, much like the moon blocks the light from the sun in a solar eclipse. This optical setup provides a totally dark background and enhances the contrast of the image to bring out fine details—bright areas at boundaries within the specimen.

The various types of light microscopy techniques include:

- **Brightfield**—This is the basic microscope configuration. This technique has very little contrast.
- **Darkfield**—This configuration enhances contrast, as mentioned above.
- **Rheinberg illumination**—This setup is similar to darkfield, but uses a series of filters to produce an "optical staining" of the specimen.

Microscopes come in two basic configurations: upright and inverted. An upright microscope has the illumination system below the stage and the lens system above the stage. An inverted microscope has the illumination system above the stage and the lens system below the stage. Inverted microscopes are better for looking through thick specimens, such as dishes of cultured cells, because the lenses can get closer to the bottom of the dish, where the cells grow.

Light microscopes can reveal the structures of living cells and tissues, as well as nonliving samples such as rocks and semiconductors. Microscopes can be simple or complex in design, and some can do more than one type of microscopy, each of which reveals slightly different information. The light microscope has greatly advanced our biomedical knowledge and continues to be a powerful tool for scientists.

Fluorescence Microscopy

A fluorescence microscope uses a mercury or xenon lamp to produce ultraviolet light. The light comes into the microscope and hits a *dichroic* mirror—a mirror that reflects one range of wavelengths and allows another range to pass through. The dichroic mirror reflects the ultraviolet light up to the specimen. The ultraviolet light causes some molecules in the specimen to fluoresce.

The objective lens collects the fluorescent-wavelength light produced. This fluorescent light passes through the dichroic mirror and a barrier filter (that eliminates wavelengths other than fluorescent), making it to the eyepiece to form the image.

Fluorescence-microscopy techniques are useful for seeing structures and measuring physiological and biochemical events in living cells. Various fluorescent indicators are available to study many physiologically important chemicals such as DNA, calcium, magnesium, sodium, pH, and enzymes. In addition, antibodies that are specific to various biological molecules can be chemically bound to fluorescent molecules and used to stain specific structures within cells.

How **GENE POOLS** Work

You hear people talk about the "gene pool" all of the time, both seriously and comically. On the serious side, animals nearing extinction can develop problems because of the shrinking gene pool for the species. On the funny side, you may hear people say things like "Get that guy out of the gene pool!"

What is a gene pool, and how is it able to grow and shrink?

In order to understand how gene pools work, it's important to know how cells work.

How Life Works: DNA and Enzymes

Using the E. coli bacteria as an example, here's a quick summary highlighting how cells work:

- A bacterium is a small single-celled organism. In the case of E. coli, the bacteria are about one-hundredth the size of a typical human cell. You can think of the bacteria as a cell wall (think of the cell wall as a tiny plastic bag) filled with various proteins, enzymes, and other molecules, plus a long strand of DNA, that are all floating in water.
- The DNA strand in E. coli contains about 4 million base pairs, and these base pairs are organized into about 1,000 genes. A gene is simply a template for a protein, and often these proteins are enzymes.
- An enzyme is a protein that speeds up a particular chemical reaction. For example, one of the 1,000 enzymes patterned in an E. coli's DNA might know how to break a maltose molecule (a simple sugar) into its two glucose molecules. That is all that particular enzyme can do, but that action is important when an E. coli is consuming maltose. After the maltose is broken into glucose, other enzymes act on the glucose molecules to turn them into energy for the cell to use.
- To make an enzyme that it needs, the chemical mechanisms inside an E. coli cell make a copy of a gene from the DNA

strand and use this template to form the enzyme. The E. coli might have thousands of copies of some enzymes floating around inside it and only a few copies of others. The collection of 1,000

Nucleotide Base Pairs (Rungs)

DNA helix.

Backbone of DNA Strand

or so different types of enzymes floating in the cell makes all of the cell's chemistry possible. This chemistry makes the cell alive—it allows the E. coli to sense food, move around, eat, and reproduce.

You can see that, in any living cell, DNA helps create enzymes, and enzymes create the chemical reactions that are life.

Creating Offspring

Bacteria reproduce asexually. This means that, when a bacteria cell splits, both halves of the split are identical—they contain exactly the same DNA. The offspring is a clone of the parent.

Higher organisms like plants, insects, and animals reproduce sexually, and this process makes the actions of evolution more interesting. Sexual reproduction can create a tremendous amount of variation within a species. For example, if two parents have multiple children, all of the children can be remarkably different. Two brothers can have

HSW Web Links

www.howstuffworks.com

How Evolution Works
How Cells Work
How Cloning Works
How Sex Works
How DNA Evidence Works

153

different hair color, different heights, different blood types, and so on. Here's why that happens:

- Instead of having a long loop of DNA, as in a bacterium, cells of plants and animals have chromosomes that hold the DNA strands. Humans have 23 chromosomes. Fruit flies have 5. Dogs have 39, and some plants have as many as 100.

meet to give the new child two copies of each chromosome.

- To form the single strand in the sperm or egg, one or the other gene from each pair is randomly chosen. One or the other gene from the pair of genes in each chromosome gets passed on to the child.

Because of the random nature of gene selection, each child gets a different mix of genes

Enzyme creation.

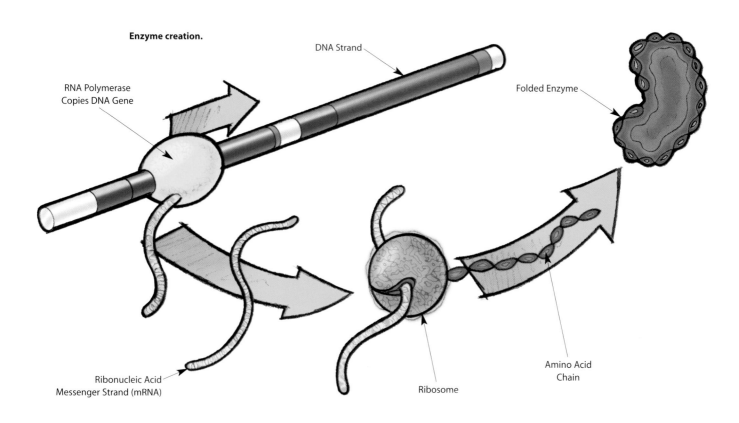

RNA Polymerase Copies DNA Gene

DNA Strand

Folded Enzyme

Ribonucleic Acid Messenger Strand (mRNA)

Ribosome

Amino Acid Chain

- Chromosomes come in pairs. Each chromosome is represented by two strands of DNA. One strand comes from the mother and one from the father.

- Because there are two strands of DNA for each chromosome, animals have two copies of every gene, rather than one copy as in an E. coli cell.

- When a female creates an egg or a male creates a sperm, the two strands of DNA for each chromosome must combine into a single strand. The sperm and egg from the mother and father each contribute one copy of each chromosome. They

from the DNA of the mother and father. This is why children from the same parents can have so many differences.

A gene is nothing but a template for creating an enzyme. This means that, in any plant or animal, there are actually two templates for every enzyme. In some cases, the two templates are the same (they're *homozygous*), but in many cases the two templates are different (they're *heterozygous*).

Sometimes a mutation (a change) can occur. It's important to note that a mutation in a single gene may have no effect on an organism, or its offspring, or its offspring's offspring. For example, imagine an animal that has two identical copies of a gene. A mutation changes one of the two genes in a harmful way. Assume that a child receives this mutant gene from the father. The mother contributes a normal gene, so the mutatant gene may have no effect on the child. The mutant gene might persist through many generations and never be noticed at all—until, at some point, both parents of a child contribute a copy of the mutant gene.

Another thing to notice is that many different forms of a gene can be floating around in a species.

Understanding the Gene Pool

The combination of all of the versions of all of the genes in a species is the gene pool of the species.

Because scientists have studied the DNA of a fruit fly and understood it very well, we'll use fruit fly DNA as an example. Here are some facts about fruit fly DNA:

- The DNA of a fruit fly is arranged on five chromosomes.
- There are about 250 million base pairs in this DNA.
- There are 13,601 individual genes.

Each gene appears at a certain location on a certain chromosome, and there are two copies of the gene. The location of a particular gene is called the *locus* of the gene. Each of the two copies of the gene is called an *allele*.

Let's say we look at locus 1 on chromosome 1 on a particular fruit fly's DNA. There are two alleles at that location, and there are two possibilities for those alleles:

- The two alleles are homozygous (the same).
- The two alleles are heterozygous (different).

If we look across a population of 1,000 fruit flies living in a jar, we might identify a total of 20 different alleles that occupy locus 1 on chromosome 1. Those 20 alleles are the gene pool for that locus. The set of all alleles at all loci is the full gene pool for the species.

One Size Doesn't Fit All

Over time, the size of a gene pool changes. The gene pool increases when a mutation changes a gene and the mutation survives. The gene pool decreases when an allele dies out. For example, let's say that we took the 1,000 fruit flies described in the previous paragraph and selected five of them. These five fruit flies might possess a total of only three alleles at locus 1. If we then let those flies breed and reproduce to the point where the population is once again 1,000, the gene pool of these 1,000 flies is much smaller. At locus 1, there are only three alleles among the 1,000 flies instead of the original 20 alleles.

This is exactly what happens when a species faces extinction. The total population dwindles down to the point where there might be just 100 or 1,000 surviving members of the species. In the process, the number of alleles at each locus shrinks and the gene pool of the species contracts significantly. If conservation efforts are successful and the species rebounds, then it does so with a much smaller pool of genes to work with than it had originally.

A small gene pool is generally bad for a species because it reduces variation. Let's go back to our fruit fly example. Let's say there are 20 alleles at locus 1, and one of those alleles causes a particular disease when a fly has two copies of that allele. Because there are 20 total alleles, the probability of a fly getting two copies of that harmful allele is relatively small. If that harmful allele survives when the gene pool shrinks down to a total of only three alleles, then the probability of flies getting the disease from that allele becomes much larger. A large gene pool provides a good buffer against genetic diseases. Some of the common genetic problems that occur when the gene pool shrinks include:

- Low fertility
- Deformities
- Genetic diseases

The two most common places to see these effects is in animals nearing extinction and in animal breeds, like a specific breed of dog.

A lot of care must be taken when breeding animals in order to avoid genetic diseases. When breeding, it is sometimes helpful to outcross. In outcrossing, an animal outside the breed is allowed to mate with an animal inside the breed. The offspring from that mating increase the size of the gene pool, decreasing the probability of genetic diseases being passed on.

How CLONING Works

Cloning has been used for many years to produce plants. Growing a plant from a cutting is a type of cloning, and people have been doing that forever. Animal cloning has been the subject of scientific experiments for about a century, but gar-nered little attention until the birth of the first cloned mammal in 1997, a sheep named Dolly. Since Dolly, several scientists have cloned other animals, including cows and mice. What is cloning and why do it?

HSW Web Links

www.howstuffworks.com

How Human
  Reproduction Works
How Cells Work
How Human Cloning Will
  Work

Cloning is the process of making a genetically identical organism through nonsexual means.

On January 8, 2001, scientists at Advanced Cell Technology, Inc. announced the birth of the first clone of an endangered animal, a baby bull gaur (a large wild ox from India and southeast Asia) named Noah. Although Noah died of an infection unrelated to the procedure, the experiment demonstrated that it is possible to save endangered species through cloning.

Send In the Clones

Nature has been cloning organisms for billions of years. For example, when a strawberry plant sends out a runner (a form of modified stem), a new plant (clone) grows where the runner takes root. Similar cloning occurs in grass (rhizomes), potatoes (tubers), and onions (bulbs).

With many plants, you can take a leaf cutting from a plant and grow it into a new plant. That's a clone too. Vegetative propagation works because the end of the cutting forms a mass of nonspecialized cells called a *callus*. With luck, the callus will grow, divide, and form various specialized cells (like those in roots and stems), and then eventually form a new plant.

More recently, scientists have been able to clone plants by taking pieces of specialized roots, breaking them up into root cells, and growing the root cells in a nutrient-rich culture. In the culture, the specialized cells become unspecialized and become calluses. These calluses can then be stimulated with the appropriate plant hormones to grow into new plants that are identical to the original plant from which the root pieces were taken. This procedure, called *tissue culture propagation*, has been widely used by horticulturists to grow prized orchids and other rare flowers.

Plants are not the only organisms that can be cloned naturally. The unfertilized eggs of some animals (small invertebrates, worms, some species of fish, lizards, and frogs) can develop into full-grown adults under certain environmental conditions (usually involving a chemical stimulus of some kind). This process is called *parthenogenesis*, and the off-spring are clones of the females that laid the eggs. Another example of natural cloning is identical twins. Although they are genetically different from their parents, identical twins are naturally occurring clones of each other.

Scientists have experimented with animal cloning, but have never been able to stimulate a specialized cell (like a skin cell) to produce a new organism directly. Instead, they rely on transplanting the genetic information from a specialized cell into an unfertilized egg cell whose genetic information has been destroyed or physically removed. In the 1970s, a scientist named John Gurdon successfully cloned tadpoles. He transplanted the nucleus from a specialized cell (a skin or intestinal cell) of one frog into an unfertilized egg of another frog in which the nucleus was destroyed by ultraviolet light. The egg with the transplanted nucleus developed into a tadpole that was genetically identical to the first frog. However, the tadpoles did not survive to grow into adult frogs. Gurdon's experiment showed that the process of specialization in

Sexual versus Asexual Reproduction

Sexual reproduction involves the merging of two sets of DNA or genetic information (one from the father's sperm and the other from the mother's egg) to produce a new off-spring that is genetically different from either parent. Asexual reproduction (without sex) produces off-spring that are genetically identical to the single parent organism.

animal cells was reversible, and his technique of nuclear transfer paved the way for later cloning successes.

In 1997, cloning was revolutionized when Ian Wilmut and his colleagues at the Roslin Institute in Edinburgh, Scotland successfully cloned a sheep named Dolly. Wilmut and colleagues transplanted a nucleus from a mammary gland cell of a Finn Dorsett sheep into the egg of a Scottish blackface ewe. The nucleus-egg combination was stimulated with electricity to fuse and to begin cell division. The new cell divided and was placed in the uterus of a blackface ewe to develop, and Dolly was born months later. Dolly was shown to be genetically identical to the Finn Dorsett mammary cells and not to the blackface ewe, which clearly demonstrated that she was a successful clone. Dolly has since grown and reproduced several offspring of her own through normal sexual means. Therefore, Dolly is a viable, healthy clone.

Since Dolly, several university laboratories and companies have used various modifications of the nuclear transfer technique to produce other cloned mammals, including cows, pigs, monkeys, and mice.

Why Clone?

The main reason to clone plants or animals is to mass-produce organisms with desired qualities, such as prize-winning orchids or genetically engineered animals (sheep, for example, have been engineered to produce human insulin). If you had to rely on sexual reproduction (breeding) alone to mass produce these animals, then you would run the risk of breeding out the desired traits, because sexual reproduction reshuffles the genetic deck of cards. Other reasons for cloning might include replacing lost or deceased family pets and repopulating endangered or even extinct species.

Whatever the reasons, the new cloning technologies have sparked many debates about ethics among scientists, politicians, and the general public. Several governments have considered or enacted legislation to slow down, limit, or ban cloning experiments outright. It is clear that cloning will be a part of our lives in the future, but the course of this technology has yet to be determined.

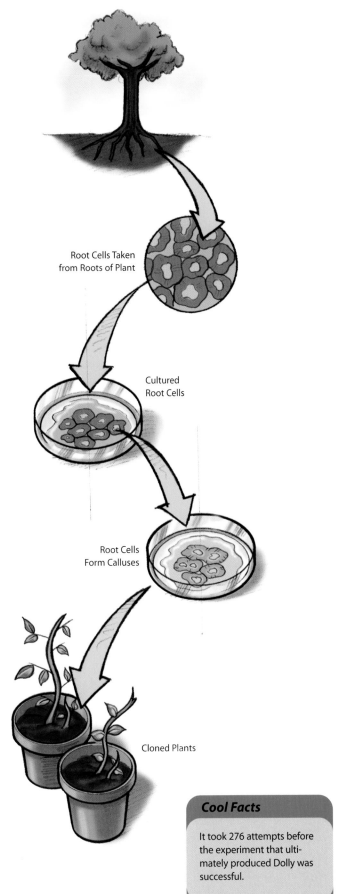

Root Cells Taken from Roots of Plant

Cultured Root Cells

Root Cells Form Calluses

Cloned Plants

Cool Facts

It took 276 attempts before the experiment that ultimately produced Dolly was successful.

How **MUMMIES** Work

The Egyptian mummy has joined the vampire and werewolf as a horror movie staple. But unlike its counterparts, the mummy is absolutely real, though mummies don't really chase after intruding tomb raiders. Mummies are, in a sense, actual tangible ghosts—humans who stick around long after death.

HSW Web Links

www.howstuffworks.com

How Cells Work

How DNA Evidence Works

How Sunburns and Suntans Work

How Food Preservation Works

How Human Cloning Will Work

Cool Facts

The term *mummy* most likely came from early Arabic travelers visiting Egypt. When the outsiders saw some mummies that had been coated with black resin material, they assumed the Egyptian embalming process involved dipping the bodies in bitumen, a dark, sticky component of tar. Based on this misconception, they dubbed the preserved bodies *mummies*, after *mummiya*, the Arabic word for bitumen.

Ordinarily, after a person dies the decomposition process reduces the body to a bare skeleton in a matter of months. The first stage of decomposition actually begins within a few hours. In this initial stage, called *autolysis*, organs that contain digestive enzymes (the intestines, for example) begin to digest themselves.

Putrefaction, the breakdown of organic matter by bacteria, follows autolysis. In warm areas, putrefaction gets going about three days after death. Within a few months, the body is reduced to a skeleton. Bacteria reproduce rapidly in hot, humid conditions, accelerating the process. The process is very slow in colder or drier environments because bacteria need warmth and water to thrive. If the conditions are cold or dry enough, or if there isn't enough oxygen, the environment is so harsh that few bacteria can survive. The body doesn't fully decompose, possibly for thousands of years. In nature, bodies have been preserved in the frozen ice of glaciers and the oxygen-depleted depths of peat bogs.

Hot sand can also preserve a corpse, because it absorbs the body's fluids quickly and dries everything out. Scientists believe this natural process created the first Egyptian mummies. Before developing the science of artificial mummification, the ancient Egyptians would simply bury their dead in the sand, with no tomb or casket. The hot sand drew the moisture out of corpses, preserving their internal organs and crisping their skin to a hard, dark shell.

Most likely, these natural mummies fueled the Egyptian's belief in an afterlife. If the body could survive long after death, it made sense that the human spirit could too.

The Egyptian Method

As their concept of the afterlife evolved, the Egyptians became concerned about the comfort of their departed family members. They began covering the bodies with long wicker baskets and later with sturdy wooden boxes. Eventually, this led to fully enclosed coffins and tomb-like housings.

Of course, with the body fully enclosed, it was not exposed to the drying properties of the sand. The fluids remained in the body, bacteria thrived, and the flesh naturally decomposed. To insure both survival and comfort in the afterlife, Egyptian scientists had to figure out a way to replicate the preservative qualities of the desert.

Through experimentation, they discovered that decomposition works largely from the inside out. Bacteria collect first in the body's internal organs and move on from there. To stop the putrefaction process, the embalmers realized, they would have to remove the internal organs. This, combined with the discovery of the natural drying agent *natron*, led to the famous Egyptian mummies we know today.

The science and theology of embalming continued to evolve over the years, so there is no single Egyptian embalming ritual. But the standard practices of the New Kingdom's 18th through 20th dynasties (in the years 1570 to 1075 B.C.), an era that produced some of the best preserved mummies, are fairly representative.

Removing Organs

Before beginning the embalming process, the Egyptians took the body to the ibu, the place where the body was purified. Here, they washed the body in water gathered from the Nile. This represented a sort of rebirth, as the person passed from one world into the next. Once the body was cleaned, the embalmers carried it to a special enclosure called the per nefer where they began the embalming process.

At the per nefer, they laid the body out on a wooden table to remove the brain. To get into the cranium, the embalmers had to hammer a chisel through the bone of the nose. Then they inserted a long iron hook into the skull and slowly pulled out the brain matter. Once they had removed most of the brain with the hook, they scooped out any remaining bits with a long spoon. Finally, they rinsed the skull with water. Surprisingly, the brain was one of the few organs the Egyptians did not try to preserve. They weren't sure what it did, but they assumed you wouldn't need it in the next world.

After they had removed the brain, the embalmers took a special blade made from obsidian (a sacred stone) and made a small incision along the left side of the body. They carefully removed the abdominal organs through this slit, setting each one aside. Next, the embalmers cut open the diaphragm to remove the lungs.

The Egyptians believed that the heart was the core of a person, the seat of emotion and the mind, so they almost always left it in the body. The other organs were washed, coated with resin, wrapped in linen strips, and stored in decorative pottery. These vessels, which Egyptologists call *canopic jars,* protected the organs for passage to the next world.

Once they removed the organs, the embalmers rinsed the empty chest cavity with palm wine in order to purify it. Then, to maintain the body's lifelike form, they filled the cavity with incense and other material. This kept the skin from shrinking down inside the cavity when the body was dried out.

Drying and Wrapping

After the embalmers removed the organs and restuffed the body, they laid the body down on a sloped board and covered it with natron powder. The Egyptians collected this powder, a mixture of sodium compounds, from the shores of Egyptian lakes in the desert west of the Nile delta. Unlike the hot sand that dried the earliest Egyptian mummies, the salty natron absorbed moisture without completely darkening and hardening the skin.

The embalmers left the body in the powder for 35 to 40 days to allow enough time for it to dry completely. During this waiting period, somebody had to stand guard, as the body's strong odor attracted desert scavengers. After the 40 days were finished, the embalmers removed the stuffing from the body cavity and refilled it with natron and resin-soaked linen. In some eras, the embalmers also stuffed material under the skin to make the desiccated body more lifelike. After stuffing the body, the embalmers closed the incisions and covered the skin with a resin layer to keep moisture out.

While the deceased was drying in the desert, his or her family gathered roughly 4,000 square feet (372 square meters) of linen and brought it to the embalmers. The wealthy typically used material that had clothed sacred statues, while the lower classes collected old clothing and other household linen. The embalmers selected the highest-quality material and stripped it into long bandages measuring 3 to 8 inches (8 to 20 cm) across.

Next, they wrapped the body in a shroud and began methodically winding the bandages around the different parts of the body. After wrapping the individual parts, the

You Get What You Pay For

Embalmers spent a lot of time working on Egyptian royalty and other members of the upper class. They cut precise incisions and carefully removed each of the organs so they could be preserved along with the outer body.

The "budget package" was scaled down considerably. The embalmers injected the body with an oil mixture, filling the entire torso cavity. Then they stopped up all the body's orifices and let the oil sit inside for several days. When they finally unstopped the body, all the oil flowed out, carrying the liquefied remains of the internal organs with it.

159

embalmers began wrapping the body as a whole. As they applied new layers, the embalmers coated the linen with hot resin material to glue the bandages in place. During this entire process, the embalmers uttered spells and laid protective amulets on the body (for protection in the next world), wrapping them up at different layers.

The bandages served three purposes:

- They kept moisture away from the body
- They let the embalmers shape the mummy to give it a more lifelike form
- They kept everything together. Without this binding system, the fragile, desiccated mummies would likely have burst or fallen apart.

After wrapping the mummy, the embalmers attached a rigid cage of *cartonnage* (linen or papyrus held together with glue) to the body and affixed a decorated mask to the head, typically depicting the deceased or an Egyptian god.

The completed mummy was then place in a *suhet*—a coffin decorated to look like a person. A procession of mourners escorted the suhet to the tomb, where a priest dressed as the jackal god Anubis performed the *ceremony of the mouth*, touching the suhet's face with sacred objects to give the deceased the powers of speech, sight, touch, hearing, and taste in the next world. The mourners propped the mummy against the wall and sealed it in the tomb with everything it would need in the next world.

Modern Mummies

In the nineteenth and twentieth centuries, there was a surge of interest in the mummies of ancient Egypt. One effect of this phenomenon was that some people began revisiting the idea of mummification—with the addition of some new technology.

The most famous modern mummies are Vladimir Ilyich Lenin, the Russian revolutionist, and Eva Peron, the revered wife of Argentinean president Juan Peron. The exact chemicals and procedure that keep Lenin's body perfectly preserved are a Russian secret, but we do know that the mummification is an ongoing process. The Russians periodically immerse him in a preservative bath and then dress him in a waterproof suit to hold the fluids inside.

Like Lenin, Eva Peron's body is so perfectly preserved that she appears to be alive. The embalmers replaced the fluid in her body with wax. Peron and similar mummies are really a lot like the wax dummies you see in a wax museum, except, of course, that they are the actual remains of a person.

In the 1970s, a group of scientists expanded on this idea to create a process called *plastination*. In the complicated plastination process, all of the water and lipids in the body's cells are replaced with polymers. The body takes on the properties of plastic: It is durable, flexible, doesn't have a strong odor, and, most importantly, doesn't decompose. Plastination is used to preserve body parts for anatomical research and education, but it's also used to create art. In one controversial exhibit, artists sculpted plastinated human bodies into wild shapes and positioned them in dynamic poses. The artwork showed all of the inner workings of the human body, in both healthy and diseased bodies.

Remarkably, thousands of years after the ancient Egyptians developed their elaborate preservation technology, people are still drawn to mummification as a means of insuring immortality. Today, a growing number of people plan to have their bodies frozen after death, in the hopes that future doctors will be able to reverse whatever killed them. The technology is very different, but the motivation is basically the same.

chapter seven

HEALTH HELP

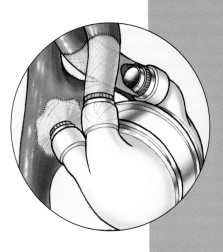

How **EXERCISE** Works

As you exercise or play a sport, you probably notice several things about your body. You breathe heavier and faster, your heart beats faster, your muscles hurt, and you sweat. These are all normal responses to exercise, whether you work out regularly or only once in a while. The body has an incredibly complex set of processes to meet the demands of working muscles. Every system in the body is involved—blood circulation, breathing, heat regulation, and the muscles themselves are all affected.

HSW Web Links

www.howstuffworks.com

How the Heart Works
How Cells Work
How Blood Works
How Performance-
  Enhancing Drugs Work
How Food Works
How Force, Power, Torque
  and Energy Work

Any sport or exercise you can imagine uses different muscle groups to generate motion. As you use your muscles, they begin to make demands on the rest of your body. When you're exercising strenuously, just about every system in your body either focuses its efforts on helping the muscles do their work or shuts down to save energy. For example, your heart beats faster during strenuous exercise so that it can pump more blood to the muscles, and your stomach shuts down during strenuous exercise so that it does not waste energy that the muscles can use.

When you exercise, your muscles act something like electric motors. Your muscles take in a source of energy and they use it to generate force. An electric motor uses electricity to supply its energy. Your muscles are biochemical motors, and they use a chemical called *adenosine triphosphate* (ATP) for their energy source. During the process of "burning" ATP, your muscles need three things:

- They need oxygen, because oxygen is consumed to produce ATP.
- They need to eliminate metabolic wastes (carbon dioxide, lactic acid) that the chemical reactions in the muscles generate.
- They need to get rid of heat. Just like an electric motor, a working muscle generates heat that it needs to get rid of.

In order for you to continue exercising, your muscles must continuously make ATP. For this to happen, your body must supply oxygen to the muscles and eliminate the waste products and heat. The more strenuous the exercise, the greater the demands are on working muscle. If these needs are not met, then exercise will cease—that is, you become exhausted and you won't be able to keep going.

ATP Is Energy!

The entire reaction that turns ATP into energy is a bit complicated, but here is a good summary:

- Chemically, ATP is an adenine nucleotide bound to three phosphates.
- A lot of energy is stored in the bond between the second and third phosphate groups. This energy can be used to fuel chemical reactions.
- When a cell needs energy, it breaks this bond to form adenosine diphosphate (ADP) and a free phosphate molecule.
- In some instances, the second phosphate group can also be broken to form adenosine monophosphate (AMP).
- When the cell has excess energy, it stores this energy by forming ATP from ADP and phosphate.

ATP is required for the biochemical reactions involved in any muscle contraction. As the work of the muscle increases, more and more ATP gets consumed and must be replaced in order for the muscle to keep moving.

Because ATP is so important, the body has several different systems to create ATP. These systems work together in phases. The interesting thing is that different forms of exercise use different systems, so a sprinter gets ATP in a completely different way than a marathon runner!

Getting Oxygen to the Cells

If you are going to be exercising for more than a couple of minutes, your body needs to get oxygen to the muscles or the muscles will stop working. Just how much oxygen your muscles will be able to use depends on two processes: getting blood to the muscles and extracting oxygen from the blood into the muscle tissue. Your working muscles can take oxygen out of the blood three times as well as your resting muscles. Your body has several ways to increase the flow of oxygen-rich blood to working muscle:

- It diverts blood flow from nonessential organs to the working muscle
- It increases the flow of blood from the heart (your *cardiac output*)
- It increases the rate and depth of breathing
- It increases the unloading of oxygen from hemoglobin in the working muscle

These mechanisms can increase the blood flow to your working muscle by almost five times what it typically is. That means that the amount of oxygen available to the working muscle can be increased by almost 15 times! The sections that follow look a little closer at how blood flow to working muscle can be increased.

Making the Pipe Bigger

As you exercise, the blood vessels in your muscles dilate and the blood flow is greater, just as more water flows through a fire hose than through a garden hose. Your body has an interesting way of making those vessels

expand. As ATP gets used up in working muscle, the muscle produces several metabolic byproducts (such as adenosine, hydrogen ions, and carbon dioxide). These by-products leave the muscle cells and cause the capillaries within the muscle to expand or dilate. This is referred to as *vasodilation*. The increased blood flow delivers more oxygen-rich blood to the working muscle.

Taking Blood from the Organs

When you begin to exercise, a remarkable diversion happens. Blood that would have gone to the stomach or the kidneys goes instead to the muscles, and the way that happens shows how the body's processes can sometimes override one another. As your muscles begin to work, the sympathetic nervous system, a part of the automatic or autonomic nervous system (that is, the brainstem and spinal cord) stimulates the nerves to the heart and blood vessels. This nervous stimulation causes those blood vessels to contract or constrict (*vasoconstriction*). This vasoconstriction reduces blood flow to tissues. Your muscles also get the command for vasoconstriction, but the metabolic byproducts produced within the muscle override this command and cause vasodilation. Because the rest of the body gets the message to constrict the blood vessels and the muscles dilate their blood vessels, blood flow from organs that are nonessential to your current activity (like your stomach, intestines, and kidneys) is diverted to working muscle. This helps to further increase the delivery of oxygenated blood to working muscle.

Making the Heart Pump Harder

Your heart, also a muscle, gets a workout during exercise, too, and its job is to get more blood out to the body's hard-working muscles. The heart's blood flow increases by about four or five times from its resting state. Your body increases blood flow by increasing the rate of your heartbeat and the amount of blood that comes through the heart and goes out to the rest of the body. The rate of blood pumped by the heart (the cardiac output) is a product of the rate at which the heart beats (heart rate) and the volume of blood that the heart ejects with each beat (stroke volume). In a resting heart, the cardiac output is about 5 liters a minute (0.07 L $\times$ 70 beats/min $= 4.9$ L/min). As you begin to exercise, sympathetic nerves stimulate the heart to beat faster and with more force; the heart rate can increase about threefold. Also, the sympathetic nerve stimulation to the veins causes them to constrict. This, along with more blood being returned from the working muscles, increases the amount of blood returned to the heart.

Breathing Faster and Deeper

So far, we've talked about getting more blood to working muscle. Your lungs and the rest of your respiratory system need to provide more oxygen for the blood, too. The rate and depth of your breathing will increase because of these events:

- Sympathetic nerves stimulate the respiratory muscles to increase the rate of breathing.
- Metabolic byproducts from muscles (lactic acid, hydrogen ions, carbon dioxide) in the blood stimulate the respiratory centers in the brainstem, which in turn further stimulates the respiratory muscles.
- Slightly higher blood pressure, caused by the increased force of each heartbeat and by the elevated cardiac output, opens blood flow to more air sacs (alveoli) in the lungs. This increases the ventilation and allows more oxygen to enter the blood.

As the lungs absorb more oxygen and the blood flow to the muscles increases, your muscles have more oxygen.

Getting Rid of Waste

Your exercising body is using energy and producing waste, such as lactic acid, carbon dioxide, adenosine, and hydrogen ions. Your muscles need to dump these metabolic wastes to continue exercise. All that extra blood that is flowing to the muscles and bringing more oxygen can also take the wastes away. Hemoglobin in the blood carries away the carbon dioxide, for example.

Getting More Oxygen Quickly

To become a world-class athlete or to get the most out of your exercise, you want your muscles to get the oxygen they need most efficiently. To make that happen, you need to increase:

- Cardiac output
- Respiration
- The amount of oxygen carried by your blood

You can increase all three of these through resistance training, possibly in combination with *cross-training* (training for more than one sport at a time or for multiple fitness components—strength, endurance and flexibility—at the same time).

The main effects of training on cardiac output appear to be an increase in stroke volume (that is, a larger heart) and a decrease in the resting heart rate. The increased stroke volume allows the heart to pump more blood with each beat. Because there is a limit to the maximum heart rate (about 180 to 190 beats/min), a slower resting heart rate (50 to 60 beats/min in a trained athlete versus the normal 70 to 80 beats/min) allows the heart to have a greater increase in heart rate during exercise. The greater increase in heart rate during exercise, along with the larger stroke volume, increases cardiac output and blood flow to working muscle.

Training can help the respiratory system by decreasing the resting rate of breathing, increasing the respiration rate during exercise, and increasing the volume of air exchanged with each breath (*tidal volume*). These changes allow the lungs to take in more air during exercise. Training can also boost the amount of oxygen that the working muscles take from the blood, which probably reflects the increases in metabolic enzymes.

Getting the Most from Muscles

If you exercise regularly or if you are an athlete in training, you're trying to make your muscles work better. You want to be stronger if you're a weightlifter, you want to be able to throw a blistering fastball if you're a baseball pitcher, or you want to be able to finish strong at the end of a 26-mile race if you're a marathon runner. Those three activities illustrate three major factors in muscle performance:

- **Strength**—Muscle strength is the maximal force that a muscle can develop. Strength is directly related to the size (that is, the cross-sectional a rea) of the muscle.

- **Power**—The power of muscle contraction is how fast the muscle can develop its maximum strength. Muscle power depends on strength and speed [power = (force × distance) × time].

- **Endurance**—Muscle endurance is the capacity to generate or sustain maximal force repeatedly.

Lots of people use resistance training, such as free weights, jump training, and isometric training to enhance their perform-ance. Resistance training is very useful in increasing the strength of muscles. Resistance training can also increase the power of muscles. Increases in muscle strength, improve-ments in a person's diet, and cardiovascu-lar fitness can increase muscle endurance, too.

Heating Up

Working muscle produces heat in two ways:

- The chemical energy used when muscles contract is not efficiently turned into mechanical energy. (It is about 20% to 25% efficient.) The excess energy is lost as heat.

- The various metabolic reactions (anaer-obic, aerobic) also produce heat.

Your body needs to remove this excess heat. The heat produced by exercising muscle causes blood vessels in the skin to dilate, which increases the blood flow to the skin. This elevated blood flow to the skin and the large surface area of the skin allows the excess heat to be lost to the surrounding air.

Also, receptors carry the message of excess heat to your body's thermostat, the hypothalamus in the brain. Nerve impulses from the hypothalamus stimulate sweat glands in the skin to produce sweat. The sweat evaporates from the skin, remov-ing heat and cooling the body. Evaporation of sweat removes fluid from the body, so it is important to maintain fluids for blood flow and sweat production by drink-ing water. Sports drinks are good if you're in training or com-petition because they also replace ions (sodium, potassium) that are lost in your sweat and provide additional glucose to fuel anaerobic and aerobic respiration.

Evaporation of sweat is an important cool-ing system that can efficiently remove heat.

How **CPR** Works

You're playing your usual weekend pick-up basketball game with friends. Without warning, one of your teammates suddenly crumples to the ground. You scream out his name, but there's no response. His face turns pale and bluish, and you can't see his chest rise and fall. You listen for a heartbeat and feel his wrist, but you can't find a pulse at all. You quickly grab your cell phone and dial 911.

HSW Web Links

www.howstuffworks.com

How Your Lungs Work
How Your Heart Works
How Heart Attacks and
    Angina Work
How Congestive Heart
    Failure Works
How Diagnosing Heart
    Disease Works

Every year, this type of scenario is played out more than 600 times a day in the United States alone. Without rapid medical intervention, the prognosis is grim. Sudden cardiopulmonary arrest is the leading cause of death for all adults, male or female.

Fortunately, modern medicine has come up with a number of tools to combat cardiopulmonary arrest. Many of these emergency procedures require medical training and/or complex equipment, but one, cardiopulmonary resuscitation (CPR), can be used in the field by almost anyone who has a little bit of training.

PLEASE NOTE: This article is not intended to be used as a method for teaching CPR. For proper CPR training, consult your local hospital or American Red Cross location for available classes.

Cardiopulmonary Arrest

Cardiopulmonary arrest simply means that your heart (*cardio*) and lungs (*pulmonary*) aren't working—your heart isn't beating and you aren't breathing. Many different things can lead to cardiopulmonary arrest, including:

• Stroke
• Drug overdose
• Heart attack
• Near drowning
• Choking
• Blood loss
• Electric shocks
• Carbon monoxide poisoning

The heart is a muscle that expands and contracts under the electrical control of a special group of pacemaking cells. The pumping action of the heart pushes blood loaded with oxygen and other nutrients out to the rest of your body. If your heart isn't beating properly or isn't beating at all, blood isn't supplied to your body, and oxygen and other vital nutrients don't get delivered to your tissues and organs (including your heart).

Cardiopulmonary arrest is an extremely dangerous situation. After 4 to 6 minutes without oxygen, your brain cells begin to die off rapidly. With each additional minute, the damage builds up. Most people cannot survive long in cardiopulmonary arrest.

CPR Basics

CPR is a first-aid technique used—while more advanced medical help is on the way—to keep victims of cardiopulmonary arrest alive and to prevent brain damage. CPR has two goals:

• Keep blood flowing throughout the body.
• Keep air flowing in and out of the lungs.

CPR is a simple technique that requires little or no equipment. What you do is pretty basic:

• Blow forcefully into the victim's mouth to push oxygenated air into the lungs. Doing so allows oxygen to diffuse through the lining of the lungs into the bloodstream.
• Compress the victim's chest to artificially re-create blood circulation.

The only catch is that CPR must be performed in a specific, timed sequence to accurately mimic your body's natural breathing pattern and the way your heart pumps.

CPR Step by Step

What should you do to help a seemingly unconscious victim? The first thing you'll want to do is to figure out whether or not the victim is really unconscious. Just like you were trying to wake someone up, you should call out to the victim and gently shake him or her to try and provoke a response. You also should check to see if the person is breathing. If you try to perform CPR on someone who is not in cardiopulmonary arrest, you can actually do damage!

If you can't rouse the victim, the very next thing to do is have someone call 911 so that professional paramedics will be on their way to the scene while you are performing CPR. This is very important, because, with the exception of choking, CPR doesn't address the underlying causes of cardiopulmonary arrest. It is only meant to buy time until the victim can get intensive medical care.

In order for CPR to work, the victim must be lying on his or her back on a flat surface. If the victim is face-down, gently roll the person toward you while making sure that you support the neck. Once the person is on their back, you can then use the American Heart Association's ABCs of CPR to guide you through the rest of process:

- **A**irway—Clear obstructed airways.
- **B**reathing—Perform mouth-to-mouth breathing.
- **C**irculation—Start chest compressions.

Here's a summary of how you might perform CPR on a nonresponsive adult (a different procedure is used to save infants and young children). *To learn all about CPR and practice it in detail, you should sign up for training from an organization like the American Red Cross.*

A is for Airway

When a person passes out, the tongue relaxes, and it can roll back in the mouth and block the windpipe. Before you can start CPR on an unconscious person, you'll probably need to move the person's tongue out of the way. Here's how to clear a blocked airway:

- Place the palm of your hand across the victim's forehead and push down gently.
- With the other hand, slowly lift the chin forward and slightly up.
- Move the chin up until the teeth are almost together, but the mouth is still slightly open.

Tilting the head back and lifting the chin moves the tongue out of the airway.

At this point, you should check again for breathing. If the victim is choking on something, you may see their chest heave as they try to breathe, but you won't be able to feel or hear air being exhaled. You'll have to take additional measures to clear out what's blocking their windpipe, including:

- Compressing the abdomen with forceful thrusts. This creates pressure that forces the object up and out of the windpipe.
- Trying to manually dislodge the object with your fingers.

Once you remove the obstruction, you have to check for signs of breathing again. Just clearing out the windpipe may sometimes be enough to allow the victim to start breathing on his or her own. If the victim starts breathing and moving around, you can stop CPR. If this doesn't happen, you'll have to help the victim breath by providing mouth-to-mouth resuscitation.

Mouth-to-mouth resuscitation.

B is for Breathing

Rescue breathing uses your lungs to force air into the victim's lungs at regular intervals. The timing of each breath (about 1.5 to 2 seconds per breath) mimics normal

Checking for a Pulse

Cardiopulmonary arrest means that both the victim's heart and lungs have stopped working properly, so it would make sense to check and see whether a victim is breathing and whether or not the heart is beating. However, current CPR guidelines don't require a layperson to check the victim's pulse before starting CPR. Why is this?

The answer is that the average person has a lot of trouble finding and determining pulse accurately. Think about how difficult it can be to find your own pulse, and then imagine trying to repeat the process on an unresponsive person. If someone is not breathing, the heart is already in danger of quitting (if it hasn't already) due to lack of oxygen. Since the first steps in CPR address the victim's respiratory state, you can try to get breathing going again right away. Then, you can check for a pulse. Skipping an initial pulse check simplifies CPR and saves valuable time.

breathing. However, the process is much more like blowing up a balloon than real breathing. You inhale deeply, form a tight seal with your mouth over the other person's mouth, and exhale strongly to push air out of your mouth into the victim's. Because you also pinch the victim's nostrils closed, the air has nowhere to go except down into the lungs, which expand as they fill with air.

Chest compression.

Mouth-to-mouth breathing is hard work. Normally, when you inhale your chest muscles drive the process. In artificial respiration, you're working against the victim's relaxed chest muscles. When the chest muscles are relaxed, the chest cavity is small, keeping the lungs in a deflated state. As a rescuer, you have to exhale forcefully into the victim's mouth to overcome this resistance. As the lungs fill with air, the victim's chest is pushed up; you can actually see it rise. When you remove your mouth from the victim's and break the air seal, their chest falls and once again deflates the lungs. As in normal breathing, this results in air being exhaled from the victim's mouth.

Does air exhaled from someone else's mouth really provide enough oxygen to save an unconscious person? Normally, the air you inhale contains about 20% oxygen by volume, and your lungs remove some of that oxygen in each breath. The air you blow into a victim's mouth actually contains about 15% to 16% oxygen, which is more than enough to take care of the person's needs.

After you've given the victim two breaths, you then check to see whether the person has a pulse and whether he or she is able to breathe without your help. This will determine what you do next.

- If the victim is breathing and has a pulse, you should stop CPR, but stay with the person until help arrives.
- If the victim is not breathing and has a pulse, you should continue rescue breathing.
- If the victim has no pulse, you should begin chest compressions, alternating with rescue breathing.

C Is for Circulation

If the victim's heart isn't beating, all your breathing efforts are for nothing; the oxygen that you're giving them isn't going anywhere! Once again, you have to take over for a failing organ. This time you essentially become a surrogate heart to

CPR and Infectious Disease

It's clear that CPR is an effective part of the emergency response to cardiopulmonary arrest, one that could potentially save thousands of lives a year. Yet, in most cases, victims aren't getting this lifesaving treatment, even when bystanders are familiar with CPR.

Why won't people perform CPR on strangers? It turns out that a large number of people are afraid of contracting some nasty disease during mouth-to-mouth resuscitation. This fear was heightened by the emergence of deadly infectious diseases, such as HIV, that are spread by bodily fluids. Even though there have been no documented cases of anyone ever catching AIDS during CPR, there is still a chance, however tiny, of this happening.

Personal protective gear.
This CPR safety kit includes a personal resuscitation mask and gloves.

To assuage the public's fear, CPR training now incorporates personal protective gear during the various steps. Some of the safety devices used include:

- **Gloves**—To prevent contact with saliva or blood.

- **Personal resuscitation masks**—These allow you to provide artificial respiration through a barrier that prevents contact with saliva and mucus membranes.

The one drawback is that you have to carry your CPR kit with you everywhere. Most people don't plan when they go into cardiopulmonary arrest, so you never know where you might need your gear.

pump oxygenated blood out to the rest of the body. How can you have any effect on another person's blood flow? All it takes is your hands and some strength. The steps are simple:

1) Kneeling by the victim, place the heel of your hands one atop the other, about 0.4 to 0.8 inches (1 to 2 cm) from tip of the breastbone.
2) Using the weight of your body, push the victim's chest down. You should compress the chest 1 to 2 inches (2.5 to 5 cm).
3) Hold this position for half a second, and then relax for half a second.
4) Repeat steps 2 and 3 fourteen more times.
5) Give the victim two rescue breaths as you did before to deliver more oxygen to the blood.

Repeat steps 1 through 5 three more times, and then check for a pulse.

In reality, all you are doing is squeezing the heart between the breastbone and the backbone to force blood out. Compressing the chest creates positive pressure inside the chest that pushes oxygenated blood out of the heart through the aorta. From here, it travels to the brain and then on to other parts of the body, delivering oxygen for cellular respiration. When you relax, the pressure inside the victim's chest subsides. Deoxygenated blood moves back into the heart from the veins.

CPR's Role in Rescue

CPR extends the window of opportunity to perform more elaborate first aid procedures. By itself, CPR cannot save the majority of victims of cardiopulmonary arrest. CPR only temporarily restores circulation, and even then the circulation is only 10% to 30% of what it would be with a healthy heart. Further, in about two out of three people in cardiopulmonary arrest, the heart goes into what's known as *ventricular fibrillation*. In this state, the heart muscle quivers rapidly, like a bowl of Jell-O, and is unable to beat properly. CPR cannot stop ventricular fibrillation, however a defibrillator (an electronic device that shocks

the heart) can. According to the American Heart Association, each minute of delay in returning the heart to its normal pattern of beating decreases the victim's chance of survival by 7% to 10%. Emergency medical practitioners now carry small portable defibrillators, called automated external defibrillators (AEDs) to handle cases of ventricular fibrillation.

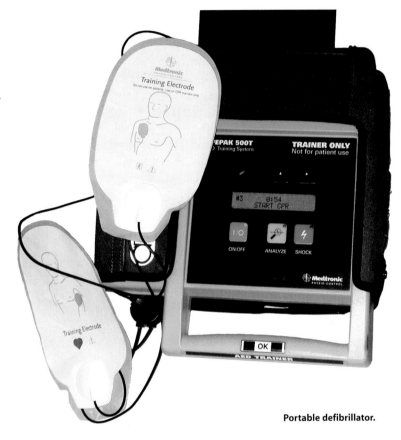

Portable defibrillator.

When someone collapses right in front of you, your first reaction is often sheer terror. But while you're panicked and unable to act, valuable minutes are slipping away. To counter this, many organizations such as the American Heart Association and the American Red Cross offer classes that train you in CPR and basic first aid and give you hands-on practice to hone your CPR skills. Then, if you are confronted with an emergency situation, you are prepared to jump into action.

169

How **EMERGENCY ROOMS** Work

A tumble down a ladder at home or work, chest pains that won't go away, or an injury during an intense rugby match could all send someone in for emergency medical treatment. In fact, each year millions of Americans find themselves visiting an emergency room. Millions more have seen the popular TV show ER. *Americans have an almost insatiable interest in the fascinating, 24-hour-a-day, nonstop world of emergency medicine.*

HSW Web Links

www.howstuffworks.com

How MRI Works
How Anesthesia Works
How Ultrasound Works
How X-Rays Work
How CAT Scans Work

A visit to the emergency room (ER) can be a stressful, scary event. First of all, there is the fear of not knowing what is wrong with you. There is the fear of having to visit an unfamiliar place filled with people you have never met. Also, you may have to undergo tests that you don't understand at a pace that discourages questions and comprehension.

A patient is moved from an ambulance into the emergency room.

Patients, Patients Everywhere!

One of the most amazing aspects of emergency medicine is the huge range of reasons that bring people to the ER. No other specialty in medicine sees the variety of conditions that an ER physician sees in a typical week. Some of the conditions that bring people to the ER include:

- Automobile and motorcycle accidents
- Sports injuries
- Cuts, broken bones, and burns
- Heart attacks or chest pain
- Strokes or loss of function/ numbness in arms or legs
- Loss of vision or hearing
- Unconsciousness
- Suicidal or homicidal thoughts
- Overdoses
- Severe abdominal pain or persistent vomiting
- Food poisoning
- Severe allergic reactions from insect bites, foods, or medications

Understanding the ER Maze

The classic emergency room scene involves an ambulance screeching to a halt, a gurney hurtling through the hallway, and five people frantically working to save a person's life with only seconds to spare. This does happen and it's not uncommon, but the majority of cases seen in a typical emergency department aren't quite this dramatic. Almost everyone that walks or is wheeled through the ER doors follows a typical path . . .

Triage

When you arrive at the emergency department, your first stop is triage. This is where each patient's condition is prioritized, typically by a nurse, into one of three general categories:

- Immediately life threatening
- Urgent but not immediately life threatening
- Less urgent

This categorization is necessary so that someone with a life threatening condition isn't kept waiting while someone with a more routine problem sees the doctor. The triage nurse records the patient's vital signs (temperature, pulse, respiratory rate, and blood pressure). He or she also gets a brief history of the patient's current medical complaints, past medical problems, medications, and allergies to determine the appropriate triage category for the patient.

Registration

The next stop is registration—not very exciting and rarely seen on TV. This is where the patient's vital statistics are recorded. A unit secretary will also obtain insurance, Medicare, Medicaid, or HMO information at this time. This step is necessary to develop a medical record so that the patient's medical history, lab tests, X-rays, and so on will all be located on one chart that can be referenced at any time. The bill will also be generated from this information. **Note**: All patients must receive a medical screening exam regardless of their ability to pay.

If the patient's condition is life threatening or if the patient arrives by ambulance, this step may be completed later at the bedside.

Examination Room

Now the patient is brought to the exam room. Here, an emergency department nurse will gather more information and may assist the patient into a gown so that he or she can be examined properly.

Some emergency departments are subdivided into separate areas to better serve their patients. These separate areas can include a pediatric ER, a chest pain ER, a fast track (for minor injuries and illnesses), a trauma center (usually for severely injured patients), and an observation unit (for patients who do not require admission but require prolonged treatment or many diagnostic tests).

After the nurse has finished his or her tasks, the next visitor is an emergency medicine physician, or another qualified person like a physician in training (an intern or resident) or a Physician Assistant (PA). The physician gets a more detailed medical history of the patient's present illness, past medical problems, family history, social history, and a complete review of all his or her body systems. The physician then formulates a list of possible causes of the patient's symptoms. This list is called a *differential diagnosis*.

The ER Team

A number of people contribute to the care a patient receives in an emergency room:

- **Emergency physician—** Emergency physicians, at the very least, must complete 4 years of college, 4 years of medical school, a 1-year internship and a 2- to 3-year residency prior to working in an emergency department.

- **Emergency nurse—** Emergency nurses complete some type of formal program, such as acquiring a bachelor or associate's degree in nursing.

After completing the program, a nurse must take a licensing exam to become an RN (registered nurse). Some ER nurses take an additional exam to become a CEN (certified emergency nurse).

- **Physician assistant—** Physician assistants (PAs) can examine, diagnose, and treat patients and review their findings with the physician. Typically, a PA has 2 to 4 years of college and some healthcare experience and has completed another 2-year program.

- **Emergency department technician—** Training varies widely, but technicians are often ambulance personnel or are trained through the hospital.

- **Unit secretary—**An essential member of the team, the unit secretary often handles the communication needs of the ER and coordinates the ordering of diagnostic tests.

- **Physicians in training—** At teaching hospitals, you may be examined by an intern or resident.

The most likely diagnosis is then determined by the patient's symptoms and a physical examination. If this information isn't sufficient to determine the diagnosis, then diagnostic tests are required.

Disposition

Depending on a patient's specific medical condition, the patient will either be:

- Admitted to the hospital
- Discharged
- Transferred to a more appropriate medical facility

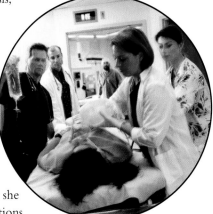

Emergency room doctors and attendants treat a patient.

If the patient is discharged, he or she will receive discharge instructions that explain medications and other treatments. The patient will also be referred for follow-up care should the condition continue or worsen.

A transfer may be necessary if the condition can be better treated at another institution.

The modern emergency department performs an important role in our society. It is one of the most important creations of modern society and it has saved countless lives.

How **ULTRASOUND** Works

Perhaps you are pregnant, and your obstetrician wants you to have an ultrasound to check on the developing baby or determine the due date. Maybe you are having problems with blood circulation in a limb or your heart, and your doctor has requested a Doppler ultrasound to look at the blood flow. Ultrasound has been a popular medical imaging technique for many years.

HSW Web Links

www.howstuffworks.com

How Bats Work
How Radar Works
How Prenatal Testing
  Works
How Human
  Reproduction Works
How Magnetic Resonance
  Imaging (MRI) Works

Dangers of Ultrasound

There have been many concerns about the safety of ultrasound. Because ultrasound is energy, the question becomes, "What is this energy doing to my tissues or my baby?" There have been some reports of low birth weight in babies who are born to mothers who had frequent ultrasound examinations during pregnancy. The two possible dangers of ultrasound are:

• Development of heat—tissues or water absorb the ultrasound energy, which increases their temperature locally.

• Formation of bubbles (*cavitation*)—dissolved gases come out of solution due to local heat caused by the ultrasound.

However, there have been no documented ill-effects of ultrasound, either in studies of humans or animals. This being said, ultrasound should still be used only when necessary (in other words, it's better to be cautious).

Ultrasound (or *ultrasonography*) is a medical imaging technique that uses high frequency sound waves and their echoes. The technique is similar to echolocation, used by bats, whales, and dolphins, and sonar, used by submarines. In ultrasound, here is what happens:

1) The ultrasound machine transmits high-frequency (1 to 5 megahertz) sound pulses into your body using a probe.

2) The sound waves travel into your body and hit a boundary between tissues.

3) Some of the sound waves get reflected back to the probe, while some travel on farther until they reach another boundary and get reflected.

4) The reflected waves are picked up by the probe and relayed to the machine.

5) The machine calculates the distance from the probe to the tissue or organ and the time of each echo's return.

6) The machine displays the distances and intensities of the echoes on the screen, forming a two-dimensional image.

In a typical ultrasound, millions of pulses and echoes are sent and received each second. The probe can be moved along the surface of the body and angled to obtain various views.

The Ultrasound Machine

A basic ultrasound machine contains several parts. If you understand each part, you can understand how the machine works as a whole. Here's a look at the separate pieces of an ultrasound machine:

Transducer Probe

The transducer probe is the main part of the ultrasound machine. It makes the sound waves and receives the echoes. You can think of the transducer probe as the mouth and ears of the ultrasound machine.

The transducer probe generates and receives sound waves using a principle called the *piezoelectric effect*. In the probe, there are one or more quartz crystals called piezoelectric crystals. When an electric current is applied to these crystals, they change shape rapidly. The rapid shape changes, or vibrations, of the crystals produce sound waves that travel outward. Conversely, when sound or pressure waves hit the crystals, the crystals emit electrical currents. Therefore, the same crystals can be used to send and receive sound waves. The probe also has a sound absorbing substance to eliminate back reflections from the probe itself and an acoustic lens to help focus the emitted sound waves.

Transducer probes come in many shapes and sizes. The shape of the probe determines its field of view, and the frequency of emitted sound waves determines how deep the sound waves penetrate and the resolution of the image. Transducer probes may contain one or more crystal elements; in multiple-element probes, each crystal has its own circuit. Multiple-element probes have an advantage in that the ultrasound beam can be "steered" by changing the timing in which each element gets pulsed; steering the beam is especially important for cardiac ultrasound. In addition to probes that can be moved across the surface of the body, some probes are designed to be inserted through various openings of the body (such as the esophagus) so that they can get closer to the organ being examined (say, the stomach) for more detailed views.

Central Processing Unit (CPU)

The CPU is the brain of the ultrasound machine. The CPU is basically a computer that contains the microprocessor, memory, amplifiers, and power supplies for the microprocessor and transducer probe. The CPU sends electrical currents to the transducer probe to emit sound waves, and also receives the electrical pulses from the probes that are created from the returning echoes. The CPU does all of the calculations that are involved in processing the data. After the raw data is processed, the CPU forms the image on the monitor. The CPU can also store the processed data and/or image on disk.

Transducer Pulse Controls

The transducer pulse controls allow the operator (called the *ultrasonographer*) to set and change the frequency and duration of the ultrasound pulses and to change the scan mode of the machine. The commands from the operator are translated into changing electric currents that are applied to the piezoelectric crystals in the transducer probe.

3-D Ultrasound Imaging

In the past few years, ultrasound machines capable of three-dimensional (3-D) imaging have been developed. In these machines, several two-dimensional images are acquired by moving the probes across the body surface or rotating inserted probes. The two-dimensional scans are then combined by specialized computer software to form 3-D images.

3-D imaging allows you to get a better look at the organ being examined and is best used for:

- Early detection of cancerous and benign tumors
- Examining the prostate gland for early detection of tumors
- Looking for masses in the colon and rectum
- Detecting breast lesions for possible biopsies

- Visualizing a fetus to assess its development, especially for observing abnormal development of the face and limbs
- Visualizing blood flow in various organs or a fetus

Doppler Ultrasound

Doppler ultrasound is based on the Doppler effect. When the object reflecting the ultrasound waves is moving, it changes the frequency of the echoes, creating a higher frequency if it is moving toward the probe and a lower frequency if it is moving away from the probe. How much the frequency is changed depends upon how fast the object is moving. Doppler ultrasound measures the change in frequency of the echoes to calculate how fast an object is moving.

Doppler ultrasound has been used mostly to measure the rate of blood flow through the heart and major arteries.

Ultrasound helps in lots of different situations, including obstetrics and gynecology, cardiology, and cancer detection. The main advantage of ultrasound is that certain structures can be observed without using radiation. Also, taking an ultrasound requires much less time than taking an X-ray or using other radiographic techniques.

How **X-RAY MACHINES** Work

X-ray machines are an invaluable tool in modern medicine—they make the human body transparent! X-ray technology lets doctors see straight through tissue to examine broken bones, cavities, and swallowed objects. Modified X-ray procedures can be used to examine softer tissue, such as the lungs, blood vessels, or the intestines.

![gear] **HSW Web Links**

www.howstuffworks.com

How Light Works
How Atoms Work
How How Magnetic Resonance Imaging (MRI) Works
How Nuclear Medicine Works
How Ultrasound Works
How Airport Security Works

X-rays are basically the same thing as visible light rays. Both are wavelike forms of electromagnetic energy, carried by particles called photons. The difference between X-rays and light rays is the energy level of the individual photons, also expressed as the wavelength of the rays.

When a photon collides with another atom, the atom may absorb the photon's energy by boosting an electron to a higher level. For this to happen, the energy level of the photon has to "match" the energy difference between the two electron positions. If not, the photon can't shift electrons between orbitals.

The atoms that make up your body tissue absorb visible light photons very well. The energy level of the photon fits with energy differences between electron positions. Radio waves don't have enough energy to match most atoms, so they pass right through your body. X-ray photons don't fit with your body's atoms either, because they have too much energy.

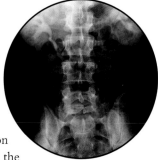

X-ray photons can knock an electron away from an atom. Some of the energy from the photon works to separate the electron from the atom and the rest sends the electron flying through space. A larger atom is more likely to absorb an X-ray photon in this way. Smaller atoms, where the electron orbitals are separated by relatively low jumps in energy, are less likely to absorb X-ray photons.

The soft tissue in your body is composed of smaller atoms, and so it does not block high-energy X-ray photons very effectively. The calcium atoms that make up your bones are much larger, so they are better at blocking X-ray photons.

The X-Ray Machine

The heart of an X-ray machine is an electrode pair—a cathode and an anode—that sits inside a glass vacuum tube. The cathode is a heated filament, like you might find in an older fluorescent lamp. The machine passes current through the filament, heating it up. The heat sputters electrons off of the filament surface. The positively-charged anode, which is a flat disc made of tungsten, draws the electrons across the tube.

The voltage difference between the cathode and anode is extremely high, so the electrons fly through the tube with a lot of force. When a speeding electron collides with a tungsten atom, the impact knocks loose an electron in one of the atom's lower orbitals. An electron in a higher orbital immediately

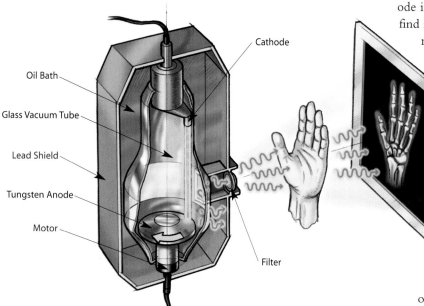

Oil Bath
Glass Vacuum Tube
Lead Shield
Tungsten Anode
Motor
Cathode
Filter

falls to the lower energy level, releasing its extra energy in the form of a photon. It's a big drop, so the photon has a high energy level. This photon is an X-ray photon.

Electrons can also emit photons without the electron hitting an atom. An atom's nucleus may attract a speeding electron just enough to alter its course. Like a comet whipping around the sun, the electron slows down and changes direction as it speeds past the atom. This "braking" action causes the electron to lose energy. The released energy takes the form of an X-ray photon.

The high-impact collisions involved in X-ray production generate a lot of heat in the X-ray machine. A motor rotates the anode to keep it from melting. This way the electron beam isn't always focused on the same area. A cool oil bath surrounding the glass vacuum tube also absorbs heat.

A thick lead shield surrounds the entire mechanism. The shield keeps the X-rays from flying out of the machine in all directions—the energy can only escape through a small window. The X-ray beam passes through a series of filters and travels on to the patient.

A camera on the other side of the patient records the pattern of X-ray light that passes through the patient's body. The X-ray camera uses the same film technology as an ordinary camera, but X-ray light sets of the chemical reaction instead of visible light.

Reading the X-Ray

Generally, doctors keep the film image as a negative. The areas exposed to more light appear darker, and the areas exposed to less light appear lighter. Hard material, such as bone, appears white, and softer material appears black or gray. Doctors can bring different materials into focus by varying the intensity of the X-ray beam.

Are X-Rays Bad for You?

In the past, a lot of doctors would expose themselves and patients to X-ray radiation for long periods of time. Eventually, doctors

and patients started developing radiation sickness, and the medical community knew something was wrong.

The problem is that X-rays are a form of ionizing radiation.

Free Electrons

Orbiting Electron

Tungsten Atom Nucleus

Displaced Electron

X-Ray Photon

X-Ray Photon

When normal light hits an atom, it can't change the atom in any significant way. But when an X-ray photon hits an atom, it can knock electrons off the atom to create an ion, an electrically-charged atom. Free electrons collide with other atoms to create more ions.

An ion's electrical charge can lead to unnatural chemical reactions inside cells. Among other things, the charge can break DNA chains in the cell. When this happens, the cell dies, or the DNA develops a mutation. If a lot of cells die, the body can develop various diseases. If the DNA mutates, a cell may become cancerous, and this cancer may spread. If the mutation is in a sperm or an egg cell, it may lead to birth defects. Because of all these risks, doctors now use X-rays sparingly.

Even with its risks, X-ray scanning is still a safer option than surgery. X-ray machines are an invaluable tool in medicine, and a terrific asset in security and scientific research. They are truly one of the most useful inventions of all time.

Did You Know?

As with many of mankind's monumental discoveries, X-ray technology was invented completely by accident. One day in 1895, a German physicist named Wilhelm Roentgen was experimenting with electron beams in a gas discharge tube. Roentgen noticed that when he turned the electron beam on, a fluorescent screen in his lab started to glow, apparently reacting to electromagnetic energy emitted by the electron activity.

Roentgen placed various objects between the tube and the screen, and the screen still glowed. Finally, he put his hand in front of the tube, and saw the silhouette of his bones projected onto the fluorescent screen.

How **CAT SCANS** Work

CAT (computerized-axial tomography) scans take the idea of conventional X-ray imaging to a new level. Instead of finding the outlines of bones and organs, a CAT scan machine forms a full three-dimensional computer model of a patient's insides. While the computer technology involved is fairly advanced, the fundamental concept at work is really very simple.

HSW Web Links

www.howstuffworks.com

How Light Works
How Atoms Work
How Magnetic Resonance
 Imaging (MRI) Works
How Nuclear Medicine
 Works
How Ultrasound Works

A conventional X-ray image is basically a shadow: You shine a "light" on one side of the body, and a piece of film on the other side registers the silhouette of the bones.

No Place to Hide

Although shadows give you useful information, they create an incomplete picture of an object's shape. Imagine that you are standing outside in front of a wall, and someone is looking at your shadow on the wall. You hold a pineapple against your chest with your right hand and a banana out to your side with your left hand. If the sun is behind you, your friend will see the outline of the banana but not the pineapple—the shadow of your torso blocks the pineapple. If you turn 90 degrees, your friend will see the outline of the pineapple, but not the banana.

The same thing happens in conventional X-ray machines. If a larger bone is directly between the X-ray machine and a smaller bone, the larger bone may cover the smaller bone on the film. In order to see the smaller bone, you would have to turn your body or move the X-ray machine.

In order to know that you are holding both a pineapple and a banana, your friend would have to see your shadow in both positions and form a complete mental image. This is the basic idea of computerized-axial tomography (CAT). In a CAT scan machine, the X-ray beam moves all around the patient, scanning from hundreds of different angles. The computer takes all this information and puts together a 3-D image of the body.

Scanning Procedure

The CAT machine looks like a giant donut tipped on its side. The patient lies down on a platform that slowly moves through the hole. The X-ray tube is mounted on a movable ring around the edges of the hole. The ring also supports an array of X-ray detectors, which are directly opposite the X-ray tube.

A motor turns the ring so that the X-ray tube and the X-ray detectors revolve around the body. Each full revolution scans a narrow horizontal slice of the body. After each revolution, the control system moves the platform farther into the hole so the tube and sensor can scan the next slice.

In this way, the machine records X-ray slices all the way across the body. The computer varies the intensity of the X-rays in order to scan each type of tissue with the optimum power. After the patient passes all the way through the machine, the computer combines all the information from each scan to form a detailed image of the body.

Since they examine the body slice by slice, from all angles, CAT scans are much more precise than conventional X-rays. Today, doctors use CAT scans to diagnose and treat a wide variety of ailments, including head trauma, cancer, and osteoporosis.

How **MAGNETIC RESONANCE IMAGING (MRI)** Works

Your back has been hurting for weeks, and you finally make an appointment with the doctor. After a thorough exam, the doctor stands in front of you with a slightly puzzled look on her face. "Mr. Smith, I'm sorry, but at this time, the tests seem to be inconclusive. You're going to need another test." She decides that an MRI will be necessary to make an accurate diagnosis. An MRI. So there you sit, wondering what that means. All kinds of questions and thoughts run through your mind— Will it require a trip to the hospital? Is it invasive? Will it hurt?

Magnetic resonance imaging (MRI) provides an unparalleled view inside the human body. The level of detail is extraordinary compared with any other imaging system. MRI is the method of choice for the diagnosis of many types of injuries and conditions because of the incredible ability to tailor the MRI exam to the particular medical question being asked. By changing exam parameters, the MRI system can cause tissues in the body to take on different appearances. This is very helpful to the *radiologist*, the person who reads the MRI, in determining if something seen is normal or not. We know that when we do "A," normal tissue will look like "B"—if it doesn't, there might be an abnormality.

Think of an MRI machine as a really big magnetic box. From the front view, an MRI machine usually resembles a giant front-loading (think laundromat-style) washing machine. Although newer models are getting smaller, a typical MRI system might be 7 feet tall x 7 feet wide x 6 to 10 feet long (2 m x 2 m x 1.8 to 3 m). MRI scanners vary in size and shape, and newer models have some degree of openness around the sides, but the basic design is the same. There is a long, tubular chamber, known as the *bore* of the magnet, running through the center of the box (where the door of the washing machine might be).

Lying on his or her back, the patient slides into the bore on a special table. Whether or not the patient goes in head first or feet first, as well as how far inside the magnet the patient will go, is determined by the type of exam to be performed. Once the body part to be scanned is in the exact center, or *isocenter,* of the magnetic field, the scan can begin.

HSW Web Links

www.howstuffworks.com

How Ultrasound Works
How Nuclear Medicine Works
How Prenatal Testing Works
How Emergency Rooms Work
How Your Brain Works
How Cancer Works

Advantages of an MRI

Why would your doctor order an MRI? Because the only way to see inside your body any better is to cut you open. MRI is ideal for situations such as:

- Diagnosing multiple sclerosis (MS)
- Diagnosing tumors of the pituitary gland and brain
- Diagnosing infections in the brain, spine, or joints
- Visualizing torn ligaments in the wrist, knee, and ankle
- Visualizing shoulder injuries
- Diagnosing tendonitis

- Evaluating masses in the soft tissues of the body
- Evaluating bone tumors, cysts, and bulging or herniated discs in the spine
- Diagnosing strokes in their earliest stages

The fact that MRI systems do not use ionizing radiation is a comfort to many patients, as is the fact that MRI contrast materials have a very low incidence of side effects. Another major advantage of MRI is its ability to image in any plane. CAT scans are limited to one plane, the *axial* plane (in the loaf-of-bread analogy, the axial plane would be how a loaf of bread is normally sliced). An MRI system can create axial images as well as images in the *sagittal* plane (think of slicing the bread side-to-side lengthwise) and *coronally* (think of the layers of a layer cake) or any degree in between, without the patient ever moving. If you have ever had an X-ray, you know that every time the technicians take a different picture, you have to move. The three gradient magnets discussed earlier allow the MRI system to choose exactly where in the body to acquire an image and how the images are oriented.

Magnetic Intensity

To understand how MRI works, let's start by focusing on the *magnetic* in *magnetic resonance imaging*. The magnet is the most essential part of an MRI system. Magnets are rated using one of two units of measure: the tesla or the gauss. One tesla is equal to 10,000 gauss. The magnets used in current MRI technology are in the 0.5-tesla to 2.0-tesla range, or 5,000 to 20,000 gauss. To give you an idea of how incredibly strong that is, consider the Earth's magnetic field—it's only 0.5-gauss! Although extremely powerful magnets—up to 60 tesla—are used in research, magnetic fields greater than 2 tesla have not been approved for use in medical imaging.

Numbers like these help you understand the magnetic strength, but everyday examples are also helpful. You might not realize it, but the MRI scan room can be an amazingly hazardous environment. If taken into the scan room, some seemingly innocuous, everyday objects can become dangerous projectiles in an instant. Without warning small metal objects like paperclips, keys, pens, and even scissors and stethoscopes have flown across the room to make contact with the super-strong MRI magnet. You don't want to take your wallet in with you, either. Anything with a magnetic coding, including your credit cards and bank cards, will be erased by most MRI systems.

Before a patient or medical staff member can enter the MRI scan room, he or she must be closely examined for metal objects. It's not uncommon for patients to have metal pins or other pieces inside their bodies from prior surgeries or procedures. Being inside the magnet of an MRI could be incredibly dangerous for these patients. In fact, even a small fragment of metal in an MRI-bound patient's eye could pose a serious threat. Your eyes do not form scar tissue as the rest of your body does. A fragment of metal in your eye that has been there for 25 years is just as dangerous today as it was then, because no scar tissue is holding it in place. Pacemakers also pose an incredible risk. The magnet in an MRI can cause the pacemaker to stop working or break down. Some patients are turned away because it's too dangerous. When this happens, there is usually an alternative method of imaging that can help them (like a PET scan or CAT scan).

There are no known biological hazards to humans from being exposed to magnetic fields of the strength used in medical imaging

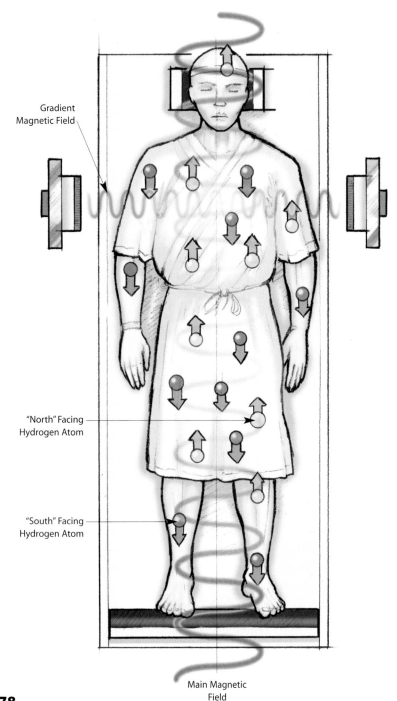

Gradient Magnetic Field

"North" Facing Hydrogen Atom

"South" Facing Hydrogen Atom

Main Magnetic Field

today. However, most facilities prefer not to image pregnant women. This is due to the fact that there has not been much research done in the area of the biological effects powerful magnetic forces have on a developing fetus.

The Magnets

Three basic types of magnets are used in MRI systems:

- **Resistive magnets** consist of many windings or coils of wire wrapped around a cylinder or bore through which an electric current is passed. The current in the wire causes a magnetic field to be generated. If the electricity is turned off, the magnetic field dies out.
- **Permanent magnets** are just that— permanent. A permanent magnet's magnetic field is always there and always on full strength, so it costs nothing to maintain the field. The major drawback is that these magnets are extremely heavy: They weigh many tons at the 0.4-tesla level.
- **Superconducting magnets** are by far the most commonly used. A superconducting magnet is somewhat similar to a resistive magnet. The important difference is that the coil of wire is continually bathed in liquid helium at −452.4°F. At this temperature, the resistance in the wire drops to zero, reducing the electrical requirement for the system dramatically and making it much more economical to operate. Superconductive systems can easily generate 0.5-tesla to 2.0-tesla fields, allowing for much finer resolution while imaging.

Another type of magnet found in every MRI system is a gradient magnet. There are three gradient magnets inside the MRI machine. These magnets are very, very low strength compared to the main magnetic field; they may range in strength from 180 gauss to 270 gauss.

The main magnet immerses the patient in a stable and very intense magnetic field, and the gradient magnets create a variable field. The rest of an MRI system consists of a very powerful computer system, some equipment that can transmit radio frequency (RF) pulses into the patient's body while they are in the scanner, and many other secondary components.

Understanding the Technology

In conjunction with radio-wave pulses of energy, the MRI scanner can pick out a very small point inside the patient's body and ask it, essentially, "What type of tissue are you?" The point might be a cube that is half a millimeter on each side. The MRI system goes through the patient's body point by point, building up a 2-D or 3-D map of tissue types. It then integrates all of this information together to create 2-D images or 3-D models.

MRI uses hydrogen atoms found in a person's body to study his or her tissue. There are many different types of atoms in the human body; hydrogen is an ideal atom for MRI because its nucleus has a single proton and a large magnetic moment. The large magnetic moment means that, when placed in a magnetic field, the hydrogen atom has a

---

Disadvantages of an MRI

Although MRI is an excellent imaging system, there are some drawbacks to using an MRI machine— such as:

- Many people cannot safely be scanned with MRI, such as a patient with a pacemaker. Also, some people are too big to be scanned (some machines only accommodate patients up to about 295 pounds).
- There are many claustrophobic people in the world, and being in an MRI machine can be a very disconcerting experience for them.
- The machine makes a tremendous amount of noise during a scan. The noise sounds like a con-

tinual, rapid hammering. Patients are given earplugs or stereo headphones to muffle the noise (in most MRI centers, you can even bring your own cassette or CD to listen to). The noise is due to the rising electrical current in the wires of the gradient magnets being opposed by the main magnetic field. The stronger the main field, the louder the gradient noise.

- MRI scans require patients to hold very still for extended periods of time. MRI exams can range in length from 20 minutes to 90 minutes or more. Even very slight movement of the body part that is being

scanned can cause very distorted images, and the scan has to be repeated.

- Orthopedic hardware (screws, plates, artificial joints) in the area of a scan can cause severe *artifacts* (distortions) on the images. The hardware causes a significant alteration in the main magnetic field. Remember, a uniform field is critical to good imaging.
- MRI systems are very, very expensive to purchase, and therefore the exams are also very expensive.

The almost limitless benefits of MRI for most patients far outweigh the few drawbacks, however.

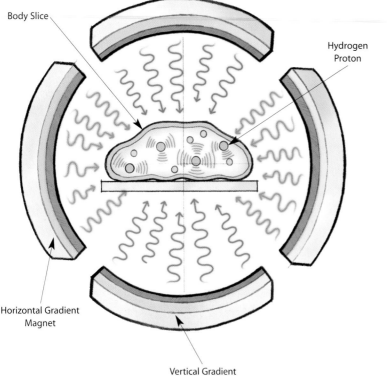

Body Slice

Hydrogen
Proton

Horizontal Gradient
Magnet

Vertical Gradient
Magnet

These RF pulses are usually applied through a coil. MRI machines come with many different coils designed for different parts of the body: a knee, a shoulder, a wrist, the head, the neck, and so on. These coils usually conform to the contour of the body part being imaged, or at least reside very close to it during the exam. At approximately the same time, the three gradient magnets come into play. They are arranged inside the main magnet so that when they are turned on and off very rapidly in a specific way, they alter the main magnetic field on a very local level. What this means is that an exact area of the body can be pinpointed. Those areas are usually referred to as *slices,* and can be as thin as a few millimeters. Using MRI, the radiologist can slice any part of the body in any direction, providing a huge advantage over any other imaging system. That also means that the patient doesn't have to move for the machine to get an image from a different direction—the machine can manipulate everything with the gradient magnets.

When the RF pulse is turned off, the hydrogen protons begin to return to their natural alignment within the magnetic field and release their excess stored energy. When they do this, they give off a signal that the coil now picks up and sends to the computer system. What the system receives is mathematical data that is converted into a picture that is put on film. That is the *imaging* part of *magnetic resonance imaging.*

The Future of MRI

This technology is still in its infancy, comparatively speaking. It has been in widespread use for not quite 20 years (compared with over 100 years for X-rays).

The future of MRI seems limited only by our imagination. There's no doubt it will be exciting for those in the field of medicine, and very beneficial to patients.

strong tendency to line up with the direction of the magnetic field.

Inside the bore of the scanner, the magnetic field runs straight down the center of the tube in which the patient is placed. This means that if a patient is lying on his or her back in the scanner, the hydrogen protons in his or her body will line up in the direction of either the feet or the head. The vast majority of these protons will cancel each other out—that is, for each one lined up toward the feet, one toward the head will cancel it out. Only a couple of protons out of every million are not canceled out. This doesn't sound like much, but the sheer number of hydrogen atoms in the body gives the MRI machine what it needs to create wonderful images.

The MRI machine applies an RF pulse that is specific only to hydrogen. The system directs the pulse toward the area of the body to be examined. The pulse causes the protons in that area to absorb the energy required to make them spin in a different direction. This is the *resonance* part of *magnetic resonance imaging.*

How **PET SCANS** Work

One problem that doctors have with the human body is that it is opaque, and looking inside generally is a painful experience for the patient. In the past, exploratory surgery was one common way to look inside the body, but today doctors can use a huge array of noninvasive techniques. Some of these techniques are X-rays, MRI scanners, CAT scans, ultrasound, and so on. Each of these techniques has advantages and disadvantages that make it useful for different conditions and different parts of the body.

PET (positron emission tomography) scans give doctors a unique way to see inside the body without opening it up. Nuclear medicine utilizes various methods of medical imaging to establish a diagnosis and treatment. PET scans are one form of nuclear medical imaging.

Nuclear Medical Imaging

Nuclear imaging techniques use radioactive substances to form images of different parts of the body. Nuclear techniques include:

- Positron emission tomography (PET)
- Single photon emission computed tomography (SPECT)
- Cardiovascular imaging
- Bone scanning

All of these techniques use different properties of radioactive elements to create an image.

Nuclear medicine imaging is useful for detecting tumors, aneurysms (weak spots in blood vessel walls), irregular or inadequate blood flow to various tissues, blood cell disorders, and inadequate functioning of organs, such as thyroid and pulmonary function deficiencies.

The use of any specific test, or combination of tests, depends upon the patient's symptoms and the disease being diagnosed.

Positron Emission Tomography (PET)

PET produces images of the body by detecting the radiation emitted from radioactive substances. These substances are injected into the body, and are usually tagged with a radioactive atom, such as carbon-11, fluorine-18, oxygen-15, or nitrogen-13, that has a short decay time. These radioactive atoms are formed by bombarding normal chemicals with neutrons to create short-lived radioactive isotopes. PET detects the gamma rays given off at the site where a positron emitted from the radioactive substance collides with an electron in the tissue.

In a PET scan, the patient is injected with a radioactive substance and placed on a flat table that moves in increments through a donut-shaped housing. This housing contains a circular gamma ray detector array, which has a series of scintillation crystals, each connected to a photomultiplier tube. The crystals convert the gamma rays, emitted from the patient, to photons of light, and the photomultiplier tubes convert and amplify the photons to electrical signals. These electrical signals are then processed by the computer to generate images. The table moves and the process repeats, resulting in a series of thin-slice images of the body over the region of interest (like the brain, breast, or liver). These images can be assembled into a 3-D representation of the patient's body.

PET provides images of blood flow or other biochemical functions, depending upon the type of molecule that is radioactively tagged. For example, PET can show images of glucose metabolism in the brain or rapid changes in activity in various areas of the body. Although this technique is very useful to many diagnoses, there are few PET centers in the country because they must be located near a particle accelerator device that produces the short-lived radioisotopes used in the technique.

If you've seen a PET scan, it was most likely a brain scan. Time-lapse PET scans let scientists see which parts of the brain are active during different thought processes, and the technique has opened a fascinating window into the functioning of the human brain.

HSW Web Links

www.howstuffworks.com

How Magnetic Resonance Imaging (MRI) Works
How Cells Work
How Nuclear Radiation Works
How Ultrasound Works
How DNA Evidence Works

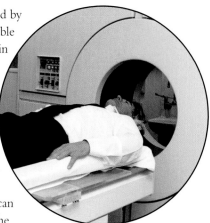

181

How **CORRECTIVE LENSES** Work

One of the most common sights almost anywhere in the world is—eyeglasses! Some people simply need reading glasses. Others might need to wear glasses all the time. Whether it's to see something close up or far away, it's incredible that two small pieces of glass or plastic can improve a person's vision in an instant.

HSW Web Links

www.howstuffworks.com

How Light Works
How Sunglasses Work
How Your Eyes Work

Since we depend so much on corrective lenses in our glasses to improve our view of the world, you might wonder just what goes into creating them. In order to understand how corrective lenses work, it's good to know something about how eyesight works, in general, first.

On the back of your eye is a complex layer of cells known as the *retina*. The retina reacts to light, and conveys information about light to the brain. The brain, in turn, translates all that activity on the retina into an image. Because the eye is a sphere, the surface of the retina is curved.

When you look at something, three things must happen:

- The image must be reduced in size to fit onto the retina.
- The scattered light must come together—that is, it must focus—at the surface of the retina.
- The image must be curved to match the curve of the retina.

To do all this, the eye uses a lens between the retina and the pupil and a transparent covering, or cornea. The lens, which would be classified as a *plus lens* because it is thickest toward the center, and the cornea work together to focus the image onto the retina. Sometimes the eye doesn't focus quite right. The reason could be:

- The surfaces of the lens or cornea may not be smooth, causing an aberration that results in a distortion.
- The lens may not be able to change its curve to properly match the image (called *accommodation*).

- The cornea may not be shaped properly, resulting in blurred vision.

Most vision problems occur when the eye cannot focus the image onto the retina. Here are a few of the most common problems:

- Myopia (nearsightedness) occurs when a distant object looks blurred because the image comes into focus before it reaches the retina. Myopia can be corrected with a *minus lens,* which moves the focus farther back.
- Hyperopia (farsightedness) occurs when a close object looks blurred because the image doesn't come into focus before it gets to the retina. Hyperopia, which can occur as we age, can be corrected with a plus lens. Bifocal lenses, which have a small plus segment, can help a farsighted person read or do close work, such as sewing.
- Astigmatism is caused by a distortion that results in a second focal point. It can be corrected with a cylinder curve.

Lenses can be made to correct all of these problems. Lenses can also correct double vision (crossed eyes). Lenses do so by moving the image to match the wayward eye. Corrective lenses are prescribed to correct for aberrations, to adjust the focal point onto the retina, or to compensate for other abnormalities.

How a Lens Works

The best way to understand the behavior of light through a curved lens is to relate it to a prism. A prism is thicker at one end, and light passing through it is bent (refracted) toward the thickest portion.

A lens can be thought of as two rounded prisms joined together. Light passing through the lens is always bent toward the thickest part of the prisms. To make a minus

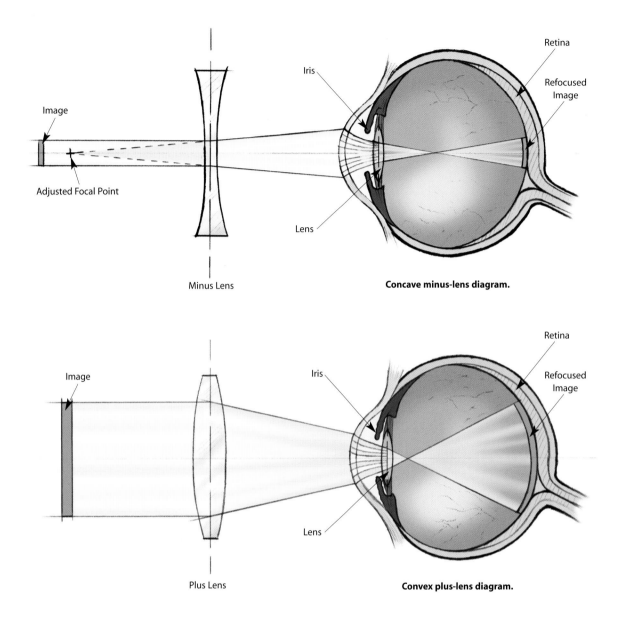

Image

Adjusted Focal Point

Iris

Retina

Refocused Image

Lens

Minus Lens

Concave minus-lens diagram.

Image

Iris

Retina

Refocused Image

Lens

Plus Lens

Convex plus-lens diagram.

lens, the thickest part—the base—of the prisms is on the outer edges and the thinnest part, the apex, is in the middle. This shape spreads the light away from the center of the lens and moves the focal point forward. The stronger the lens, the farther the focal point is from the lens.

To make a plus lens, the thickest part of the lens is in the middle and the thinnest part on the outer edges. The light is bent toward the center and the focal point moves back. The stronger the lens, the closer the focal point is to the lens.

Placing the correct type and power of lens in front of the eye will adjust the focal

point to compensate for the eye's inability to focus the image on the retina.

Determining Lens Strength

The strength of a lens is determined by the lens material and the angle of the curve that is ground into the lens. Lens strength is expressed in diopters (D), which indicates how much the light is bent. The higher the diopter, the stronger the lens. A plus (+) or minus (–) sign before the diopter strength indicates the type of lens.

Plus and minus lenses can be combined, with the total lens type being the algebraic sum of the two. For example, a +2.00D lens added to a –5.00D lens yields:

[+2.00] + [–5.00] = –3.00 or a 3.00D minus lens

How to Read the Prescription

Most prescriptions have four parts:

- The base (spherical) strength and type (plus or minus).
- The cylinder strength and type.
- The cylinder axis orientation.
- The strength of the bifocal segment (*plus* indicating *in addition*) and type.

A short form prescription from the optometrist or ophthalmologist might read:

2.25 –1.50 x 127 plus +2.00

This means:

- A +2.25D spherical base curve (a plus lens).
- A –1.50D cylinder at 127 degrees (a minus cylinder lens is added to the base curve).
- An additional bifocal segment of +2.00D.

Total power of the lens with the cylinder is +2.25 + (–1.50) = +0.75D. At the bifocal segment, the power is (+0.75) + (+2.00) = +2.75D. And in case you've ever wondered, OD means right eye and OS means left eye.

Lens Shapes

Two basic lens shapes are commonly used in optometry:

- **A spherical lens** looks like a basketball cut in half. The curve is the same all over the surface of the lens.
- **A cylindrical lens** looks like a pipe cut lengthwise. The direction of a cylinder curve's spine (axis) defines its orientation. It will only bend light along that axis. Cylinder curves are commonly used to correct astigmatism, as the axis can be made to match the axis of the aberration on the cornea.

Making a Lens

To make a lens, the first thing you need is a lens blank. Blanks are made in factories and shipped to individual labs to be made into eyeglasses. The raw lens material is poured

into molds that form discs about 4 inches in diameter and between 1 and 1½ inches thick. The bottom of the mold forms a spherical curve on the front face. A small segment with a stronger curve may be placed in the mold to form the segment for bifocals or progressive lenses.

Corrective lenses can be made with glass or plastic, but nowadays plastic is more common. While several different types of plastic are used in making lenses, all of them follow the same general manufacturing procedures.

In the lab the patient's full prescription gives these exact details:

- The total power (in diopters) that the finished lens must have
- The strength and size of the segment (if needed)
- The power and orientation of any cylinder curves
- Details such as the location of the optical center and any induced prism that may be needed

The lab technician selects a lens blank that has the correct segment (called an *add*) and a base curve that is close to the prescribed power. Then to make the power match the prescription exactly, another curve is ground on the back of the lens blank.

In most labs, the equipment is designed to grind minus curves, so a strong, plus lens blank is usually selected.

If the base curve is too strong, then a minus curve is ground in the back of the lens, which reduces the total power of the lens.

Glass lenses are ground and polished much the same way as plastic lenses are, except that diamond cutting surfaces are used, and some details may vary. The blanks are made of relatively soft glass and must be tempered, either by chemicals or heat, to strengthen them before they're inserted into the frame.

Advances in automation are rapidly changing how lenses are made. For example, the vast majority of labs now use computers to determine curve parameters and lens choice, and equipment is available that will combine several steps or even do the entire operation automatically.

Tinted Lenses

To make tinted lenses, special dyes are kept in heated containers and the lenses are immersed. The density of the tint is determined by how long the lenses are left in the dye. Lenses may be only partially tinted, tinted different colors at the top and bottom, or tinted a custom color by combining different colors. Also, for sunglasses, special UV blocking dyes may be applied in the same way.

How **ARTIFICIAL HEARTS** Work

Your heart is the engine inside your body that keeps everything running. Basically, the heart maintains oxygen and blood circulation through your lungs and body. In a day, your heart pumps about 2,000 gallons of blood. Like any engine, if the heart is not well taken care of it can break down—a condition called heart failure.

Between 2 million and 3 million Americans are currently living with heart failure, and 400,000 new cases are diagnosed annually. Heart failure causes 39,000 deaths per year, according to the National Heart, Lung, and Blood Institute (NHLBI). Until recently, the only option for severe heart failure patients has been heart transplants. However, only slightly more than 2,000 heart transplants are performed in the United States annually, meaning that tens of thousands of people die waiting for a donor heart.

On July 2, 2001, heart failure patients were given new hope as surgeons at Jewish Hospital in Louisville, Kentucky, performed the first artificial heart transplant in nearly two decades. The AbioCor Implantable Replacement Heart, developed by ABIOMED, used in this surgery is the first completely self-contained artificial heart and is expected to at least double the life expectancy of heart patients.

A Hydraulic-Driven Heart

The average adult human heart pumps blood at a rate of 60 to 100 beats per minute. The heart contracts in two stages:

1) In the first stage, the right and left atria contract at the same time, pumping blood to the right and left ventricles.
2) In the second stage, the ventricles contract together to propel blood out of the heart.

The heart muscle then relaxes before the next heartbeat. This allows blood to fill up the heart again.

Patients with an implanted AbioCor heart still have their natural atria, but the artificial heart replaces both ventricles. It can only force blood out one ventricle at a time; therefore it alternately sends blood to the lungs and then to the body, instead of both at the same time as a natural heart does. The AbioCor is able

to pump more than 10 liters per minute, which is enough for everyday activities.

The AbioCor is a very sophisticated medical device, but the core mechanism is simple: a hydraulic pump that shuttles hydraulic fluid from side to side. Here's how the AbioCor works:

- **The hydraulic pump**—in theory, this pump is similar to the hydraulic pumps used in heavy equipment: Force that is applied at one point is transmitted to another point using an incompressible fluid. A gear inside the pump spins at 10,000 revolutions per minute to create pressure.
- **The porting valve**—This valve opens and closes to let the hydraulic fluid flow

HSW Web Links

www.howstuffworks.com

How Your Heart Works
How Congestive Heart
 Failure Works
How Heart Attacks Work
How Hydraulic Machines
 Work

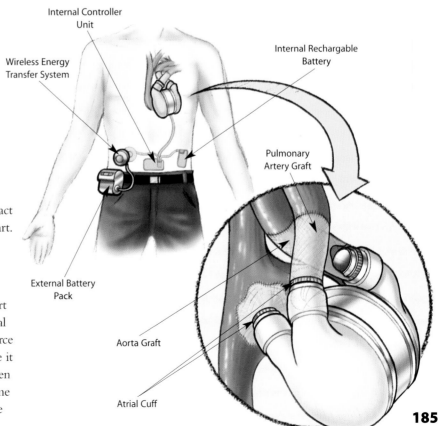

Internal Controller Unit

Wireless Energy Transfer System

Internal Rechargable Battery

Pulmonary Artery Graft

External Battery Pack

Aorta Graft

Atrial Cuff

185

from one side of the artificial heart to the other. When the fluid moves to the right, blood gets pumped to the lungs through an artificial ventricle. When the fluid moves to the left, blood gets pumped to the rest of the body.

- **The wireless energy-transfer system—** Also called the transcutaneous energy transfer (TET), this system consists of two coils, one inside the body and one outside the body, that transmit power via magnetic force from an external battery across the skin without piercing the surface. The internal coil receives the power and sends it to the internal battery and controller device.
- **The internal battery—**A rechargeable battery is implanted inside the patient's abdomen. This gives a patient 30 to 40 minutes to perform activities such as showering while being disconnected from the main battery pack.
- **The external battery—**This battery is fastened to a belt that's to be worn around the patient's waist. Each rechargeable battery offers about 4 to 5 hours of power.
- **The controller—**This small electronic device is implanted in the patient's abdominal wall. It monitors and controls the pumping speed of the heart.

The AbioCor heart, made of titanium and plastic and weighing about 2 pounds (0.9 kg), connects to the right atrium, the left atrium, the aorta and the pulmonary artery.

The Seven-Hour Surgery

The surgery to implant a AbioCor artificial heart is extremely delicate. Not only are the surgeons cutting off and extracting the natural heart's right and left ventricles, but they are also placing a foreign object into the patient's chest. The patient must be placed on, and later removed from, a heart–lung machine. The surgery also requires hundreds of stitches, to properly secure the remaining natural heart to the artificial ventricles. Grafts connect the AbioCor to remaining parts of the natural heart. Grafts are a kind of synthetic tissue used to connect the artificial device to the patient's natural tissue.

Here is the procedure, as described by University of Louisville surgeon Robert Dowling:

1) Surgeons implant the energy-transfer coil in the abdomen.
2) The breast bone is opened and the patient is placed on a heart-lung machine.
3) Surgeons remove the right and left ventricles of the native heart. They leave in the right and left atria, the aorta, and the pulmonary artery. This part of the surgery alone takes two to three hours.
4) Atrial cuffs are sewn to the native heart's right and left atria.
5) A plastic model is placed in the chest to determine the proper placement and fit of the heart in the patient.
6) Grafts are cut to an appropriate length and sewn to the aorta and pulmonary artery.
7) The AbioCor is placed in the chest. Surgeons use *quick connects*—which are sort of like little snaps—to connect the heart to the pulmonary artery, aorta, and left and right atria.
8) All of the air in the device is removed.
9) The patient is taken off the heart-lung machine.
10) The surgical team ensures that the heart is working properly.

The Right Candidate

The U.S. Food and Drug Administration (FDA) cleared ABIOMED to perform five initial implants (one patient died during surgery, so there were actually six patients.) The FDA plans to review the results of these transplants on a case-by-case basis to determine the future of the AbioCor device.

One requirement is that the grapefruit-sized device fit inside the patient's chest. To determine the fit of the device, the patient undergoes a CAT scan and chest X-ray. Using a computer-aided design (CAD) program, the natural heart is virtually removed and the AbioCor heart is virtually placed in the patient's chest. If the computer program shows that the device will fit, doctors can proceed with the operation to implant the artificial heart.

Most people diagnosed with heart failure can expect to live about five years, and will usually need to have a heart transplant to extend their life. Doctors still encourage the public to become organ donors, but the AbioCor may save many of those who don't have the option of a natural transplant or of waiting for an available heart.

AROUND THE HOUSE

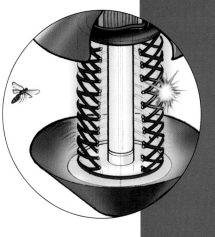

How **AEROSOL CANS** Work

Whether you're spraying paint, bug repellant, or hair spray, you probably use an aerosol can just about every day. It's one of those technologies that we don't think about often, but it makes life a little bit easier every time you use it.

Why a Curved Bottom?

In most aerosol cans, the bottom curves inward. This serves two functions:

- **The shape strengthens the structure of the can.** If the can had a flat bottom, the force of the pressurized gas might push the metal outward. A curved bottom has greater structural integrity, just like an architectural arch or dome. With this shape, most of the force applied at the top of the curved metal is distributed to the sturdy edges of the can.

- **The shape makes it easier to use up all the product.** Draining a flat-bottom can would be like sucking up the last little bit of milkshake through a straw: You would have to tilt the can to one side so the product would collect under the plastic tube. With a curved bottom design, the last bit of product collects in the small area around the edges of the can. This makes it easier to empty almost all of the liquid.

The basic idea of an aerosol can is very simple: One fluid stored under high pressure is used to propel another fluid out of a can. To understand how this works, you need to know a little about fluids and fluid pressure.

A Few Words About Fluids

A fluid is any substance made up of free-flowing particles. This includes substances in a liquid state, such as the water from a faucet, as well as substances in a gaseous state, such as the air in the atmosphere.

The particles in a liquid are loosely bound together, but they move about with relative freedom. In a gas, the particles are completely separate from each other, and they are constantly in motion. If you want to think about it in simple terms, you can think of gas particles as tiny BBs moving at high speed. They constantly run into each other and the walls of any container that holds them.

The force of the individual, moving particles can add up to considerable pressure. Because the particles aren't bound together, a gas doesn't have a set volume like a liquid; the particles keep pushing outward in all directions. In this way, a gas expands to fill any open space.

Propellant and Product

An aerosol can applies these basic principles toward one simple goal: pushing a liquid substance out of the can. The can contains one fluid that boils well below room temperature (called the propellant) and one that boils at a much higher temperature (called the product). The product is the substance you actually use—the hair spray or insect repellent, for example—and

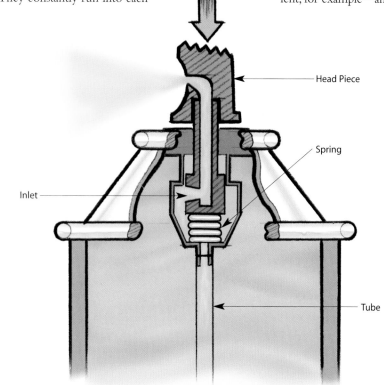

Head Piece

Spring

Inlet

Tube

the propellant is the way to get the product out of the can.

In most modern aerosol cans, the propellant and the product are both stored in a liquid state. Since the product is a liquid at room temperature, it can be poured directly into the can. The propellant, on the other hand, must be pumped in at high pressure after the can is sealed. Even though the propellant is hot enough to boil, it remains a liquid in the can because there is no room to expand.

The diagram shows a typical can design. A long plastic tube runs from the bottom of the can up to a valve system at the top of the can. The valve consists of a small, depressible head piece, with a narrow channel running through it. The channel runs from an inlet near the bottom of the head piece to a nozzle at the top. A spring pushes the head piece up, so the channel inlet is blocked by a tight seal.

When you push the head piece down, opening a passage from the inside of the can to the outside, the pressure on the liquid propellant is instantly reduced. With less pressure, it can begin to boil. Particles break free, forming a gas layer at the top of the can. This pressurized gas layer pushes the liquid product, as well as some of the liquid propellant, up the tube to the nozzle.

Some cans, such as spray-paint cans, have a ball bearing inside. If you shake the can, the rattling ball bearing helps to mix up the propellant and the product. When the liquid flows through the nozzle, the propellant rapidly expands into a gas. In some designs, this helps atomize the product—it breaks the product up into tiny drops to form a fine spray. In other designs, the evaporating propellant forms bubbles in the product, creating a foam (as in shaving cream). The consistency of the expelled product depends on several factors, including:

- The chemical makeup of the propellant and product
- The ratio of propellant to product
- The pressure of the propellant
- The size and shape of the valve system

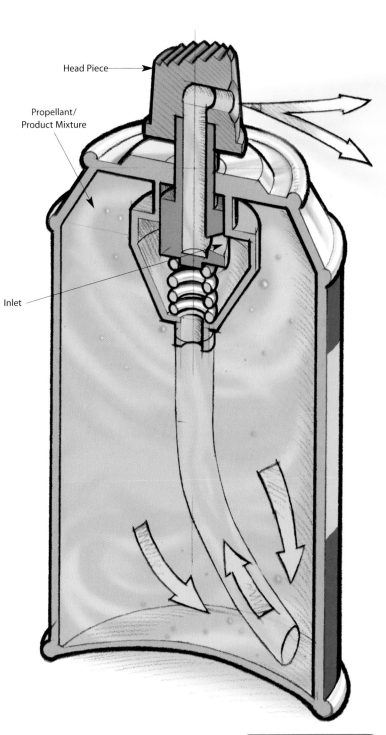

Head Piece

Propellant/ Product Mixture

Inlet

Manufacturers produce a wide variety of aerosol devices by configuring these factors in different combinations. But whether the can shoots out foamy whipped cream, thick shaving gel, or a fine mist of deodorant, the basic mechanism at work is the same: One fluid pushes another.

Cool Facts

You've probably never heard of Eric Rotheim, but you're definitely familiar with his work. Rotheim, a Norwegian engineer and inventor, came up with the first aerosol-can design more than 75 years ago. Today's can is nearly identical to his original invention.

189

How **FIRE EXTINGUISHERS** Work

A fire extinguisher is an absolute necessity in any home or office. While there's a good chance that the extinguisher will sit on the wall for years, collecting dust, it could end up saving your property and your life at some point.

HSW Web Links

www.howstuffworks.com

How Fire Engines Work
How Wildfires Work
How Aerosol Cans Work
How Smoke Detectors Work
How Flamethrowers Work
How Gasoline Works

Under Pressure

Most dry-chemical fire extinguishers have a built-in pressure gauge. If the gauge indicator is pointing to "recharge," the pressure in the extinguisher may be too low to expel the contents. The National Fire Protection Association recommends having dry extinguishers inspected every 6 years, even if the gauge indicates correct pressure.

Fire is almost always a reaction between oxygen in the atmosphere and some sort of fuel (wood or gasoline, for example). Of course, wood and gasoline don't spontaneously catch on fire just because they're surrounded by oxygen. For the combustion reaction to take place, the fuel has to be heated to its ignition temperature.

In a typical wood fire, when the wood reaches about 500°F (260°C), the heat decomposes some of the cellulose material that makes up the wood and turns it into volatile gases made up of hydrogen and carbon. These gases burn quite easily, and that's where the flame comes from.

Putting Out Fires

As you can see, there are three essential elements involved in the firemaking process:

- Fuel
- Heat
- Oxygen (or similar gas)

If you eliminate any one of the three, the fire goes out. So, fire extinguishers are designed to remove at least one of these elements.

In most fires, removing the fuel isn't a practical solution. In a house fire, for example, the entire house is fuel, which you obviously can't remove.

The best way to remove heat in many cases is to dump water on the fire. This cools the fuel below the ignition point, interrupting the combustion cycle. Water can be dangerous in the wrong situation, however. Water can put out things like burning wood, paper, or cardboard, but it does not work well on electrical fires or fires involving flammable liquids. In an electrical fire, water may conduct the current, which can electrocute you. Water will only spread out a flammable liquid, making the fire worse.

As a result, most fire extinguishers work by cutting off the supply of oxygen to the fire. In other words, they smother the fire. One popular smothering material is pure carbon dioxide. Carbon dioxide gas is heavier than oxygen, so it displaces the oxygen surrounding the burning fuel. Carbon dioxide fire extinguishers are common in restaurants because they don't contaminate cooking equipment or food.

The most common extinguisher material is dry chemical foam or powder that is typically made of sodium bicarbonate (normal baking soda), potassium bicarbonate (nearly identical to baking soda), or monoammonium phosphate. Baking soda starts to decompose at only 158°F (70°C), and when it decomposes, it releases carbon dioxide. In addition to the chemical reaction, the foam or powder forms a blanket to smother the fire.

Inside an Extinguisher

A fire extinguisher is a sturdy metal cylinder filled with water or a smothering material. When you depress a lever at the top of the cylinder, the material is expelled by gas kept under high pressure (typically carbon dioxide).

In one common extinguisher design, the compressed gas is housed in a smaller cylinder within the larger cylinder. The cylinder is sealed so the gas won't escape. To use the extinguisher, you pull out the safety pin and depress the operating lever. The lever pushes on an actuating rod, which presses a spring-mounted valve down to open up the passage to the nozzle. The bottom of the actuating rod has a sharp point that pierces the seal on the gas cylinder.

The compressed gas escapes, applying downward pressure on the fire-suppressant material. This drives the material up the siphon and out the nozzle with considerable force. The proper way to use the extinguisher

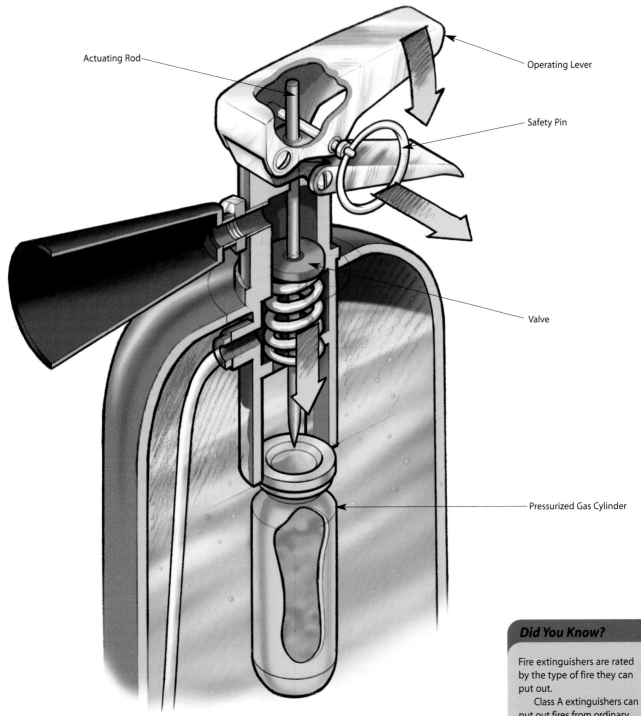

Actuating Rod

Operating Lever

Safety Pin

Valve

Pressurized Gas Cylinder

Did You Know?

Fire extinguishers are rated by the type of fire they can put out.

Class A extinguishers can put out fires from ordinary combustibles such as wood, plastic, or paper. Class B can put out burning liquids such as gasoline or grease. Class C can put out electrical fires. Extinguishers marked A, B, and C can put out all types. Class D extinguishers, designed to put out burning metal, are rare. Check the letters and pictures on your extinguisher to find out what it can handle.

is to aim it directly at the fuel, rather than the flames themselves, and move the stream with a sweeping motion.

Most fire extinguishers contain only a little bit of fire-suppressant material—you can use it all up in a matter of seconds. For this reason, extinguishers are only effective on relatively small, contained fires. To put out a larger fire, you need much bigger equipment—a fire engine, for example—and the professionals who know how to use the equipment. But for the dangerous flames that can pop up in your house or car, a fire extinguisher is an invaluable lifesaver.

How **WATER HEATERS** Work

You probably don't even think about it, but you use your water heater three or four times every day. Every time you take a hot shower, wash the dishes, or run a load of clothes through the washing machine you're depending on the technology of a hot water heater.

HSW Web Links

www.howstuffworks.com

How Water Blasters Work

How Thermoses (Vacuum Flasks) Work

How Clothes Dryers Work

How Hair Dryers Work

How Car Cooling Systems Work

In most houses you'll find one of two kinds of storage-type water heaters: gas or electric. A gas-powered water heater is a lot like a pot of water on a stove. A burner under the tank heats the water. An electric water heater, on the other hand, is like an electric teakettle. Heater elements inside the tank heat the water.

The cool thing about a water heater is that you can use hot water from the tank while the tank heats incoming cold water. A water heater uses the "heat rises" principle to separate hot water from cold water in the tank. Cold water is fed into the bottom of the tank by a pipe, and since cold water is denser than hot water it stays at the bottom until it's heated. The outlet for the hot water is at the top of the tank, so only the hottest water in the tank makes it up to the outlet.

Inside a Water Heater

A water heater consists of the following parts:

- A heavy inner tank made of steel that holds the hot water. Typically this tank holds 40 to 60 gallons. It has to be able to hold the pressure of a residential water system, which typically runs at 50 to 100 psi (the tank is tested to handle 300 psi). The steel tank normally has a bonded glass liner to keep rust out of the water.
- Insulation surrounding the tank
- A drain valve at the bottom to drain the tank—handy when you need to remove or replace the water heater
- A dip tube to let cold water into the tank. This tube normally lets cold water in at the top of the tank and releases it at the bottom of the tank.
- A pipe at the top to let hot water out of the tank
- A thermostat to control the temperature of the water inside the tank. Many electric water heaters have a separate thermostat on each heating element.
- A pressure relief valve—an important safety feature that keeps the tank from exploding
- An anode rod to help keep the steel tank from corroding

If your water heater is electric, two heating elements heat the water. These are thick electric elements similar to the elements you see inside an electric oven. If it's a gas water heater, a burner at the bottom of the tank heats the water.

The thermostat controls the temperature of the water inside the tank. Normally you can set the temperature between 120°F

Caveat Emptor...

The majority of homeowners in the U.S. use storage-type water heaters. Unfortunately, water heaters don't last forever. That means if you're a homeowner, chances are you're going to have to purchase a new water heater sometime. In addition to the upfront cost, there are other things you will want to consider, such as:

- **Size of the tank**—this can range anywhere from 20 to 80 gallons
- **Fuel source**—such as electric, oil, and propane or natural gas
- **Energy efficiency rating**—a higher number indicates a more energy-efficient appliance

Check the EnergyGuide label for information regarding a water heater's energy efficiency rating. The annual cost of operation and the first hour rating (FHR) should also be included on the label. The FHR can be important in your selection, too. It is an estimate of the maximum amount of hot water the water heater will provide during a peak hour. It's important to note that a larger tank doesn't necessarily reflect a higher FHR.

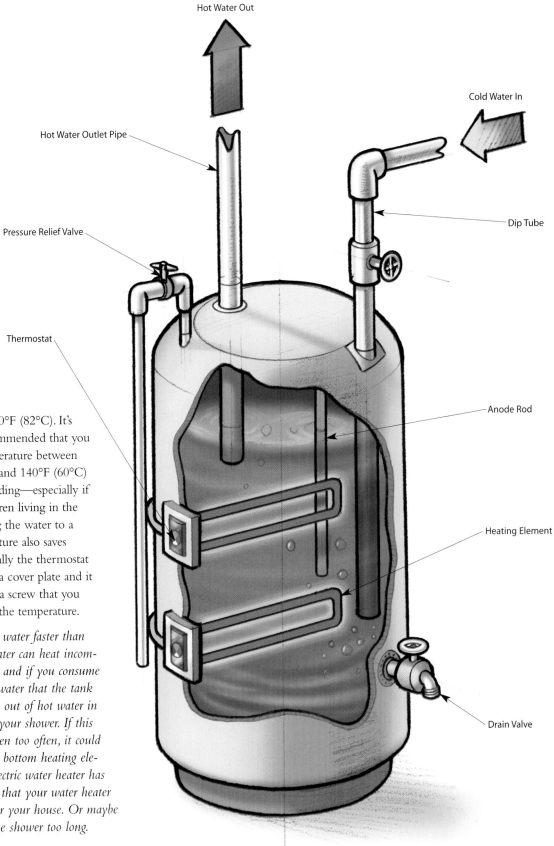

Hot Water Out

Cold Water In

Hot Water Outlet Pipe

Dip Tube

Pressure Relief Valve

Thermostat

Anode Rod

Heating Element

Drain Valve

(48°C) and 180°F (82°C). It's generally recommended that you keep the temperature between 120°F (48°C) and 140°F (60°C) to prevent scalding—especially if there are children living in the house. Heating the water to a lower temperature also saves energy. Normally the thermostat is underneath a cover plate and it has a knob or a screw that you can use to set the temperature.

If you use hot water faster than your water heater can heat incoming cold water, and if you consume all of the hot water that the tank holds, you run out of hot water in the middle of your shower. If this seems to happen too often, it could mean that the bottom heating element in an electric water heater has burned out or that your water heater is too small for your house. Or maybe you stay in the shower too long.

How **WASHING MACHINES** Work

It's washed your clothes hundreds of times, and using it is definitely a lot easier than beating your clothes on a rock by the river. But have you ever wondered what's inside your trusty washing machine? How does it spin the clothes so fast without shaking the house apart or leaking water? And why is a washing machine so heavy?

HSW Web Links

www.howstuffworks.com

Inside an Electric Motor
How Home Dry Cleaning
 Works
How Clothes Dryers Work
How Gears Work
How Electric Motors Work

Operating a washing machine is pretty simple. You fill the tub with clothes, throw in some soap, and hit a button. The machine does the rest—it fills the tub with water, then stirs the clothes around using an agitator, rinses them, and spins the water out. What's inside this machine to help it do all this work?

Doing the Dirty Work

If you take a look under a washing machine, you'll see one reason why it's so heavy—a block of concrete! The concrete is there to balance the equally heavy electric motor, which drives a very heavy gearbox. The weight of the motor and the concrete also give the machine stability during the spin cycle.

A washing machine contains two steel tubs, one inside the other. The outer one is stationary and sealed so it can hold water. It bolts to the body of the washer. The inner tub holds the clothes. It has an agitator in the middle of it, and the sides are perforated with holes so that when the tub spins, the water can leave.

Because the inner tub vibrates and shakes during the wash cycle, it has to be mounted in a way that lets it move around without banging into other parts of the machine. The inner tub attaches to the gearbox, which attaches to a metal frame. This frame holds the motor, gearbox, and the concrete weight. A cable and pulley system supports the weight of the heavy components, letting them move in a way that does not shake the entire machine.

If all of these parts are just hanging by cables, why don't they swing around all the time? Because a washing machine has a damping system that uses friction to absorb some of the force from the vibrations.

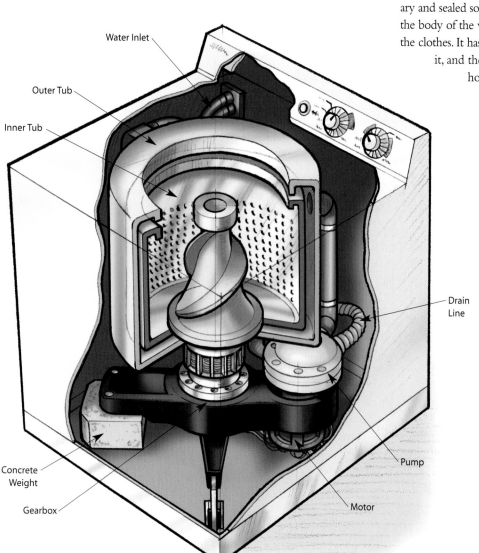

Water Inlet

Outer Tub

Inner Tub

Concrete
Weight

Gearbox

Drain
Line

Pump

Motor

In each of the four corners of the machine is a mechanism that works a little like a disc brake. It squeezes two pads against a metal plate that is attached to the frame. As the frame vibrates and rocks, the friction of these pads absorbs a lot of the motion.

Plumbing the Depths

The plumbing inside a washing machine has several jobs:

- It fills the washing machine with water.
- It recirculates the wash water from the bottom of the wash tub back to the top during the wash cycle.
- It pumps water out of the machine and into a drain between cycles.

There are hookups for two water lines on the back of a washing machine, one for hot water and one for cold. These lines are hooked up to the body of a solenoid (electromagnetic switch controlled by an electrical current) valve. So depending on the water temperature that you select, either the hot valve, the cold valve, or both valves will open.

Before the hose releases water into the washtub, it sends the water through an anti-siphon device. This device prevents wash water from being sucked back into the water supply lines and possibly contaminating the water for your house or even your neighborhood.

Pump It Up

The rest of the plumbing system—the part that recirculates the water and the part that drains it—involves the pump. The pump is actually two separate pumps in one: The bottom half of the pump is hooked up to the drain line, while the top half recirculates the wash water. So how does the pump decide whether to pump the water out through the drain line or back into the washtub?

This is where one of the neat tricks of the washing machine comes in: The motor that drives the pump can reverse direction. So, the motor spins one way when the washer is running a wash cycle and recirculating the water; and it spins the other way when the washer is doing a spin cycle and draining the water.

- If the pump spins clockwise, the bottom pump sucks water from the bottom of the washtub and forces it out through the drain hose. And the top pump tries to suck air from the top of the washtub and force it back up through the bottom, so that no water recirculation takes place.
- If the pump spins counter-clockwise, the top pump sucks water from the bottom of the tub and pumps it back up to the top. And the bottom pump tries to pump water from the drain hose back into the bottom of the tub. There is actually a little bit of water in the drain hose, but the pump doesn't have the power to force much of it back into the tub.

Drive Mechanism

The drive mechanism, or gearbox, on a washing machine is one of the coolest parts of the washing machine. It has two jobs:

- To agitate the clothes, moving them back and forth inside the washtub
- To spin the entire washtub, forcing the water out

To handle these two jobs, the gearbox uses the same trick as the pump does. If the motor spins in one direction, the gearbox agitates; if it spins the other way, the gearbox goes into spin cycle.

If you were to look inside the gearbox, you would see a gear with a link attached to it. This link is just like the one attached to an old steam-train wheel—as the gear (and the link) turns, it pushes another pie-shaped piece of gear back and forth. This pie-shaped gear engages a small gear on the agitator. In addition to rotating the inner shaft in alternating directions, there are other gears within the system that provide the gear reduction to slow the rotation. Because the motor spins only at one speed, spin-cycle speed, gear reduction lets the agitator move at a more sedate pace.

When the washer goes into spin cycle, the whole mechanism locks up, causing everything to spin at the same speed as the motor.

When you put it all together, you have a machine that takes what used to be a day of hard work and turns it into a two-minute task. Isn't technology great!

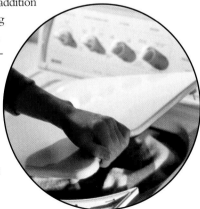

Agitator gearbox assembly.

How **CLOTHES DRYERS** Work

A clothes dryer can be found in just about every house in the United States, and millions more of them are manufactured each year. They are reliable, and cheap to build. You'll be surprised at how simple this common machine is.

HSW Web Links

www.howstuffworks.com

How Toasters Work
How Washing Machines Work
How Dry Cleaning Works
How Home Dry Cleaning Works
How Hair Dryers Work

You probably do it once or twice a week. After the washing machine makes its final spin, you lift the lid, grab the wet stuff, and cram it in the clothes dryer. You push a button or twist a dial and voilà—about 45 minutes later that soggy pile of material is now a fluffy, fresh bundle of dry clothes. You know air and tumbling is involved somehow, but what really happens inside this magic box? The basic idea behind a clothes dryer is this:

- You have a box.
- You want to force warm air inside the box.
- You want to get rid of the resulting moist air.
- You want to keep the clothes moving so that all surfaces of the fabric get exposed to the warm air.

You may know that the warm, moist air leaves the dryer through a hole in the back, which is usually hooked up by a pipe to a vent on your house. But where does the air enter the dryer?

Where the Air Goes

In the example here, the dryer is the type that has the lint screen in the door. In brief, this is how the air makes its way through the dryer:

- Air enters the body of the dryer through the bottom or back of the dryer.
- Air is sucked past the heating element and into the tumbler.
- Air enters the door and is directed down through the lint screen.
- Air passes through a duct in the front of the dryer and into the fan.
- The fan forces air into the duct leading out the back of the dryer, at which point it exits your house.

The first thing that the air hits is the heating element. In an electric dryer the heating element is a standard nichrome-wire affair similar to what you would find in a toaster or a space heater. This type of heating

element consumes lots of power—4,000 to 6,000 watts on most dryers. In a gas dryer, burning gas provides the heat.

The hot air now makes its way through the clothes in the tumbler and then into the holes in the door. The air passes through the holes in the door and out through the big slot in the bottom of the door that leads to the lint screen.

The air is drawn through the lint screen and down a duct in the front of the dryer, where it enters the fan. The fan is a centrifugal type of device—as it spins, it flings the air to the outside, sucking air from the center and forcing it out the duct at the back of the dryer.

Tumbling the Clothes

If you take the outer sheet of metal off your dryer, one surprising thing is the lack of any gears on the tumbler. It turns out that the tumbler is one giant gear (actually a pulley), and the motor drives a tiny pulley. Because of the ratio between the huge diameter of the tumbler and the tiny diameter of the motor pulley, no other gears are needed!

The same electric motor drives both the fan and the tumbler. The pulley for the tumbler belt is hooked up to one output shaft of the motor and the fan is hooked to the other.

Another funny thing about most dryers is that the tumbler has no bearings to help it spin smoothly. So what supports the weight of the clothes? The tumbler rides on two hard, slippery nylon pads that are mounted to the top of the support structure. This arrangement lets the tumbler hold a lot of weight in wet clothing, but is relatively low-friction.

Controls

Most dryers have some type of cycle switch and/or heat setting buttons. By turning a knob to various positions and/or pushing a few buttons, you can control the type of cycle, the length of time the dryer runs, and the heat setting.

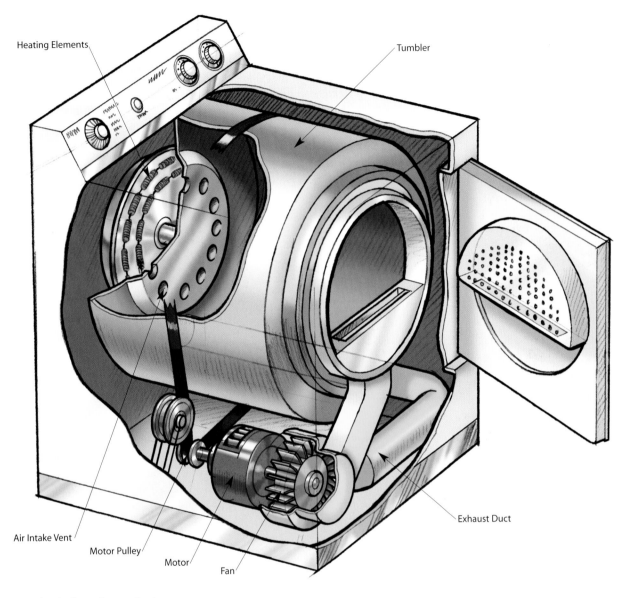

Heating Elements

Tumbler

Air Intake Vent

Motor Pulley

Motor

Fan

Exhaust Duct

A mechanical or electronic timer controls how long the cycle will last. Some dryers have a moisture switch that can sense when the clothes are dry and end the cycle early.

The cycle switch and the heat setting buttons together control which heating elements are on at a given time:

- If none of the heating elements are on, only cool air blows through the clothes.
- If one heating element is on, the air is warm.
- If both heating elements are on, the air is hot.

Safety Systems

Clothes dryers have safety features that help to prevent overheating. In many dryers,

there are two temperature shut-off sensors. When these sensors reach certain preset temperatures, they break contact, which shuts the dryer off.

The first sensor monitors the lint screen and the tumbler. If the temperature in the tumbler gets too hot, the sensor cuts the power, shutting the dryer off.

But what happens if the belt breaks? Or if the fan gets clogged and no air is coming out of the tumbler? This is where the second temperature sensor comes in.

The second sensor is located close to the heating elements. If airflow is shut off for any reason, the air near this sensor will quickly heat up to the temperature that triggers this sensor, and it will switch off the power.

How **SEPTIC TANKS** Work

It's invisible, so we don't even notice it. But each of us generates lots of wastewater every day. For example, you flush the toilet three or four times a day, you take a shower, you brush your teeth, wash clothes and dishes, and so on. A family of four can easily produce 100 to 200 gallons of sewage a day. Where does it all go?

HSW Web Links

www.howstuffworks.com

How Toilets Work
How House Construction
 Works
How Composting Works
How Landfills Work

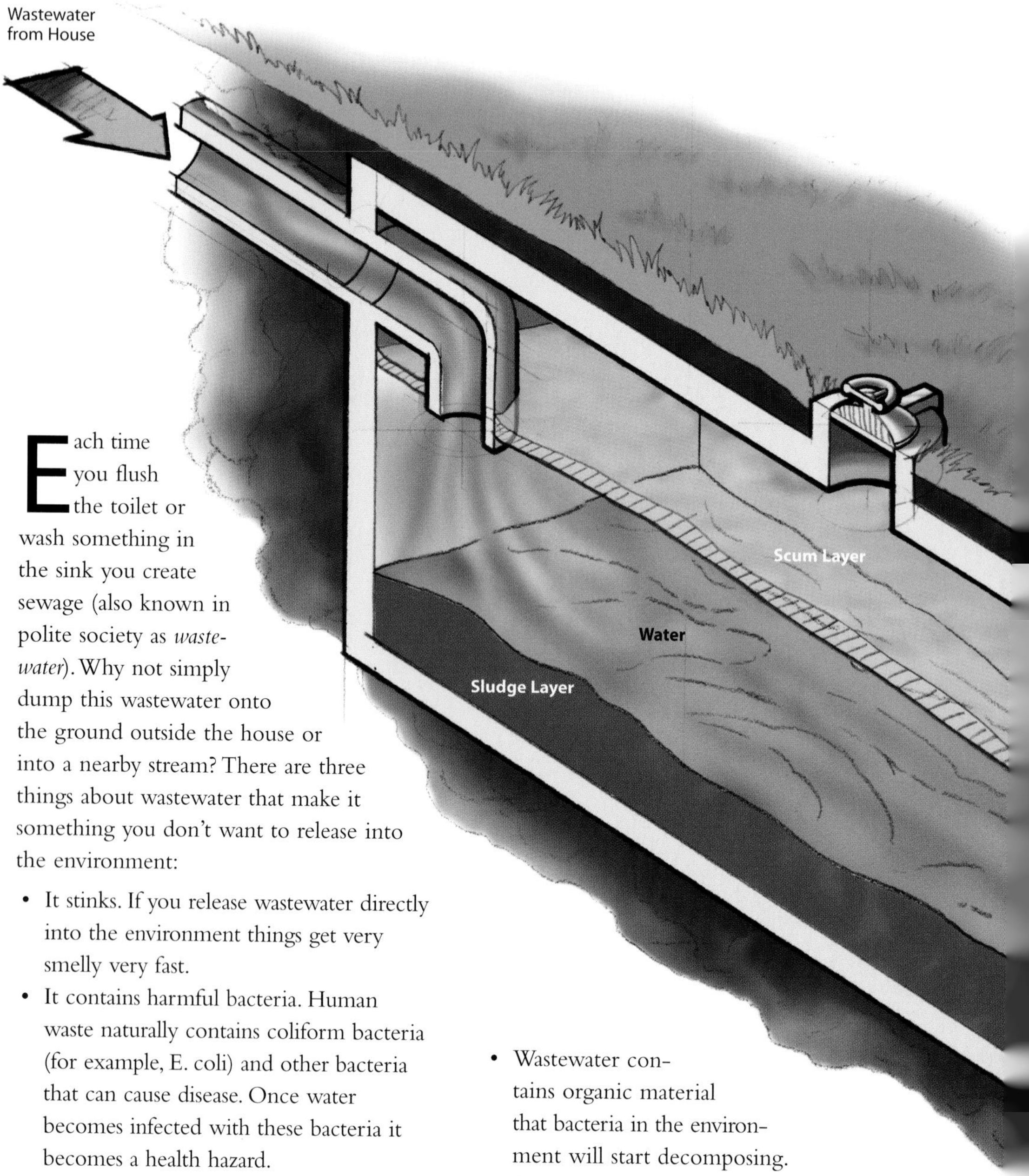

Wastewater
from House

Scum Layer

Water

Sludge Layer

Each time you flush the toilet or wash something in the sink you create sewage (also known in polite society as *waste-water*). Why not simply dump this wastewater onto the ground outside the house or into a nearby stream? There are three things about wastewater that make it something you don't want to release into the environment:

• It stinks. If you release wastewater directly into the environment things get very smelly very fast.

• It contains harmful bacteria. Human waste naturally contains coliform bacteria (for example, E. coli) and other bacteria that can cause disease. Once water becomes infected with these bacteria it becomes a health hazard.

• It contains suspended solids and chemicals that affect the environment. For example:

  • Wastewater contains nitrogen and phosphates that, being fertilizers, encourage the growth of algae. Excessive algae growth can block sunlight and foul the water.

  • Wastewater contains organic material that bacteria in the environment will start decomposing. The bacteria consume oxygen in the water, and the lack of oxygen kills fish.

  • The suspended solids in wastewater make the water look murky and can affect the ability of many fish to breath and see.

The increased algae, reduced oxygen, and murkiness destroy the ability of a stream or

198

lake to support wildlife; all of the fish, frogs, and other life forms quickly die.

No one wants to live in an place that stinks, is full of deadly bacteria, and cannot support aquatic life. That's why communities build wastewater-treatment plants and enforce laws against the release of raw sewage into the environment. Not everyone lives in a community with a wastewater-treatment plant. Those people must rely on a backyard system.

Private Treatment: The Septic Tank

In rural areas where houses are spaced so far apart that a sewer system would be too expensive to install, people install their own private sewage-treatment plants. They are called septic tanks.

A septic tank is simply a big concrete or steel tank buried in the yard. The tank might hold 1,000 gallons (4,000 liters) of water. Wastewater flows into the tank at one end and leaves the tank at the other.

There are three layers to a typical private septic system. Anything that floats rises to the top and forms a layer known as the scum layer. Anything heavier than water sinks to form the sludge layer. In the middle is a fairly clear water layer. This body of water contains bacteria and chemicals like nitrogen and phosphorous, but it is largely free of solids.

Waste Not, Want Not

The sludge layer gets deeper and deeper over time. Eventually you call a septic tank pumping service, and it sends a truck that sucks the sludge out.

As new water enters the tank, it displaces water already there. The displaced water flows out of the septic tank to a drain field. The drain field is made of perforated pipes buried in trenches filled with gravel.

A typical drain field pipe is 4 inches (10 centimeters) in diameter. It is perforated with holes, and buried in a trench that is 4 to 6 feet (about 1.5 meters) deep and 2 feet (0.6 meters) wide. The gravel fills the bottom 2 to 3 feet of the trench and dirt covers the gravel. Water leaks out of the holes in the pipe into the gravel, and seeps from the gravel into the soil.

The water is slowly absorbed and filtered by the ground in the drain field. The size of the drain field is determined by how well the ground absorbs water. In places where the ground is hard clay that absorbs water very slowly, the drain field has to be much bigger.

A septic system is normally powered by nothing but gravity. Water flows down from the house to the tank, and down from the tank to the drain field. It is a completely passive system.

You may have heard the expression, "The grass is always greener over the septic tank." Actually, it's the drain field, and the grass really is greener. The grass takes advantage of the moisture and nutrients in the drain field and it grows better.

To Drain Field

How **SEWING MACHINES** Work

Like the automobile, the cotton gin, and countless other innovations from the past three hundred years, the sewing machine takes a job that was time-consuming and laborious and makes it fast and easy. With the invention of the mechanized sewing machine, manufacturers could suddenly produce piles of high-quality clothing at minimal expense. Because of this technology, the vast majority of people in the world can now afford the sort of sturdy, finely stitched clothes that were a luxury only a couple hundred years ago.

HSW Web Links

www.howstuffworks.com

How Gears Work
How Electric Motors Work
How Gear Ratios Work
How Pendulum Clocks
  Work
Inside an Electric
  Screwdriver
How Chain Saws Work

The automated stitching mechanism at the heart of a sewing machine is incredibly simple, but the machinery that drives it involves an assembly of gears, pulleys, and motors that looks incredibly complicated. When you get down to it, the sewing machine is among the most elegant and ingenious tools ever created.

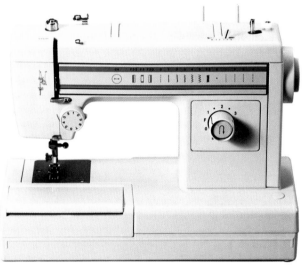

Sew What?

Sewing machines are something like cars: There are hundreds of models on the market, and they vary considerably in price and performance. But just like cars, most sewing machines are built around one basic idea. Where the heart of a car is the internal combustion engine, the heart of a sewing machine is the loop stitching system.

The loop stitch approach is very different from ordinary hand-sewing. In the simplest hand stitch, a length of thread is tied to a small eye at the end of a needle. The sewer passes the needle and the attached thread all the way through two pieces of fabric, from one side to the other and back again. The needle runs the thread in and out of the fabric pieces, binding them together.

While this is easy enough to do by hand, it is extremely difficult to pull off with a machine. The machine would have to release the needle on one side of the fabric just as it grabbed the needle again on the other side. Then it would have to pull the entire length of loose thread through the fabric, turn the needle around, and do the whole thing in reverse. This process is way too complicated and unwieldy for a simple machine, and

even by hand it only works well with short lengths of thread.

Instead, sewing machines pass the needle only partway through the fabric. On a machine needle, the eye is right behind the sharp point, rather than at the end. The

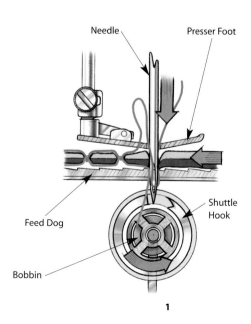

Needle   Presser Foot

Feed Dog

Shuttle
Hook

Bobbin

1

200

needle is fastened to the needle bar, which is driven up and down by the motor through a series of gears and cams (more on this later).

When the tip of the needle passes through the fabric, it pulls a small loop of thread from one side to the other. A mechanism underneath the fabric grabs this loop and wraps it around another piece of thread or another loop in the same piece of thread.

Lock and Chain

There are several different types of loop stitches, and they all work a little differently. The simplest loop stitch is the *chain stitch*. To sew a chain stitch, the sewing machine loops a single length of thread back on itself.

The fabric, sitting on a metal plate underneath the needle, is held down by a presser foot. At the beginning of each stitch, the needle pulls a loop of thread through the fabric. A looper mechanism, which moves in sync with the needle, grabs the loop of thread before the needle pulls up. Once the needle has pulled out of the fabric, the feed dog mechanism (which we'll examine later) pulls the fabric forward.

When the needle pushes through the fabric again, the new loop of thread passes directly through the middle of the earlier loop. The looper grabs the thread again and loops it around the next thread loop. In this way, every loop of thread holds the next loop in place.

The main advantage of the chain stitch is that it can be sewn very quickly. It's not especially sturdy, however, since the entire seam can come undone if one end of the thread is loosened. Most sewing machines use a sturdier stitch known as the *lockstitch*.

The most important element of a lock-stitch mechanism is the shuttle hook and bobbin assembly. The bobbin is just a spool of thread positioned underneath the fabric. The shuttle is a round housing for the spool, with a built-in hooking mechanism. The machine motor rotates the shuttle in sync with the moving needle.

Just as in a chain-stitch machine, the needle pulls a loop of thread through the fabric, rises again as the feed dogs move the fabric along, and then pushes another loop in. But instead of looping the thread around itself, the shuttle loops it around a second length of thread. The shuttle grips the loop with a hook and pulls it all the way around the unspooling bobbin. This makes for a very sturdy stitch.

The General Assembly

The conventional electric sewing machine is a fascinating piece of engineering. If you were to take the outer casing off, you would see a mass of gears, cams, cranks, and belts, all driven by a single electric motor. The exact configuration of these elements varies a good deal from machine to machine, but they all work on a similar idea.

Did You Know?

One important addition to the basic sewing machine is the ability to sew different sorts of stitches. The typical stitch options for a conventional sewing machine are variations on the zig-zag stitch, a stitch where the needle moves back and forth instead of creating a single line of stitches.

To sew a zig-zag stitch, the machine simply moves the needle assembly from side to side at the same time that it is moving up and down. In a conventional electric machine, the needle bar is attached to an additional linkage, which follows a cam on the main drive shaft. When the linkage is engaged, the rotating cam shifts the linkage from side to side. In an electronic machine, a small motor moves the needle bar.

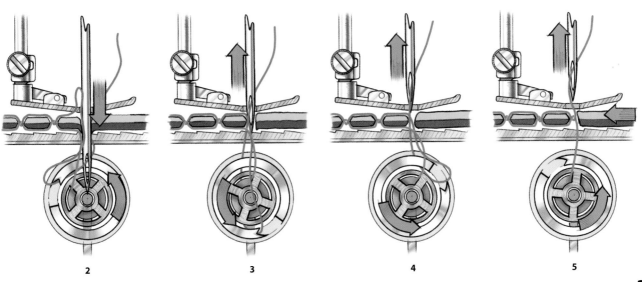

2   3   4   5

The electric motor connects to a drive wheel by way of a drive belt. The drive wheel rotates the long upper drive shaft, which is connected to several different mechanical elements:

- The shaft turns a belt driving the lower shaft.
- The end of the shaft turns a crank, which pulls the needle bar up and down.
- The crank also moves the thread-tightening arm.

Moving in sync with the needle bar, the tightening arm lowers to create enough slack for a loop to form underneath the fabric, then pulls up to tighten the loop after it's released from the shuttle hook.

The thread runs from a spool on the top of the machine, through the tightening arm and through a tension disc assembly. By turning the disc assembly, the sewer can tighten the thread feeding into the needle. The tension must be tighter when sewing thinner fabric and looser when sewing thicker fabric.

The end of the lower drive shaft connects to a set of bevel gears that rotate the shuttle assembly. Since both are connected to the same drive shaft, the shuttle assembly and the needle assembly always move in unison.

The lower drive shaft also moves linkages that operate the feed dog mechanism. One linkage slides the feed dog forward and backward with each cycle. At the same time, another linkage moves the feed dog up and down. The two linkages are synchronized so that the feed dog presses up against the fabric, shifts it forward, and then moves down to release the fabric. The feed dog then shifts backward before pressing up against the fabric again to repeat the cycle.

The person using the sewing machine controls the motor with a foot switch. The switch is pressure sensitive, so the sewer can adjust the speed of the motor by varying the pressure on the switch.

The cool thing about this design is that everything is linked together, so when you press on the pedal, the motor speeds all of the processes up at the same rate. The process is always perfectly synchronized, no matter how fast the motor is turning.

Computer Crafts

Today's high-end home sewing machines have built-in computers, as well as small monitor displays for easier operation. In these models, the computer directly controls several different motors, which precisely move the needle bar, the tensioning discs, the feed dog, and other elements in the machine. With this fine control, it's possible to produce hundreds of different stitches. The computer drives the motors at just the right speed to move the needle bar up and down and from side to side in a particular stitch pattern. Typically, the computer programs for different stitches are stored in removable memory disks or cartridges. The sewing-machine computer may also hook up to a PC in order to download patterns directly from the Internet.

Some electronic sewing machines also have the ability to create complex embroidery patterns. These machines have a motorized work area that holds the fabric in place underneath the needle assembly, and a series of sensors that tell the computer how all of the machine components are positioned. By precisely moving the work area forward, backward, and side-to-side while adjusting the needle to vary the stitching style, the computer can produce all kinds of elaborate shapes and lines. The sewer simply loads a pattern from memory or creates an original one and the computer does almost everything else. The computer prompts the sewer to replace the thread or make any other adjustments when necessary.

Obviously, this sort of high-tech sewing machine is a lot more complex than the fully manual sewing machines of two hundred years ago, but they are both built around the same simple stitching system: A needle passes a loop of thread through a piece of fabric, where it is wound around another length of thread. This ingenious method was one of those rare, inspired ideas that changed the world forever.

How **VACUUM CLEANERS** Work

Just about every home or apartment in America has one—a vacuum cleaner. If you have carpet, a vacuum is a necessity. They come in all shapes and sizes, and mutant strains of vacuum cleaners now even spray the carpet with water and suck it back up to really get things clean!

When you sip soda through a straw, you are utilizing the simplest of all suction mechanisms. Sucking on the end of the straw causes a pressure drop between the bottom of the straw and the top of the straw. With greater fluid pressure at the bottom than the top, the soda gets pushed up to your mouth.

This pressure differential is the same basic mechanism that's at work in a vacuum cleaner, though the execution is a bit more elaborate.

The Basic Cleaner

It may look like a complicated machine, but the conventional vacuum cleaner has only six essential elements:

- An intake port
- An exhaust port
- An electric motor
- A fan
- A porous bag
- A housing that holds all the other elements

When you plug the vacuum cleaner in and turn it on, the motor turns the fan. As the fan blades turn, they force air toward the exhaust port. The density of particles (and therefore the air pressure) decreases on one side of the fan and increases on the other. In a vacuum cleaner, it is the side with the pressure decrease—the vacuum—that does the work. The pressure drop creates suction inside the vacuum cleaner. The ambient air pushes itself into the vacuum cleaner through the intake port because the air pressure inside the

vacuum cleaner is lower than the pressure outside.

As long as the fan is running and the passageway through the vacuum cleaner remains open, a constant stream of air moves through the intake port and out the exhaust port. This stream of air acts just like a stream of water.

HSW Web Links

www.howstuffworks.com

How Washing Machines Work
How Clothes Dryers Work
How Air Conditioners Work
How Hot Air Balloons Work
How Snow Makers Work
How Water Blasters Work
How SCUBA Works

Porous Dust Bag

Filter

Exhaust Port

Electric Motor

Fan

Intake Port

The moving air particles rub against any loose dust or debris, and if the debris is light enough and the suction is strong enough, the friction carries the material inside the vacuum cleaner. Most vacuums have rotating brushes at the intake port, which kick dust and dirt particles loose from the carpet so the air stream can more easily pick them up.

As the dirt-filled air makes its way toward the exhaust port, it passes through the vacuum-cleaner bag. Vacuum bags are made of porous woven material (typically cloth or paper) that acts as an air filter. The tiny holes in the bag are large enough to let air particles pass by, but are too small for most dirt particles to fit through. When the air current streams into the bag, all the air moves on through the material, but the dirt and debris collect in the bag.

You can put the vacuum-cleaner bag anywhere along the path between the intake tube and the exhaust port, as long as the air current flows through it. In upright vacuum cleaners, the bag is typically the last stop on the path: Immediately after it is filtered, the air flows back to the outside. In canister vacuums, the bag may be positioned before the fan, so the air is filtered as soon as it enters the vacuum.

Suction Power

The power of the vacuum cleaner's suction depends on the speed of the fan, the blockage of the air passageway, and the size of the intake port opening.

To generate strong suction, the motor has to turn the fan very quickly. If the motor is damaged, the bag is full, or the power supply is limited, the fan will slow down and so will the suction.

Blockage is a regular problem in most vacuums. When debris builds up in the vacuum bag, the air faces greater resistance on its way out. Each particle of air moves more slowly because of the increased drag. This is why a vacuum cleaner works better when you've just replaced the bag than when you've been vacuuming for a while.

The intake port opening is the most intriguing power variable. Since the speed of the vacuum fan is constant, the amount of air passing through the vacuum cleaner per unit of time is also constant. No matter what size

you make the intake port, the same number of air particles will have to pass into the vacuum cleaner every second. If you make the port smaller, the individual air particles will have to move much more quickly in order for them all to get through in that amount of time.

At the point where the air speed increases, pressure decreases (this is known as Bernoulli's principle). The drop in pressure translates to a greater suction force at the intake port. Because they create a stronger suction force, narrower vacuum attachments can pick up heavier dirt particles than wider attachments.

Variations on a Theme

Most vacuum cleaners work on the same basic principles, but they vary a good deal in design. For heavy-duty cleaning jobs, a lot of people use wet/dry vacuum cleaners, models that can pick up liquids as well as solids. In these bagless cleaners, the air stream passes through a wide area positioned over a bucket. When it reaches this larger area, the air stream slows down, for the same reason that the air speeds up when flowing through a narrow attachment. This drop in speed effectively loosens the air's grip, so the liquid droplets and heavier dirt particles can fall out of the air stream and into the bucket. After you're done vacuuming, you simply dump out whatever has collected in this bucket.

Another bagless variation is the cyclone vacuum. This machine, developed in the 1980s by James Dyson, sends the air stream through one or more cylinders along a high-speed spiral path. The spiraling motion works something like a clothes dryer, a roller coaster, or a merry-go-round: As the air stream whips around in a circle, all of the dirt particles experience a powerful centrifugal force; they are thrown outward, away from the air stream. In this way, the dirt is extracted from the air without using any sort of filter. It simply collects at the bottom of the cylinder.

In the future, we are sure to see more improvements on the basic vacuum-cleaner design, with new suction mechanisms and collection systems. But the basic idea, using a moving air stream to pick up dirt and debris, is here to stay for some time.

Central Suction

The central vacuum system of the early 1900s turned your whole house into a cleaner. A motorized fan in the basement or outside the house created suction through a series of interconnected pipes in the walls. To use the cleaner, you turned on the fan motor and attached a hose to any of the various pipe outlets throughout the house. The dirt was sucked into the pipes and deposited in a large canister, which you only had to empty a few times a year. These cleaners were once very popular, went nearly extinct, but have seen something of a resurgence in recent years in more expensive homes.

How **ESPRESSO MACHINES** Work

Espresso consumption in the United States has grown tremendously in the last decade or two. In Seattle, for instance, you can't walk more than a couple of blocks without seeing a cafe or espresso cart—they're in bookstores, grocery stores, laundromats, gas stations, and movie theaters. There are even drive-through espresso shops in parking lots. Sometimes an espresso cart will just park on the sidewalk, like the hot-dog vendors in New York City.

Espresso has become so popular that a lot of folks want to be able to make it in their own kitchen, so they've been buying espresso machines for their homes. These machines are smaller than the commercial machines found in cafes, but they work on the same principles. Before we get to the machines, let's see what espresso is.

What Is Espresso?

If you go to a cafe or espresso bar and ask for an espresso, what you will get is a shot-sized glass holding a small amount of very strong coffee. There are many different types of espresso drinks, including cappuccino, cafe latte, and cafe mocha. All are made with one or more shots of espresso.

A shot of espresso is made by forcing about 1.5 ounces (42 grams) of hot water through tightly packed, finely ground espresso coffee. If everything goes well, what comes out is a dark brown, slightly thick liquid with a small amount of *crema* (a foam, sort of like the head on a beer) on top.

There are many variables in the process of making a shot of espresso. The temperature

HSW Web Links

www.howstuffworks.com

How Drip Coffee Makers Work
How Caffeine Works
How Are Coffee, Tea and Colas Decaffeinated?

Did You Know?

Espresso coffee is a blend of several different types of coffee beans from different countries. The beans are roasted until they are dark and look oily.

The beans are ground very finely—much finer than for drip coffee makers. The consistency is almost like powdered sugar. The more finely the coffee is ground, the slower the espresso comes out. Generally, for the best shot of espresso, the water should take about 25 seconds to pass through the coffee. Sometimes the grind is adjusted to control the brewing time.

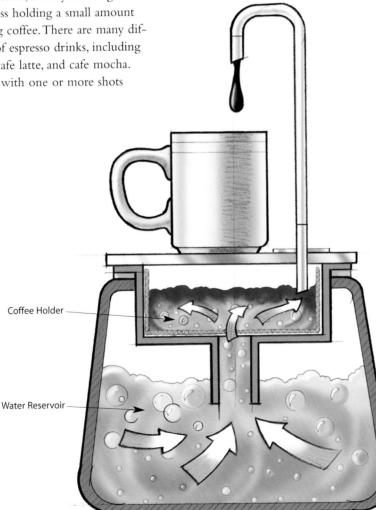

Coffee Holder

Water Reservoir

A simple espresso machine.

of the water, the pressure of the water, the fineness of the ground coffee, and how tightly the coffee is packed are just a few. The skilled espresso maker, or *barista*, controls all of these variables to produce a quality shot of espresso.

A Simple Machine

To force the water through the coffee, the simplest espresso machines use pressure that comes from heating water inside a sealed vessel. These types of machines can be bought for around $50, and there's even one that's made especially to take on a camping trip. Water-pressure espresso machines all work on the same principle, so we'll take a look at one of the camping-style machines as an example.

The coffee is packed into a funnel-shaped piece of metal that has a tube extending to the bottom of the reservoir. A few ounces of water are put into the reservoir and the top is screwed on.

When the water is heated over a fire, pressure builds inside the vessel. The only way for the pressure to escape is up the tube, through the coffee and out of the tube in the top.

There are some disadvantages to a machine like this. The pressure in the system depends on

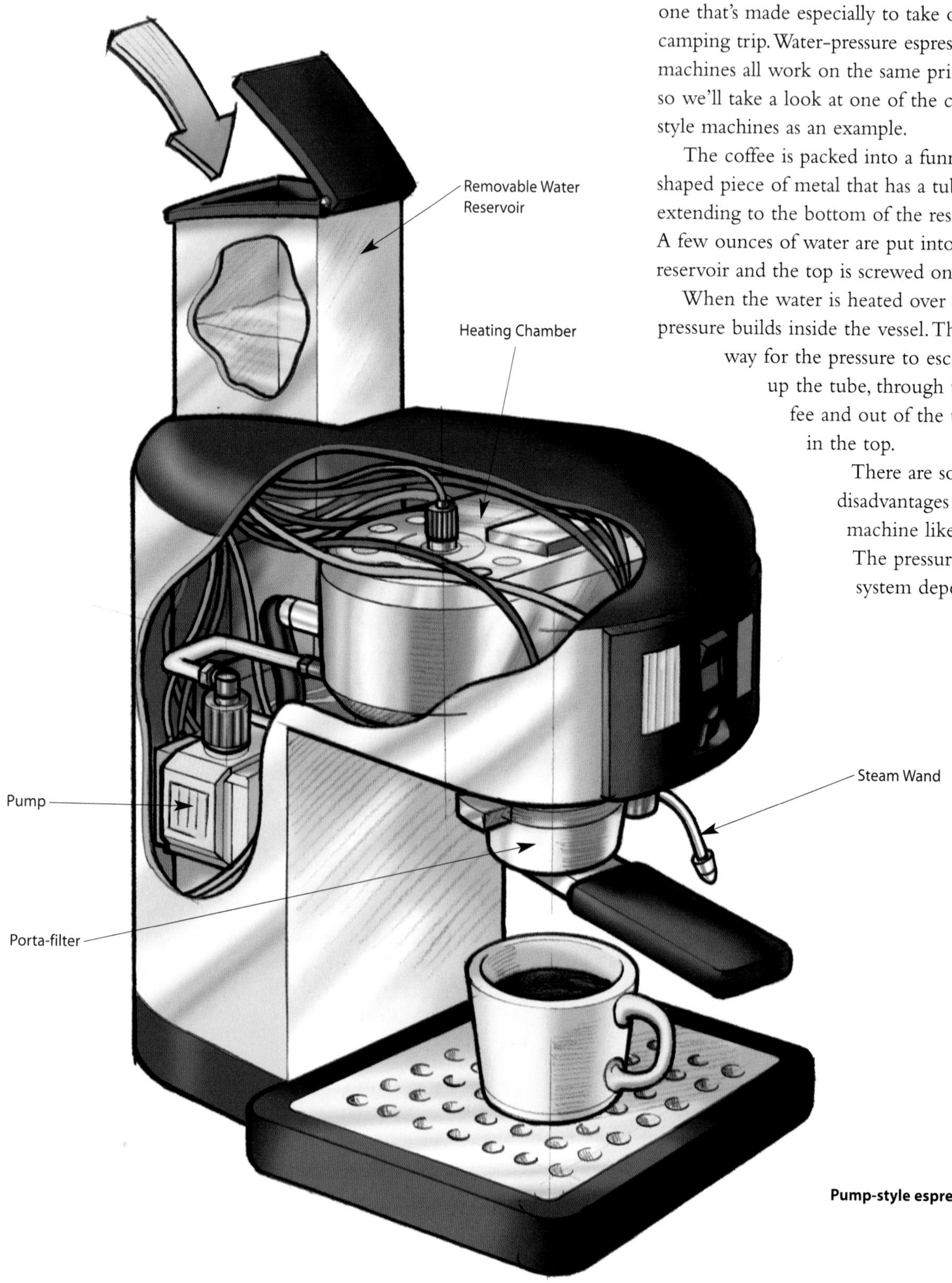

Removable Water Reservoir

Heating Chamber

Steam Wand

Pump

Porta-filter

Pump-style espresso machine.

the temperature of the water. The temperature required to build up enough pressure to force the water through the coffee might exceed the ideal brewing temperature.

This is why some home machines incorporate a pump.

Pump-Style Espresso Machines

This type of machine is a little fancier, but is still fairly simple to operate. It has the following elements:

- **A reservoir**—The reservoir holds the cold water used in the espresso machine. It is not pressure tight or heated, and it is removable.
- **A pump**—The pump draws water out of the reservoir and pumps it into the heating chamber at high pressure.
- **A heating chamber**—The heating chamber is a sturdy, stainless steel structure with a heating element built into a groove in the bottom. The resistive heating element is simply a coiled wire, very similar to the filament of a light bulb or the element in an electric toaster, which gets hot when you run electricity through it. In a resistive element like this, the coil is embedded in plaster to make it more rugged. The heating chamber also contains a one-way valve that lets water into the chamber from the pump but not back into the pump from the chamber.

- **A porta-filter**—The porta-filter is the removable part of the machine that holds the ground coffee. Inside the basket is a small removable screen into which the ground coffee is packed. On the bottom of the basket are two spouts where the espresso comes out.
- **A steam wand**—The steam wand is used to heat and froth milk for use in various espresso drinks. This wand is connected to the heating vessel. When the user puts the valve in the steam position, steam from the heating vessel is released out of the wand and into the milk.
- **A control panel**—The control panel in this machine contains the on/off switch, two indicator lights, and a control valve. One of the lights indicates that the machine is on, and the other indicates if the heating chamber is up to the proper temperature. The valve is used to start the flow of water through the coffee in the porta-filter or to start the flow of steam from the steam wand. It also engages one of two micro-switches that control the pump and heating element.

When you turn on the pump, cold water heats up in the heating chamber and flows through the coffee at high pressure. The result is a perfect shot of espresso!

How ICEMAKERS Work

Only a century ago, ice was hard to come by in most parts of the world. In hotter climates you had to buy your ice from a delivery service, which imported hefty blocks from a colder climate or from an industrial refrigeration plant. In an equatorial country, you might live your whole life and never even see a piece of ice. This all changed in the early twentieth century.

HSW Web Links

www.howstuffworks.com

How Refrigerators Work
How Snow Makers Work
How Air Conditioners
 Work
How Home Thermostats
 Work

Compact, affordable refrigerators brought economical food preservation and ice production into the home and corner store. In the 1960s, new automatic icemaker machines made life even easier. These days, most Americans take ice completely for granted, even during the hottest days of summers.

On the Rocks

The home icemaker's predecessor was the plastic ice tray. It's fairly obvious how this device works: You pour water into a mold, leave it in the freezer until it turns to a solid, and then extract the ice cubes. An icemaker does exactly the same thing, but the process of pouring water and extracting cubes is fully automated—it's an ice-cube assembly line.

Most icemakers use an electric motor, an electrically operated water valve, and an electrical heating unit. To provide power to all these elements, the icemaker is hooked up to the electrical circuit powering your refrigerator. You also have to hook the icemaker up to the plumbing line in your house to provide fresh water for the ice cubes. The power line and the water-intake tube both run through a hole in the back of the freezer.

When everything is hooked up, the icemaker begins its cycle. A simple electrical circuit and a series of switches keep everything going.

At the beginning of the cycle, a timed switch in the circuit briefly sends current to a solenoid water valve. The valve is only open for a few seconds. It lets in just enough water to fill the ice mold, which is a plastic well with several curved cavities.

After the mold is filled, the machine waits for the water in the mold to freeze. The cold air in the freezer does the actual work of freezing the water, not the icemaker itself. The icemaker has a built-in thermostat, which monitors the temperature level of the water in the molds. When the temperature dips to a particular level—say, 9°F (−13°C)—the thermostat closes a switch in the electrical circuit.

Closing this switch lets electrical current flow through a heating coil underneath the icemaker. As the coil heats up, it warms the bottom of the ice mold, loosening the ice cubes from the mold surface.

The electrical circuit then activates the icemaker's motor. The motor rotates a long plastic shaft with a series of ejector blades extending out from it. As the blades revolve, they scoop the ice cubes up out of the mold, pushing them to the front of the icemaker. Since the cubes are connected to one another, they move as a single unit.

Plastic notches in the housing match up with the ejector blades. The blades pass through these notches, and the cubes are pushed out to a collection bin underneath the icemaker.

The revolving shaft has a notched plastic cam at its base. Just before the cubes are pushed out of the icemaker, the cam catches hold of a shut-off arm, lifting it up. After the cubes are ejected, the shut-off arm falls down again. When the shut-off arm reaches its lowest resting position, it throws a switch in the circuit to activate the water valve to begin another cycle. If the arm can't reach its lowest position, because stacked-up ice cubes are in the way, the cycle is interrupted. This keeps the icemaker from filling your entire freezer with ice; it will only make more cubes when there is room in the collection bin.

This system is effective for making ice at home, but it doesn't produce enough ice to be used for commercial purposes, such as in restaurants and self-service hotel ice machines.

A Slippery Slope

There are many ways to configure a large, free-standing icemaker—all you need is a refrigeration system, a water supply, and some way of collecting the ice that forms.

One of the simplest professional systems uses a large metal ice-cube tray, positioned vertically.

The icemaker has a normal compressor-based refrigeration system. Heat-exchange tubes line the back of the ice tray and drop its temperature to 0°F or so.

The icemaker has a water pump, which draws water from a collection sump and pours it over the chilled ice tray. As the water flows over the tray, it gradually freezes, building up ice cubes in the well of the tray. When you freeze water layer by layer this way, it forms clear ice. When you freeze it all at once, as in the home icemaker, you get cloudy ice.

After a set amount of time, the icemaker triggers a solenoid valve connected to the heat-exchanging coils. Switching this valve changes the path of the refrigerant. The pipes and the ice tray heat up rapidly, which loosens the ice cubes.

Typically, the individual cube cavities are slanted so the loosened ice cubes will slide out on their own into a collection bin below. Some systems have a cylinder piston that gives the tray a little shove, knocking the cubes loose.

This sort of system is popular in restaurants and hotels because it makes ice cubes with a standard shape and size. Other businesses, such as grocery stores and scientific research firms, need smaller ice flakes for packing perishable items. Flake icemakers

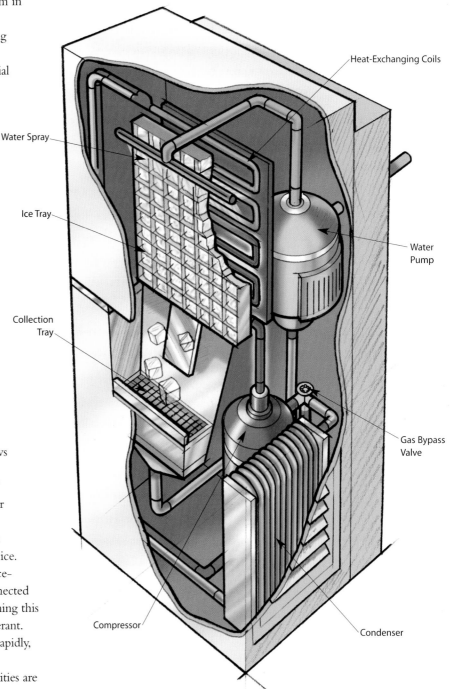

form a large block of ice and then grind it up into small pieces with a crusher.

There are dozens of different icemaker designs, but the basic idea in all of them is the same. A refrigeration system builds up a layer of ice, and a harvesting system ejects the ice into a collection bin. At the most basic level, this is all there is to any icemaker.

How **GRILLS** Work

"Fire up the grill!" That's something you hear a lot during summer holidays and gatherings. In fact, about 75% of house-holds across the United States have a grill. There are several different types of grills, but gas and charcoal grills are by far the most common.

HSW Web Links

www.howstuffworks.com

How Refrigerators Work
How Microwave Ovens
 Work
How Toasters Work
How Oil Refining Works
How Air Conditioning
 Works
How Food Works

A grill consists of a cooking surface, typically made of parallel metal bars or a porcelain-covered metal grid, that's placed over a fuel source capable of generating intense heat, usually up to temperatures of 500°F (260°C) or more.

Fire Up the Charcoal

The fuel source for charcoal grills has been around for at least 5,000 years. No one is certain who discovered charcoal or even what civilization first used it. Evidence of charcoal has been found all over the world. It was even used in the embalming process for Egyptian mummies! (see "How Mummies Work.") You may not realize it, but charcoal is not a rock or even some type of coal. It's actually wood.

Charcoal is created by heating wood to high temperatures in a sealed container that's free of oxygen. That is, you take wood, put it in a sealed box of steel or clay, and heat it to about 1000°F (538°C).

Why would you go through such a tedious process instead of just burning the wood as it is? Freshly cut wood contains a lot of water—sometimes more than half its weight is water. Seasoned wood (wood that has been allowed to sit for a year or two) or kiln-dried wood contains a lot less water, but it still contains some. Watery wood does not make for very efficient cooking. Also, when the tree was alive it contained sap and a wide variety of volatile hydrocarbons in its cells. *Volatile* means that these compounds evaporate when heated.

When you put a fresh piece of wood or paper on a hot fire, the smoke you see is those volatile hydrocarbons evaporating from the wood. They start vaporizing at a temperature of about 300°F (149°C). If the temperature gets high enough, these compounds burst into flame.

You don't see any smoke from a charcoal fire because the charcoal-making process drives off all of the volatile organic compounds and leaves behind pure carbon and the nonburnable minerals in the tree's cells (ash). When you light the charcoal, what is burning is the pure carbon. It combines with oxygen to produce carbon dioxide, and what is left at the end of the fire is the ash—the minerals. This produces a very intense heat with very little smoke, making charcoal very useful as a cooking fuel because it does not overwhelm the flavor of the food with the elements found in normal woodsmoke.

Grilling enthusiasts passionately argue the merits of charcoal grilling versus gas grilling, citing especially the difference in flavor. Charcoal does provide a distinctive flavor that's not easily reproduced. It's a tough decision for many people: the convenience of a gas grill against the flavor of charcoal.

Cooking with Gas

Even the simplest gas grill is more complex than a typical charcoal grill. Common components of a gas grill include the following:

- Gas source
- Hoses
- Valves
- Regulators
- Burners
- Starter
- Cooking surface
- Grill body
- Grill hood

Grill starter and
temperature control.

The grill body houses all of the other components except the hood. The hood covers the cooking surface and serves to trap the heated air inside, which increases the temperature inside the grill.

The gas source is connected to the valve via the main hose. The valves let you determine how much gas reaches the burner. Most grills have two main burners, with a valve for each one. Each burner has a series of tiny holes along its length that the gas exits through.

Three things are required for a gas grill to ignite properly:

- Gas
- Oxygen
- Spark

The gas is supplied from the propane tank or the natural-gas pipeline, and oxygen comes from the air. But where does the spark come from?

Making Sparks Fly

The grill starter, sometimes called the igniter, usually supplies the spark. The grill starter is a push-button or rotating knob that creates a spark of electricity to ignite the gas. The starter uses piezoelectricity to generate a nice spark that lights the grill.

Certain crystalline materials (like quartz and some ceramics) have piezoelectric behavior. When you apply pressure to them, you get a charge separation within the crystal and a voltage across the crystal that is sometimes extremely high. For example, in a grill starter, the popping noise you hear is a little spring-loaded hammer hitting a crystal and generating thousands of volts of energy across the faces of the crystal.

A voltage this high is identical to the voltage that drives a spark plug in a gasoline engine. The crystal's voltage generates a spark large enough to light the gas in the grill.

The burner is where all the actual burning occurs. It mixes the gas with oxygen and spreads it out over a large surface area to burn. Each burner has a pair of electrodes connected to the starter. When the starter's hammer is tripped, the resulting surge of electricity causes a spark to arc across these electrodes and ignite the gas/oxygen mixture.

It's a Gas!

The majority of gas grills use metal tanks full of liquid propane (LP) gas. LP gas has the advantage of coming in a portable tank, whereas natural gas uses a pipe connected to the gas main at your home.

Why can you get LP gas in a tank but not natural gas? Propane has the nice property that when you compress it, it condenses into a liquid at a reasonable pressure (around 100 psi) and will stay that way until it is uncompressed. This means that propane is much easier to store in a tank than natural gas, which doesn't liquify until it reaches 2,800 psi. Because natural gas doesn't compress well into a liquid form, it is typically delivered as a gas via a dedicated pipeline to your home.

Another benefit of LP is that it contains much more energy than natural gas. A grill's cooking capability is rated in British thermal units (Btu). A Btu is the amount of heat required to raise the temperature of 1 pound (0.45 kg) of water 1°F (0.56°C). A cubic foot of natural gas contains something like 1,000 Btu of energy. One cubic foot (1 ft^3) of propane contains perhaps 2,500 Btu. Grills typically range from 20,000 Btu to about 50,000 Btu. A higher Btu rating normally indicates a larger grill with a greater cooking surface.

Even though LP-burning grills are the norm, natural gas grills are avaible. You normally would buy a natural-gas grill if you plan to connect it directly to a gas pipe in a permanent location (like your house) and if natural gas is available in your area. You can see the difference between a natural-gas grill and an LP grill most easily by looking at the pipes connecting to the burners. The pipe on a natural-gas grill is about twice as big as the one on a propane grill. You can see the same difference in the jets.

Grill burners.

Piezoelectric igniters.

What Is Propane?

Propane is created from petroleum and contains aliphatic hydrocarbons—hydrocarbons composed of nothing but hydrogen and carbon atoms. When petroleum is processed in a refinery, hydrocarbon chains of different lengths are produced. These different chain lengths can then be separated from each other and blended to form different fuels. For example, you may get methane, propane, and butane. All three of these fuels are hydrocarbons. Methane has just a single carbon atom and four hydrogen atoms (CH_4); butane has four carbon atoms and ten hydrogen atoms (C_4H_{10}); propane has three carbon atoms chained together with eight hydrogen atoms (C_3H_8).

How **BUG ZAPPERS** Work

While you have fun outdoors, many insects get to enjoy a good meal. Either they're eating your food or they're eating you. To clear your yard of these insects, you can try a variety of devices, ranging from simple citronella candles to elaborate traps to pesticides to electronic bug zappers.

HSW Web Links

www.howstuffworks.com

How Mosquitoes Work
How Venus Flytraps Work

Transformer

Fluorescent
Light Bulb

Wire Mesh

A bug zapper, officially known as an electronic insect-control system, lures bugs into the casing and kills them with electricity.

Inside a Bug Zapper

The first bug zapper was patented in 1934 by William F. Folmer and Harrison L. Chapin (U.S. patent 1,962,439). Although there have been many improvements, mostly in the areas of safety and lures, the basic design of the bug zapper hasn't changed much since then.

Bug zappers are incredibly simple. The basic parts of the bug zapper are:

- **Housing**—The exterior casing that holds the parts is called the housing. The housing is usually made of plastic or electrically-grounded metal and may be shaped liked a lantern, a cylinder, or a big rectangular cube. The housing usually has a grid design so that insects can get through, but children and larger animals aren't able to touch the electrified wire inside.
- **Light bulb(s)**—Light bulbs provide the fluorescent light that attracts insects. This can be either mercury, neon, or ultraviolet (black light).
- **Wire grids or screens**—Wire mesh grids or screens (usually two) surround the light bulb and are electrified to kill ("zap") the insects.
- **Transformer**—The transformer is the device that electrifies the wire mesh, changing the 120-volt (V) electrical-line to 2,000 V or more.

The voltage supplied by the transformer (at least 2,000 V) is applied across the two wire-mesh grids. These grids are separated by a tiny gap that is about the size of a typical insect (a couple of millimeters across). The light inside the wire-mesh network lures the insects to the device—many insects see ultraviolet light better than visible light, and are more attracted to it than they are to visible light because the patterns that attract insects to flowers are revealed in ultraviolet light.

As the bug flies toward the light, it penetrates the space between the wire-mesh grids and completes the electric circuit. High-voltage electric current flows through the insect and vaporizes it. You often hear a loud "ZZZZ" sound when this happens. Bug zappers can lure and kill more than 10,000 insects in a single evening.

Other Bug-zapping Strategies

There are lots of different ways to control insects, particularly mosquitoes. In fact, traditional electronic bug zappers don't do much to mosquitoes, because mosquitoes aren't particularly fond of ultraviolet light. Some electronic bug zappers compensate for this by emitting octenol, a nontoxic, pesticide-free pheromone mosquito attractant.

Mosquitoes are attracted to the carbon dioxide emitted by humans in our breath and sweat, so several types of mosquito zappers try to take advantage of this. One such product emits a steady stream of carbon dioxide, octenol attractant, and moisture. Mosquitoes are attracted to this mixture, get sucked into a net, dehydrate, and die. The device is powered by a propane tank, so no electricity is required. One manufacturer claims that entire mosquito populations collapse in six to eight weeks as egg-laying females are destroyed.

Another device uses a chemical that the manufacturer claims blocks the mosquitoes' olfactory receptors. The makers of this product say that blocking the insect's ability to smell carbon dioxide reduces the number of mosquito landings and bites.

How **LOCK PICKING** Works

Your key ring is a clear demonstration of just how ubiquitous lock technology is. You probably have a couple keys for your house, one or two more for the car, and a few for the office or a friend's house. Most people use locks dozens of times every week.

The main reason we install locks everywhere is that they provide us with a sense of security. But in the movies, there are spies, detectives, and burglars who can open a lock very easily, sometimes using only a couple of paper clips. This is sobering, to say the least. Is it really possible for someone to open a lock so easily?

The Pin Is Mightier than the Sword

Locks come in all shapes and sizes, with many innovative design variations. But you can get a good idea of the process of lock picking by examining one simple, representative type of lock, called the *pin-and-tumbler design.*

A lock like this is, essentially, a puzzle. The job of the key is to "know" the answer to the puzzle instantly, so that the person who owns the lock doesn't have to spend any time working the puzzle. A person who is trying to pick the lock has to work the puzzle out the hard way.

The pin-and-tumbler design is one of the most common puzzles that you find today. This sort of lock is found on almost every door lock in America. Homes and offices with doorknobs and deadbolt locks almost always use the pin-and-tumbler approach. The design has three basic parts:

- An outer, hollow cylinder
- An inner cylinder (the tumbler)
- A set of pins

The pins come in pairs of varying lengths. Each pin pair sits in a shaft that runs through the tumbler and into the outer cylinder around the plug. Springs at the top of the shafts keep the pin pairs in position in the plug. When no key is inserted, the bottom pin in each pair is completely inside the plug, while the upper pin is halfway in the plug and halfway in the housing. The position of these upper pins keeps the plug from turning—the pins bind the plug to the housing.

HSW Web Links

www.howstuffworks.com

How Power Door Locks
   Work
Inside a Combination
   Lock
How Anti-Shoplifting
   Devices Work
How Burglar Alarms Work

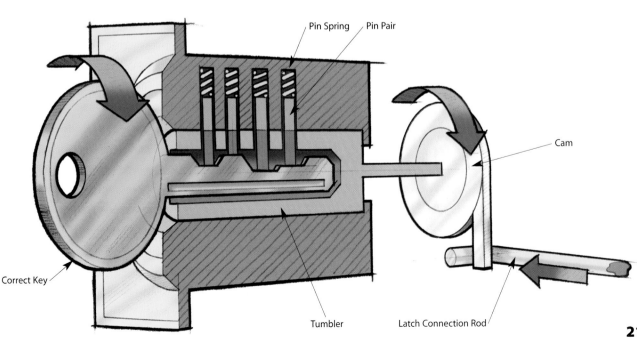

Pin Spring   Pin Pair

Cam

Correct Key

Tumbler   Latch Connection Rod

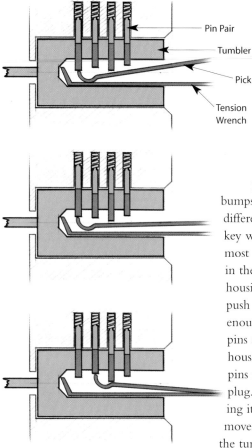

Pin Pair

Tumbler

Pick

Tension
Wrench

The puzzle is this: How do you push the pins up, each one by just the right amount, so that the tumbler can turn inside the outer cylinder? When you insert a key to solve the puzzle, its notches and bumps push the pin pairs up to different levels. The incorrect key will push the pins so that most of the pins are still partly in the plug and partly in the housing. The correct key will push each pin pair up just enough so that all of the upper pins are completely in the housing while all of the lower pins rest completely in the plug. Without any pins binding it to the housing, the plug moves freely, and you can turn the tumbler. The rotation of the tumbler normally pushes the bolt of the deadbolt lock in and out, or it can release the knob of a doorknob lock.

Feeling Lock-Key?

To solve the puzzle of a pin-and-tumbler lock without a key, you have to manually move each pin pair into the correct position, one by one. This wouldn't be too hard, except the keyhole is incredibly narrow—you can't see what's going on inside. An experienced lock-picker can move the pins blindly, by touch alone, with only two simple tools: a pick and a tension wrench.

A pick is a long, thin piece of metal that curves up at the end (like a dentist's pick). A tension wrench is a sturdy flat piece of metal. The simplest sort of tension wrench is a thin flathead screwdriver.

The first step in picking a lock is to insert the tension wrench into the keyhole and turn it in the same direction that you would turn the key. This turns the plug so

that it is slightly offset from the housing around it, creating a slight ledge in the pin shafts and tension on the pins.

While applying pressure on the plug, you insert a pick into the keyhole and begin lifting the pins. The object is to lift each pin pair up to the level at which the top pin moves completely into the housing, as if it were pushed by the correct key.

When you do this while applying pressure with the tension wrench, you feel or hear a slight click when the pin falls into position. This is the sound of the upper pin falling into place on the ledge in the shaft. The ledge keeps the upper pin wedged in the housing, so it won't fall back down into the plug. In this way, you move each pin pair into the correct position until all of the upper pins are pushed completely into the housing and all of the lower pins rest inside the plug. At this point, the plug rotates freely and you can open the lock.

Another technique for picking a lock is called *raking*. Raking is much less precise than actually picking. To rake a lock, you insert a pick with a wider tip, called a rake, all the way to the back of the plug. Then you pull the rake out quickly so that it bounces all of the pins up on its way out. As the rake exits, you turn the plug with the tension wrench. As they're moving up and down, some of the upper pins will happen to fall on the ledge created by the turning plug. Often, locksmiths will start by raking the pins, and then pick any remaining pins individually.

Conceptually, the lock-picking process is quite simple, but it is a very difficult skill to master. Locksmiths have to learn exactly the right pressure to apply and what sounds to listen for. They also must hone their sense of touch to the point where they can feel the slight forces of the moving pins and plug. Additionally, they must learn to visualize all the pieces inside the lock. Successful lock-picking depends on complete familiarity with the lock's design.

Locks and the Law

Lock picking is an essential skill for locksmiths because it lets them get past a lock without destroying it. Lock-picking skills are not particularly common among burglars, mainly because there are much easier ways of breaking into a house (throwing a brick through a back window, for example). For the most part, only intruders who need to cover their tracks, such as spies and detectives, will bother to pick a lock.

How NAIL GUNS Work

If you're hanging pictures or putting together a bookcase, a hammer is a perfect tool: simple, cheap, and entirely effective. But if you're building a two-story house, installing hardwood floors, or running your own furniture repair shop, you may want to buy a nail gun. These high-powered machines fully embed nails in a piece of wood in only a fraction of a second. Obviously, such a machine can save you hours of hammering.

The most popular sort of nail gun is the pneumatic nailer. In these machines, the hammering force comes from compressed air generated by a separate air compressor.

Types of Nail Guns

The design of a pneumatic nailer is incredibly simple and has been around for years. Compressed air drives a large piston that's connected to a rod or a blade. If a nail is in the chamber, the blade drives the nail out of the gun. When you pull the trigger, this opens a valve that lets in a pulse of compressed air. When you release the trigger, this cuts off the flow of compressed air into the piston, and also opens a valve that let's the high-pressure air out of the piston chamber so that the piston can move back and reset.

One of the newest nail gun machines to hit the market is the combustion nailer. These portable guns work in the same basic way as the piston engine in a car.

At the most basic level, combustion guns are a lot like pneumatic nailers. The main difference is the source of the pressure that drives the piston. Just like a car engine, combustion guns have a reservoir filled with a flammable gas. A sliding plate releases a little of this gas into the combustion chamber just above the piston head. A small fan in the combustion chamber vaporizes the gas, mixing it up with the air particles. A spark plug ignites the gas, and the explosion drives the piston downward.

As the piston slides back up, a valve opens to release the exhaust gases. The fan also helps to move the exhaust out of the chamber. The gun is now ready to fire the next nail.

This is an amazing mutation of engine technology. The combustion nailer does pretty much the same thing as an engine, except it uses a sliding plate, instead of conventional valves, to draw gas and expel exhaust.

Playing It Safe

Most modern nail guns have a second safety trigger to keep people from accidentally shooting nails through the air. To drive a nail, you have to pull down the main trigger and press the secondary trigger against the nailing surface.

HSW Web Links

www.howstuffworks.com

How Engines Work
How Electromagnets Work
How Force, Power, Torque and Energy Work
How Paintball Works
How Flintlock Guns Work

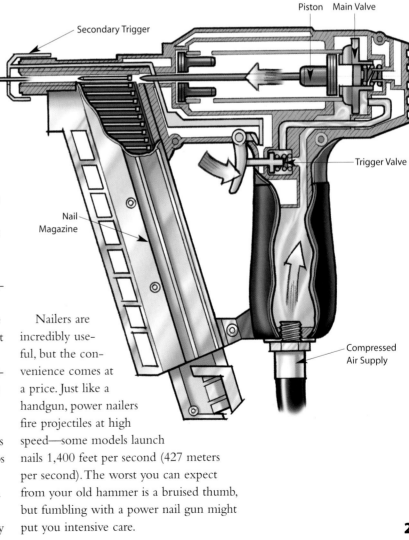

Nailers are incredibly useful, but the convenience comes at a price. Just like a handgun, power nailers fire projectiles at high speed—some models launch nails 1,400 feet per second (427 meters per second). The worst you can expect from your old hammer is a bruised thumb, but fumbling with a power nail gun might put you intensive care.

215

How **HUMIDIFIERS** Work

One thing that makes winter uncomfortable for humans, even when we're inside a nice warm building, is low humidity. People need a certain level of humidity to be comfortable. A humidifier can help make things more comfortable, and can even save a little wear and tear on your house by adding moisture to the air. It's surprising how big a difference a little water can make!

HSW Web Links

www.howstuffworks.com

How Sweat Works
What Is the Heat Index?
How Air Conditioners Work
How Snow Makers Work
How Ice Rinks Work

Weather and Humidity

Here's what happens in winter to make it feel so dry in our houses. Let's say that the outdoor temperature is 32°F (0°C). Suppose that the relative humidity outside is 100%—the maximum amount of water that a cubic meter of air can hold at this temperature is 5 grams. Now you bring this cubic meter of air inside and heat it to 77°F (25°C). The relative humidity of the air is now only 23%:

5 grams of water in the air ÷ 22 grams possible = 23% relative humidity

The situation gets worse as the temperature outside falls lower. This is why the air inside any heated building in the winter feels so dry. Any time the temperature outside is below freezing, relative humidity inside will be below 20% unless you do something to increase the humidity. During the dry months, a humidifier can help maintain a comfortable level of humidity.

In the winter, indoor humidity can be extremely low and the lack of humidity can dry out your skin and mucous membranes. Low humidity also makes the air feel colder than it actually is. Dry air can even dry out the wood in the walls and floors. As the drying wood shrinks, it can cause creaks in floors and cracks in drywall and plaster.

Relative Humidity

The relative humidity of the air affects how comfortable we feel. But what is humidity, and what is "relative humidity" relative to?

Humidity is the amount of moisture in the air. If you're standing in the bathroom after a hot shower and can see the steam hanging in the air, or if you're outside after a heavy rain, then you are in an area of high humidity. If you're standing in the middle of a desert that hasn't seen rainfall for two months, or if you're breathing air out of a SCUBA tank, then you are experiencing low humidity.

Air contains a certain amount of water vapor. The maximum amount of water vapor any mass of air can contain depends on the temperature of that air: The warmer the air is, the more water it can hold. A low relative humidity means that the air is dry and could hold a lot more moisture at that temperature.

For example, at 77°F (25°C), air can hold 22 grams of water per cubic meter. If the temperature is 77°F and a cubic meter of air contains 22 grams of water, then the relative humidity is 100%. If it contains 11 grams of water, the relative humidity is 50%. If it contains 0 grams of water, relative humidity is 0%.

Relative humidity plays a big role in determining our comfort level. If the relative

A humidifier.

humidity is 100%, water will not evaporate—the air is already saturated with moisture. Our bodies rely on the evaporation of moisture from our skin for cooling. So, we feel much hotter than the actual temperature indicates because our sweat doesn't evaporate at all. The lower the relative humidity, the easier it is for moisture to evaporate from our skin and the cooler we feel. In low relative humidity, air feels cooler than it would at the same temperature and higher relative humidity because our sweat evaporates easily. The drawback to low relative humidity is that we can feel extremely dry.

Low humidity has at least three effects on human beings:

- It dries out your skin and mucous membranes. If your home has low humidity, you will notice things like chapped lips, dry and itchy skin, and a dry, sore throat when you wake up in the morning. (Low humidity also dries out plants and furniture.)
- It increases static electricity, and most people dislike getting sparked every time they touch something metallic.
- It makes it seem colder than it actually is.

Since humidifying air costs a lot less than heating it, a humidifier can save you a lot of money!

For best indoor comfort and health, a relative humidity of about 45% is ideal. At temperatures typically found indoors, this humidity level makes the air feel approximately what the temperature indicates, and your skin and lungs do not dry out and become irritated.

Most buildings cannot maintain this level of humidity without help, so they require some kind of humidifier to balance out the relative humidity.

Types of Humidifiers

You can raise the humidity in your home many different ways. For example, you can put a pan of water on the stove or on the radiator, or you can hang wet towels near a heater duct. But most people use a mechanical humidifier to do the job. The four most popular technologies are:

- **Steam**—Often referred to as a *vaporizer,* a steam humidifier boils water and releases the warm steam into the room. This is the simplest, and therefore the least expensive, technology for adding moisture to the air. You can find inexpensive steam models for less than $10 at discount stores. Another advantage of this technology is that you can use a medicated inhalant with the unit to help reduce coughs.
- **Impeller**—In this humidifier, a rotating disc flings water at a comb-like diffuser. The diffuser breaks the water into fine droplets that float into the air. You normally see these droplets as a cool fog exiting the humidifier.
- **Ultrasonic**—An ultrasonic humidifier uses a metal diaphragm vibrating at an ultrasonic frequency, much like the element in a high-frequency speaker, to create water droplets. An ultrasonic humidifier is usually silent, and like an impeller it produces a cool fog.
- **Wick/evaporative system**—The wick system uses a paper, cloth, or foam wick or sheet to draw water out of the

reservoir. A fan blowing over the wick lets the air absorb moisture. The higher the relative humidity, the harder it is to evaporate water from the filter, which is why this type of humidifier is self-regulating—as humidity increases, the humidifier's water-vapor output naturally decreases.

Humidifiers can be installed as small portable room units, or they can be integrated into your furnace for full-house humidity control. If you're interested in tracking your home's humidity, an inexpensive hygrometer will show you the relative humidity in your house. You may be surprised to learn how low it is!

> ### Pros and Cons
>
> There are some things to keep in mind as you are weighing the advantages and disadvantages of the different technologies:
>
> - Steam vaporizers can be dangerous around children because they can cause burns. They also have the highest energy costs. However, there are no bacterial or mineral concerns with this technology.
> - Impeller and ultrasonic designs have low energy costs but raise two concerns:
>
> - First, if the water gets stagnant, these designs will spray the stagnant water, and any bacteria it contains, into your home. This is why it's important to clean the tank regularly and refill it with clean water when you haven't been running it. Many high-end ultrasonic units have antibacterial features built in. For example, some units use ultraviolet light to kill bacteria.
> - The second concern is minerals in the water.
>
> Impeller and ultrasonic designs send these minerals into the air. If the water in your area contains a lot of minerals, you will notice them as dust. The EPA does not issue health warnings about minerals in the air, but does recommend using low-mineral water (such as distilled water) in your humidifier. Many ultrasonic models feature a demineralization cartridge that filters minerals out of the water to prevent the dust.

Fan mechanism.

How **PLASMA DISPLAYS** Work

In the 1990s, a new sort of television started popping up on store shelves—the flat panel display. These television sets produce bright, vivid images, but are generally less than 6 inches thick. Plasma displays, the coolest of the new flat panel technologies, form a picture using a plasma—a gas made up of free-flowing ions and electrons.

HSW Web Links

www.howstuffworks.com

How Television Works
How Digital Television
  Works
How Projection Television
  Works
How THX Works
How Movie Sound Works

Tuning In

Most plasma displays aren't technically televisions, because they have no television tuner. The television tuner is the device that takes a television signal (the one coming from a cable wire, for example) and interprets it to create a video image.

Like LCD displays, Plasma displays are just monitors that display a standard video signal. To watch television on a plasma display, you have to hook it up to a separate unit that has its own television tuner, such as a standard VCR.

The essence of a plasma display is the same technology that runs a fluorescent light. A plasma display contains millions of tiny colored fluorescent lights that collectively form images. Each pixel in the display includes three separate fluorescent lights—a red light, a green light, and a blue light. Just as in a fluorescent lamp, an electrical current energizes a gas (in this case, a mixture of xenon and neon) to create plasma. The atoms in the energized plasma release invisible ultraviolet light photons, and these ultraviolet photons stimulate phosphorescent material to release visible light. (See "How Fluorescent Lights Work" to find out more.)

Phosphors, Pixels, and Photons

Each pixel in a plasma display is made up of tiny cells positioned between two plates of glass. Long electrodes are also sandwiched between the glass plates, on both sides of the cells. Address electrodes sit behind the cells, along the rear glass plate. Transparent display electrodes run along the front glass plate.

Both sets of electrodes extend across the entire screen. The display electrodes are arranged in horizontal rows along the screen and the address electrodes are arranged in vertical columns. They form a basic grid.

To ionize the gas in a particular cell, the plasma display's computer charges the electrodes that intersect at that cell. It does this thousands of times in a small fraction of a second, charging each cell in turn.

When the intersecting electrodes are charged (with a voltage difference between them), an electric current flows through the gas in the cell, producing ultraviolet light photons.

The phosphors in a plasma display give off colored light when they are excited by the photons. Every pixel is made up of three separate subpixel cells, each with different colored phosphors. One subpixel has a red phosphor, one subpixel has a green phosphor, and one subpixel has a blue phosphor. These colors blend together to create the overall color of the pixel.

By varying the pulses of current flowing through the different cells, the control system can increase or decrease the intensity of each subpixel color to create hundreds of different combinations of red, green, and blue. In this way, the control system can produce colors across the entire spectrum.

Pros and Cons

The main advantage of plasma display technology is that you can make a very wide screen using extremely thin materials. And because each pixel is lit individually, the image is bright and looks good from almost every angle.

The biggest drawback of this technology is the cost. With prices starting at $4,000 and going all the way up past $20,000, these sets aren't exactly flying off the shelves. But as prices fall and technology advances, plasma display TVs may start to edge out the old CRT sets. In the near future, setting up a new TV might be as easy as hanging a picture!

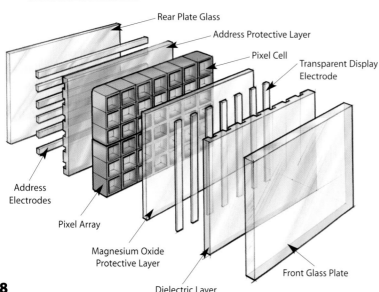

Rear Plate Glass
Address Protective Layer
Pixel Cell
Transparent Display Electrode
Address Electrodes
Pixel Array
Magnesium Oxide Protective Layer
Dielectric Layer
Front Glass Plate

chapter nine

AROUND THE OFFICE

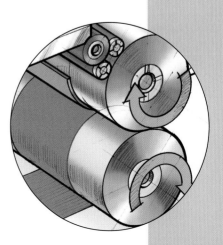

How **SCANNERS** Work

Scanners have become an important part of the office over the last few years. Whether you're scanning documents or photos, a scanner gives you an easy way to move an image into your computer so that you can manipulate it or email it to someone else.

⚙ HSW Web Links
www.howstuffworks.com

How USB (Universal Serial Bus) Ports Work
How Parallel Ports Work
How Photocopiers Work
How Digital Cameras Work

If you think about it, there's no real difference between a scanner and a digital camera. Both devices take a picture and store it as an electronic file. They simply accomplish that in different ways. A digital camera takes a picture of the 3-D world in an instant, and normally does so at fairly low resolution. A scanner creates an image of a 2-D object, like a sheet of paper or a photograph, and can do so at very high resolution. Depending on the resolution, a scanner might take up to a minute to form the image. The resolution of high-end scanners can be startling. A single square inch of a scanner photograph can have as much detail as the entire image from most digital cameras.

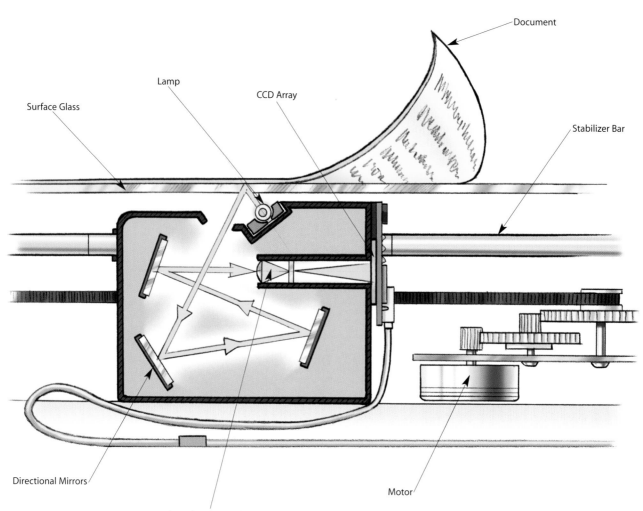

Document

Lamp

CCD Array

Surface Glass

Stabilizer Bar

Directional Mirrors

Lens Arrangement

Motor

220

Anatomy of a Scanner

The core component of any scanner is a linear CCD (charge coupled device) array that captures the image. A CCD is a collection of tiny light-sensitive diodes, which convert photons (light) into electrons (electrical charge). These diodes are called *photosites*. Each photosite is sensitive to light—the brighter the light that hits a single photosite, the greater the electrical charge that accumulates at that site. The number of photosites on the CCD controls the horizontal resolution of the scanner.

The image of the document that you scan reaches the CCD array through a series of mirrors, filters, and lenses. The exact configuration of these components depends on the model of the scanner, but the basics are pretty much the same.

Scanning

After you place a document on the glass plate and close the cover, a lamp illuminates the document. The lamp in newer scanners is either a cold cathode fluorescent lamp (CCFL) or a xenon lamp, while older scanners may have a standard fluorescent lamp (see "How Fluorescent Lamps Work" for more information).

The scan head (composed of mirrors, the lens, the filter, and the CCD array) moves slowly across the document—a belt that is attached to a stepper motor controls the motion. The scan head is attached to a stabilizer bar to ensure that there's no wobble or deviation in the pass (a single complete scan of the document). An angled mirror reflects the image of the document to another mirror. In some scanners, there are only two mirrors, while others use a three-mirror approach. Each mirror is slightly curved to focus the image it reflects onto a smaller surface. The last mirror reflects the image onto a lens. The lens focuses the image through a filter onto the CCD array.

The filter-and-lens arrangements vary based on the scanner. Some scanners use a three-pass scanning method. Each pass uses a different color filter (red, green, or blue) between the lens and CCD array. After the three passes are completed, the scanner software assembles the three filtered images into a single full-color image.

Most scanners today use the single-pass method. The lens splits the image into three smaller versions of the original. Each smaller version passes through a color filter (either red, green, or blue) onto a different section of the CCD array. The scanner combines the data from the three parts of the CCD array into a single full-color image.

If the scanner has a 600 dot-per-inch resolution, then the CCD array will have 5,100 photosites. The stepper motor is able to move the scan head in $1/600$th of an inch increments. Many high-end scanners today have resolutions exceeding 2,000 dots per inch.

Never the TWAIN Shall Meet

Scanning the document is only one part of the process. For the scanned image to be useful, it must be transferred to your computer. There are three common connections used by scanners:

- Parallel port
- Small computer system interface (SCSI)
- Universal Serial Bus (USB)

On your computer, you need to have software, called a *driver,* that knows how to communicate with the scanner. Most scanners speak a common language, TWAIN. The TWAIN driver acts as an interpreter between any application that supports the TWAIN standard and the scanner. This means that the application doesn't need to know the specific details of the scanner in order to access the scanner directly. For example, you can choose to acquire an image from the scanner from Adobe Photoshop because Photoshop supports the TWAIN standard.

The great thing about scanner technology today is that you can get exactly what you need. You can find a decent scanner with good software for less than $100 or a fantastic scanner with incredible software for less than $1,000. It all depends on your needs and budget.

Did You Know?

TWAIN is not an acronym. It actually comes from the phrase "Never the twain shall meet," because the driver is the go-between for the software and the scanner. Because computer people feel a need to make an acronym out of every term, TWAIN is known as *Technology Without An Interesting Name!*

How **PHOTOCOPIERS** Work

Walk into almost any business office, and you'll find a photocopier with a line of people waiting to use it. For most businesses, small or large, the copier is standard equipment, much like having desks to work at and chairs to sit in.

HSW Web Links

www.howstuffworks.com

How Light Works
How Atoms Work
How Photographic Film
  Works
How Van de Graaff
  Generators Work
How Lasers Work

What if you had to resort to making carbon copies of important documents, as many people did before copiers came along? Or worse, imagine how tedious it would be if you had to recopy everything by hand! Most of us don't think about what's going on inside a copier while we wait for copies to shoot neatly out into the paper tray, but it's pretty amazing to think that, in mere seconds, you can produce an exact replica of what's on a sheet of paper!

The Basic Principles of Photocopying

The human part of making a copy begins with a few basic steps:

1) Open the copier lid.
2) Place the document to be photocopied face-down on the glass.
3) Select the options you want (number of pages, enlargements, lighter/darker).
4) Press the start button.

What happens inside the copier at this point is amazing. At its heart, a copier works because of one basic physical principle: Opposite charges attract.

As a kid, you probably played with static electricity and balloons. On a dry winter day, you can rub a balloon on your sweater and build up enough static electricity in the balloon to create a noticeable force. For example, a balloon charged with static electricity will attract small bits of paper or particles of sugar very easily. A photocopier is a machine that puts the forces of static electricity to good use.

Inside a copier there is a special drum. The drum acts a lot like a balloon—you can charge it with a form of static electricity. Inside the copier there is also a very fine black powder known as *toner*. The drum, once it's charged with static electricity, can attract the toner particles. Three things about the drum and the toner let a copier perform its magic:

- The drum can be selectively charged, so that only parts of it attract toner.
- The toner is heat sensitive, so the loose toner particles are attached (fused) to the paper with heat as soon as they come off the drum.
- The drum, or belt, is made out of photoconductive material.

The copier makes an image—in static electricity—on the surface of the drum. Where the original sheet of paper is black, static electricity is created on the drum. Where it's white, no static electricity is created. What you want is for the white areas of the original sheet of paper to NOT attract toner. The way this selectivity is accomplished in a copier is with light—the *photo* in *photocopier* is from the ancient Greek word for light!

Inside a Photocopier

If you take a photocopier apart, you might be overwhelmed by how many different parts there are. The actual photocopying process relies on only a few key pieces:

- **Photoreceptor drum or belt**—A drum is basically a metal roller that's covered by a layer of photoconductive material that's made out of a semiconductor such as selenium, germanium, or silicon.
- **Corona wires**—These wires carry a high voltage, which they transfer to the drum and paper in the form of static electricity.
- **Lamp and lenses**—Photocopiers use a plain old incandescent or fluorescent bulb to flash light onto the original document. A mirror attached to the lamp assembly directs reflected light through a lens onto the rotating drum below.
- **Toner**—Toner is a fine, negatively charged, plastic-based powder.
- **Fuser**—The fuser melts and presses the toner image onto the paper.

Putting It All Together

For the photocopier to work its magic, the surface of the photoconductive material must first be coated with a layer of positive charge by the corona wire. When you hit the start button, a strong lamp casts light onto the paper you're copying and the drum starts to rotate. As light reflects off the white areas of the paper, mirrors direct it onto the drum surface. Like dark clothing on a hot sunny day, the dark areas of the original absorb the light, and the corresponding areas on the drum's surface are not illuminated.

In the places that light strikes the rotating drum, the energy of the photons kicks electrons away from the photoconductive atoms.

Opposite charges attract, so the positively charged ions coating the photoconductive layer attract the freed electrons. The marriage of one ion and one electron produces a neutral particle. Charged particles remain only in places where light didn't hit the drum because it wasn't reflected from the original—the dark spaces taken up by text and pictures on the page!

The exposed areas of the drum rotate past rollers that are coated with beads of toner. Tiny particles of toner are pressed against the drum's surface. The plastic-based toner particles have a negative charge and are attracted to the areas of positive charge that remain on the drum's surface.

The corona wire passes over a sheet of paper so that the paper's surface becomes electrically charged.

The area of the drum that's been freshly coated with toner spins into contact with a positively charged sheet of paper. The electric field surrounding the paper exerts a stronger pull than the ions coating the drum's surface, and so the toner particles stick to the paper as the drum passes by.

To fix the toner image in place on the paper's surface, the entire sheet rolls through the fuser's heated rollers. The heat melts the plastic material in the toner and fuses the pigment to the page.

By the time you reach for your copy in the collection tray, the photocopier has already prepared for the next go-round by again cleaning off the drum's surface and applying a fresh coat of positively charged ions to it.

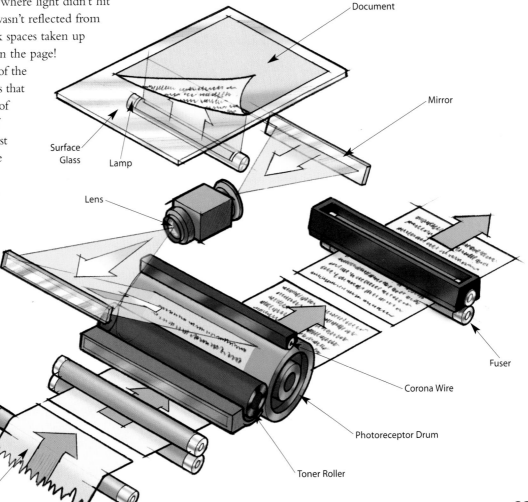

Document

Mirror

Surface Glass

Lamp

Lens

Fuser

Corona Wire

Photoreceptor Drum

Toner Roller

Paper Feed

How **FAX MACHINES** Work

Walk into nearly any office in the United States today, big or small, high-tech or low-tech, and you will find a fax machine. Once connected to a normal phone line, a fax machine allows you to transmit pieces of paper to someone else—almost instantly! Even with FedEx and e-mail, it is nearly impossible to do business without one of these machines today.

HSW Web Links

www.howstuffworks.com

How Modems Work
How Telephones Work
How Photocopiers Work
How Scanners Work
How Inkjet Printers Work
How Laser Printers Work

Although fax machines didn't become widespread until the late 1980s, they've been around in one form or another for over a century (the first patent on a fax process was in 1843). If you look back at some of the early designs, you can get a very good idea of how they work today.

Most of the early fax machine designs involved a rotating drum. To send a fax, you would attach the piece of paper to the drum with the printed side facing outward. The rest of the machine worked something like this:

- There was a small photo sensor with a lens and a light.
- The photo sensor was attached to an arm and faced the sheet of paper.
- The arm could move downward over the sheet of paper from one end to the other as the sheet rotated on the drum.

In other words, an early fax machine worked something like a lathe, with the photosensor on the lathe's arm.

The photo sensor was able to focus in and look at a very small spot on the piece of paper—perhaps an area of 0.01 inches square (0.25 millimeters square). That little patch of paper would be either black or white. The drum would rotate so that the photo sensor could examine one line of the sheet of paper and then move down a line. It did this either in steps or in a very long spiral.

To transmit the information through a phone line, early fax machines used a very simple technique: If the spot of paper that the photo cell was looking at was white, the fax machine would send one tone; if it was black, the machine would send a different tone. For example, the fax might have sent an 800-hertz tone for white and a 1,300-hertz tone for black. At the receiving end, there would be a similar rotating-drum mechanism and some sort of pen to mark on the paper. When the receiving fax machine heard a 1,300-hertz tone it would apply the pen to the paper, and when it heard an 800-hertz tone it would take the pen off the paper.

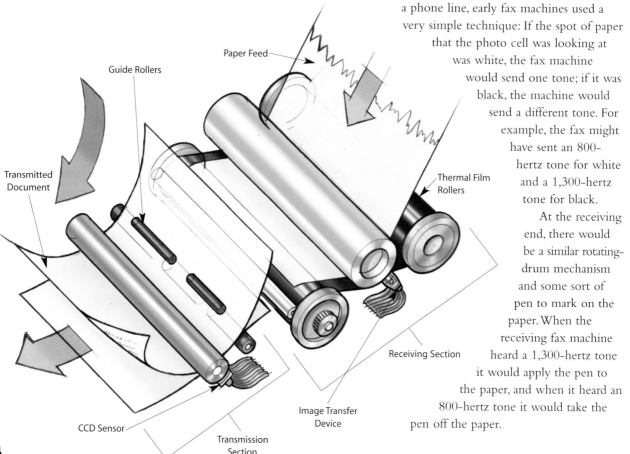

Paper Feed

Guide Rollers

Transmitted Document

Thermal Film Rollers

Receiving Section

CCD Sensor

Transmission Section

Image Transfer Device

Modern Fax Machines

A modern fax machine does not have the rotating drums and is a lot faster, but it uses the same basic mechanics to get the job done.

At the sending end, some sort of sensor reads the paper. A modern fax machine also usually has a paper-feed mechanism so that it is easy to send multi-page faxes. There is some standard way to encode the white and black spots that the fax machine sees on the paper so that the fax can send information that travels through a phone line.

At the receiving end, there is a mechanism that marks the paper with black dots.

A typical fax machine that you find in an office is officially known as a *CCITT* or *ITU-T (standardized organization) Group 3 Facsimile machine.* The Group 3 designation tells you four things about the fax machine:

- It will be able to communicate with any other Group 3 machine.
- It has a horizontal resolution of 203 pixels per inch (8 pixels/mm).
- It has two different vertical resolutions:
  - Standard: 98 lines per inch (3.85 lines/mm)
  - Fine: 196 lines per inch (7.7 lines/mm) (There is also a super fine resolution. It's not officially a Group 3 standard, but it's fairly common, and is 391 lines per inch (15.4 lines/mm)
- It can transmit at a maximum data rate of 14,400 bits per second (bps), and will usually fall back to 12,000 bps, 9,600 bps, 7,200 bps, 4,800 bps, or 2,400 bps if there is a lot of noise on the line.

The fax machine typically has a CCD (charge coupled device) or photodiode sensing array. It contains 1,728 sensors, 203 pixels per inch, so it can scan an entire line of the document at one time. The paper is lit by a small fluorescent tube so that the sensor has a clear view. The image sensor looks for black or white. Therefore, a single line of the document can be represented in 1,728 bits. In standard mode, there are 1,145 lines to the document. The total document size is:

1,728 pixels per line × 1,145 lines = approximately 2,000,000 bits of information.

To reduce the number of bits that have to be transmitted, Group 3 fax machines use three different compression techniques:

- Modified Huffman (MH)
- Modified Read (MR)
- Modified Modified Read (MMR)

The basic idea in these schemes is to look for a run of same-color bits. For example, if a line on the page is all white, the modem can transmit a dozen or so bits rather than the full 1,728 bits scanned for the line. This sort of compression can cut transmission time by a factor of at least two, and for many documents much more. A document containing a significant amount of white space can transmit in just a few seconds.

Receiving the Fax

The bits for the scanned document travel through the phone line and arrive at a receiving fax machine. The bits are decoded, uncompressed, and reassembled into the scanned lines of the original document. There are five common ways to print the fax, depending on the type of machine that receives it:

- **Thermal paper**—When fax machines started infiltrating offices en masse in the 1980s, most of them used thermal paper. The paper is coated with chemicals that react to heat by turning black.
- **Thermal film**—Thermal film uses a page-width ribbon that contains ink that melts onto paper when heated. This is more complicated mechanically than thermal paper, but less complicated than an inkjet.
- **Inkjet**—This technique uses the same mechanism as an inkjet printer.
- **Laser printer**—This technique uses the same mechanism as a laser printer.
- **Computer printer**—The fax is actually received by a fax modem (a modem that understands the Group 3 data standards), stored on the computer's hard disk as a graphics file, and then sent to the computer's usual printer.

How **OFFSET PRINTING** Works

The next time you read your favorite magazine or go through the latest catalog that arrives in your mailbox, stop for a moment and think about how that publication came to be. First, writers, editors, and designers create the magazine. Then printers take that creative work and put it on paper. Printing is a fascinating process involving huge high-speed machines, 1-ton rolls of paper, computers, metal plates, rubber blankets, and sharp knives.

HSW Web Links

www.howstuffworks.com

HowStuffWorks Express
How Stereolithography
(3-D Layering) Works
How Digital Cameras
Work
How Photographic Film
Works
How Newspapers Work

Lithography

To get a better idea of how lithography works in the abstract, imagine the following:

1. Take a piece of paper and paint a stick figure on it with linseed oil (or common vegetable oil).

2. Now moisten the rest of the paper by misting it with water.

3. Put some oil paint on a cotton ball and dab it onto the paper.

4. The parts of the paper moistened with water will not pick up any of the paint (oil and water repel).

5. The parts of the paper coated with linseed oil will pick up the paint. This sheet is now "inked."

6. If you now press another piece of paper onto the inked sheet, the painted portions will transfer to the new sheet of paper and create a print. You can re-ink the original sheet to create multiple prints.

The essence of the technique is the affinity of oil for oil and the repulsion of oil and water.

When it comes to magazine printing, once the content and layout of the magazine has been finalized, offset lithography is the most commonly used printing process. It includes three production steps: pre-press, press run, and bindery. To illustrate these steps, we'll follow the publication of our magazine, *HowStuffWorks Express,* from start to finish.

The Creative Process

Every print piece starts with the creative process. Writers, editors, graphic designers, and artists are the initial step in the creation of magazines, newspapers, brochures, flyers, catalogs, and other print pieces.

The *HowStuffWorks Express* creative team begins work months in advance of each edition's publication date. Topics for articles are identified and writers are assigned. Editors help focus copy and keep the whole process moving.

After each article is written, edited, and approved with final art, the pieces are sent electronically to the director of graphic design for page layout. The director determines what page a story will appear on, where art will be in relation to words and, in some publications, where advertising will appear. Often, there are difficult decisions to make about how to fit the pieces of art and text into very limited space. As in the making of a movie, some material gets left "on the cutting room floor."

Finally, after the layout of every page has been completed, edited, and proofread, a digital *printer's file* is created for the entire document. The printer's file contains an exact image of how every page of the magazine will look once it comes off the press. The printer's file is usually made by burning a CD.

The Printing Process

Offset lithography works on a very simple principle: ink and water don't mix. Images (words and art) are put on plates, which are dampened first by water, then ink. The ink adheres to the image area, the water to the nonimage area. Then the image is transferred to a rubber blanket, and from the rubber blanket to paper. That's why the process is called *offset*—the image doesn't go directly to the paper from the plates, as it does in gravure printing.

Let's look at the steps in the printing process.

Step One: Pre-Press Production

Before the job can be printed, the document must be etched onto plates made out of aluminum. On the plates, a chemical reaction occurs that activates an ink-receptive coating.

Each of the primary colors—black, cyan (blue), magenta (red), and yellow—has a separate plate. Even though you see many, many colors in the finished product, only these four colors are used (you'll also hear this called the four-color printing process).

Step Two: The Press Run

The printing process used to print *HowStuffWorks Express* is called *web offset lithography.* The paper feeds through the press as one continuous stream pulled from rolls of paper. Each roll can weigh as much as 1-ton (2,000 pounds). The paper flows through four separate presses—one for each plate (see "The Inking Process" that follows), and then through an oven to cure the ink. The paper is cut to size after printing. Offset lithography can also be done with precut paper in sheet-fed presses.

Web presses print at very high speeds and use very large sheets of paper. Press

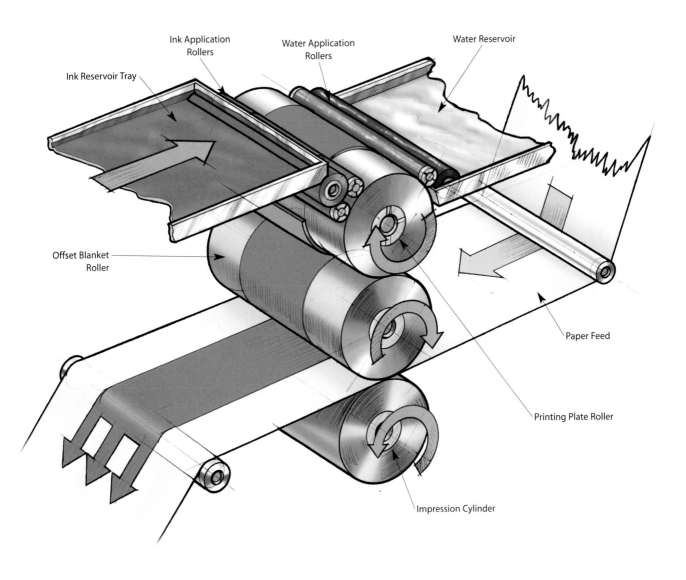

Ink Reservoir Tray

Ink Application Rollers

Water Application Rollers

Water Reservoir

Offset Blanket Roller

Paper Feed

Printing Plate Roller

Impression Cylinder

speeds can reach up to 50,000 impressions per hour. An impression is equal to one full press sheet (38 inches x 22³/4 inches), which equals 12 pages of *How Stuff Works Express.*

Even when a 1-ton roll of paper runs out, the presses don't stop rolling. Rolls can be spliced together as the web press is running by using *festoons.* Festoons are a series of rollers that extend up into a tower. Right before the roll runs out of paper, the festoons will move up into the tower, pulling in large amounts of paper. At the moment the splice occurs, the roll of paper stop rotating for a split second. At that point the paper is taped together automatically. As the

newly spliced roll begins to pick up speed, the festoons begin to drop out of the tower at a rate determined by the speed of the press. The press operator never has to adjust the press controls during this operation—everything is automatic.

The Inking Process

As we mention earlier, ink and water don't mix—this is the underlying principle of offset lithography. During the press run, the ink is distributed to the plates through a series of rollers. On the press, the plates are dampened, first by water rollers, then ink rollers. The rollers distribute the ink from the ink fountain onto the plates.

Printing Processes

There are nine main types of printing processes:

- **Offset lithography**—the workhorse of commercial printing

- **Engraving**—used in creating fine stationery

- **Thermography**—raised printing, used in stationery

- **Reprographics**—copying and duplicating

- **Digital printing**—limited now, but the technology is exploding

- **Letterpress**—the original Guttenberg process (hardly done anymore)

- **Screen-printing**—used for T-shirts and billboards

- **Flexography**—usually used on packaging, such as can labels

- **Gravure**—used for huge runs of magazines and direct-mail catalogs

The image area of the plate picks up ink from the ink rollers. The water rollers keep the ink off the nonimage areas of the plate. Each plate then transfers its image to a rubber blanket that in turn transfers the image to the paper. All of this occurs at an extremely high speed.

The paper is left slightly wet by all of the ink and water being applied. Obviously, there's a risk of the ink smudging. Passing the paper through an oven helps to avoid this smudging. The oven is gas fired, and the temperature inside runs at 350 to 400°F (176 to 206°C). Immediately after leaving the oven, the paper is run through a short series of large metal rollers that have refrigerated water flowing through them. These *chill rollers* cool the paper down instantly and set the ink into the paper.

Color and Registration Control

Four separate presses put the ink onto the paper. If the plates from each of the presses don't line up perfectly, the image looks out of focus and the color is wrong. *Registration* is the process of lining up the four presses;

registration marks on the pages help line things up. A computer takes a video image of the registration marks printed onto the press sheet. Each plate has its own individual mark. The computer reads each of these marks and makes adjustments to the position of each plate in order to achieve perfect alignment. All of this occurs many times per second while the press is running at full speed.

Color control looks at the way the inks blend together on the paper and adjusts things so the color is perfect. A computer can adjust the ink flow onto each press to get the look right, and the human who is operating the machine can also adjust things.

Step Three: Bindery

The bindery is where the printed pages get stapled or glued together to create the final publication. Imagine having six sheets of paper for each issue and a million copies of all those sheets, and you want to put them all together in the right order and staple them. And you want to do all that in less than a day. That's what's happening in the bindery.

In the case of *How Stuff Works Express,* a machine called a *stitcher* takes the folded printed papers (called *press signatures*) and collates them. Then the machine inserts stitches (staples) into the signatures, binding them together.

The final components in the stitcher machine are the knives, which trim the paper to the final. The magazine is then ready to be addressed and sent off to the readers.

When you get a magazine in the mail, it looks so simple. Now you understand all the steps that really go into the process!

How **VIRTUAL PRIVATE NETWORKS** Work

The world shrinks every day. Instead of simply dealing with local or regional concerns, many businesses now have to think about global markets and logistics. Many companies have facilities spread out across the country or even around the world. But there is one thing that all of them need: a way to maintain fast, secure, and reliable communications wherever their offices are.

A private network is a network made up of private wiring. For example, if you have three computers in your house that all talk to one another on a small network you've installed, that's a private network. If you want to privately connect your home network to your brother's home network in his house, you would either string a wire across the yard (if you're neighbors) or lease a dedicated phone line from the phone company (if the two houses are in different states, say). Because the network is private, made up of all private wiring and dedicated phone lines, no one can tap into it.

Basically, a virtual private network (VPN) is a private network that uses a public network (usually the Internet) to connect remote sites or users together. Instead of using a dedicated, real-world connection such as a leased line, a VPN uses virtual connections routed through the Internet from the company's private network to the remote site or employee.

There are two common VPN types:

- **Remote-access**—Also called a virtual private dial-up network (VPDN), this is a user-to-LAN (local area network) connection used by a company that has employees who need to connect to the private network from various remote locations.
- **Site-to-site**—Through the use of dedicated equipment and large-scale encryption, a company can connect multiple fixed sites over a public network such as the Internet. Site-to-site VPNs can be either:
  - **Intranet-based**—If a company has one or more remote locations that they wish to join in a single private network, they can create an intranet VPN to connect LAN to LAN.
  - **Extranet-based**—When a company has a close relationship with another company (for example, a partner, supplier or customer), they can build an extranet VPN that connects LAN to LAN, and that allows all of the various companies to work in a shared environment.

Analogy: Each LAN Is an IsLANd

Imagine that you live on an island in a huge ocean. There are thousands of other islands all around you, some very close and others farther away. The normal way to travel is to take a ferry from your island to whichever island you wish to visit. Of course, traveling on a ferry means that you have almost no privacy. Anything you do can be seen by someone else.

Let's say that each island represents a private LAN and the ocean is the Internet. Traveling by ferry is like connecting to a Web server or other device through the Internet. You have no control over the wires and routers that make up the Internet, just like you have no control over the other people on the ferry. You're left susceptible to security issues if you're connecting two private networks using a public resource.

Continuing with our analogy, your island decides to build a bridge to another island so that there's an easier, more secure, direct way for people to travel between the two. Building and maintaining the bridge is

HSW Web Links

www.howstuffworks.com

How Network Address
 Translation Works
How LAN Switches Work
How Web Servers Work
How Encryption Works
How Routers Work

229

A well-designed VPN can greatly benefit a company. For example, it can:

- Extend geographic connectivity
- Improve security
- Reduce operational costs versus a traditional wide area network (WAN)
- Reduce transit time and transportation costs for remote users
- Improve productivity
- Simplify network topology
- Provide global networking opportunities
- Provide telecommuter support
- Provide broadband networking compatibility
- Provide faster return on investment than a traditional WAN

expensive, even though the island you're connecting with is very close. But the need for a reliable, secure path is so great that you do it anyway. Your island would like to connect to a second island that is much farther away, but decides that the cost is simply too much to bear.

This is very much like having a leased line. The bridges (leased lines) are separate from the ocean (the Internet), yet are able to connect the islands (LANs). Many companies have chosen this route because of the need for security and reliability in connecting their remote offices. However, if the offices are very far apart, the cost can be prohibitively high—just like trying to build a bridge that spans a great distance.

So how does VPN fit in? Using our analogy, we could give each inhabitant of our island their own small submarine. Let's assume that these submarines have some amazing properties:

- They're fast.
- They're easy to take with you wherever you go.
- They're able to completely hide you from any other boats or submarines.
- They're dependable.
- They're inexpensive. It costs little to add additional submarines to your fleet once the first is purchased.

Although they are traveling in the ocean along with other traffic, the inhabitants of your island could travel back and forth between islands whenever they wanted to with privacy and security. That's essentially

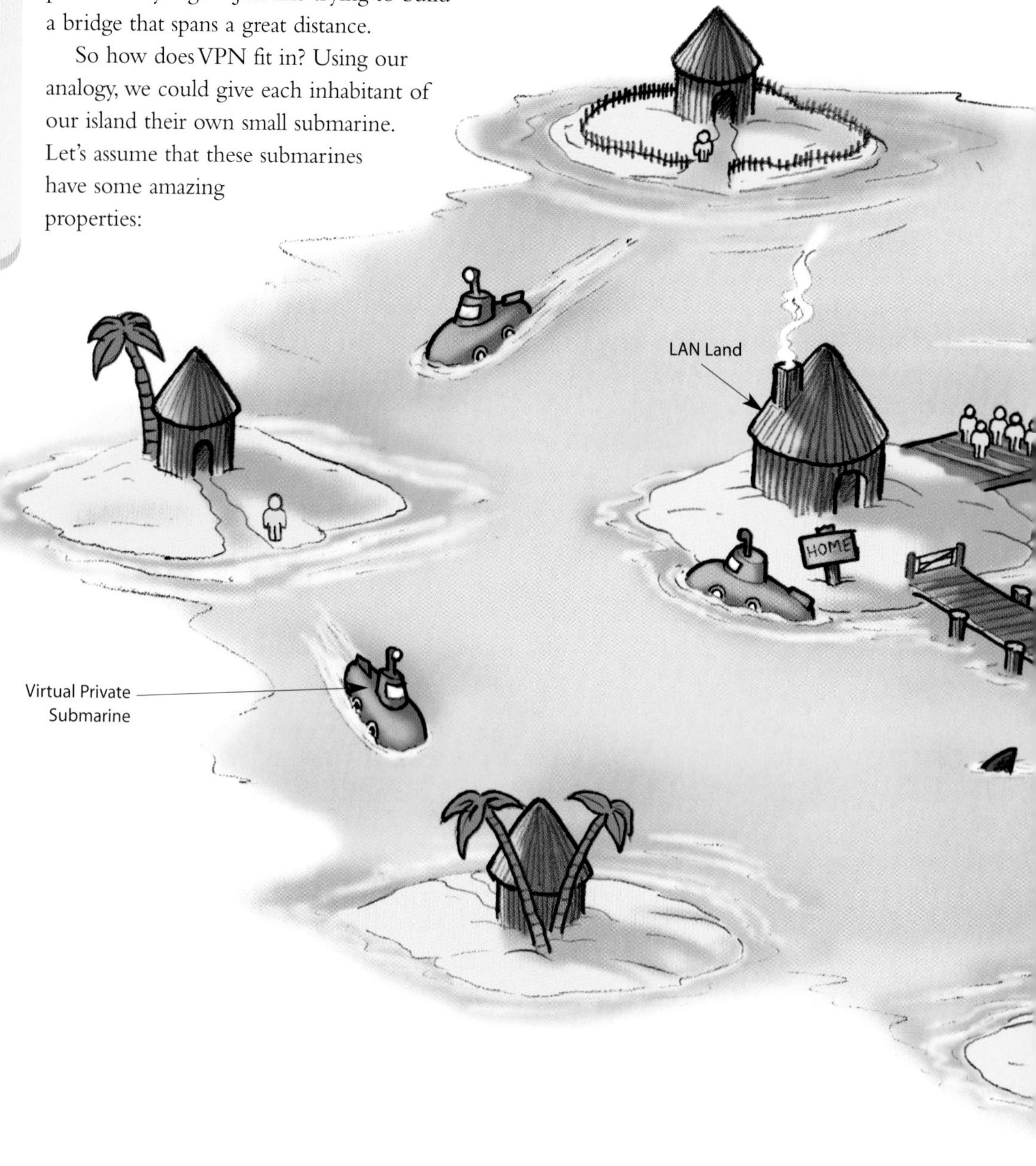

LAN Land

Virtual Private Submarine

HOME

230

how a VPN works. Each remote member of your network can communicate in a secure and reliable manner using the Internet as the medium to connect to the private LAN. A VPN can easily grow to accommodate more users and different locations.

Security Measures

A well-designed VPN uses several methods to keep your connection and data secure:

- **Firewalls**—A firewall provides a strong barrier between your private network and the Internet. You can set firewalls to restrict the number of open ports, what type of packets are passed through, and which protocols are allowed through.

- **Encryption**—This is the process of taking all the data that one computer is sending to another and encoding it into a form that only the other computer will be able to decode. (See "How Encryption Works" in the chapter on the Internet and radio for more information.)

- **IPSec**—Internet protocol security protocol (IPSec) provides enhanced security features such as better encryption algorithms and more comprehensive authentication.

- **AAA server**—AAA servers (the three 'A's are *authentication, authorization,* and *accounting*) are used for more secure access in a remote-access VPN environment.

Tunneling

Most VPNs rely on tunneling to create a private network that reaches across the Internet.

Think of tunneling like having a computer delivered to you by UPS. The vendor packs the computer (passenger protocol) into a box (encapsulating protocol), which is then put on a UPS truck (carrier protocol) at the vendor's warehouse (entry tunnel interface). The truck (carrier protocol) travels over the highways (the Internet) to your home (exit tunnel interface) and delivers the computer. You open the box (encapsulating protocol) and remove the computer (passenger protocol). Tunneling works the same way!

As you can see, VPNs are a great way for a company to inexpensively keep its employees and partners connected, no matter where they are.

Internet Ferry

Leased Line Bridge

Building a VPN

Depending on the type—remote-access or site-to-site—you will need to put in place certain components to build your VPN. These might include:

- Desktop software client for each remote user
- Dedicated hardware such as a VPN concentrator or secure PIX firewall
- Dedicated VPN server for dial-up services
- Network access server used by service provider for remote-user VPN access
- VPN network and policy-management center

How **INKJET PRINTERS** Work

Since their introduction in the latter half of the 1980s, inkjet printers have grown in popularity and performance while dropping significantly in price. You now find them in professional and home offices alike because they are an affordable way to print color images.

HSW Web Links

www.howstuffworks.com

How Laser Printers Work
How Parallel Ports Work
How USB Ports Work
How Photocopiers Work

An inkjet printer is any printer that places extremely small droplets of ink onto paper to create an image. If you ever look at a piece of paper that has come out of an inkjet printer, you know that:

- The dots are extremely small (usually between 50 and 60 microns in diameter), so small that they are tinier than the diameter of a human hair (70 microns).
- The dots are positioned very precisely, with resolutions of up to 1440 x 720 dots per inch (dpi).
- The dots can take on different colors (through combining inks) and can create photo-quality images.

Heat Wave or Good Vibrations?

Different types of inkjet printers form their droplets of ink in different ways. There are two main inkjet technologies currently used by printer manufacturers.

Thermal bubble printers (commonly referred to as *bubble jets*) use tiny resistors to create heat, and this heat vaporizes ink to create a bubble. As the bubble expands, some of the ink is pushed out of a nozzle onto the paper. When the bubble pops (collapses), a vacuum is created. This pulls more ink into the print head from the cartridge. A typical bubble-jet print head has 300 or 600 tiny nozzles, and all of them can fire a droplet simultaneously.

Piezoelectric printers use piezoelectric crystals. A crystal is located at the back of the ink reservoir of each nozzle. The crystal receives a tiny electric charge that causes it to vibrate. When the crystal vibrates, it forces a tiny amount of ink out of the nozzle and pulls more ink into the reservoir to replenish the ink.

Anatomy of an Inkjet

The print-head assembly is the prime mover in any inkjet printer. The print-head assembly consists of the following:

- Print head
- Ink cartridges
- Print-head stepper motor
- Belt
- Stabilizer bar

The print head contains a series of nozzles that spray drops of ink from the ink cartridges. The stepper motor moves the print-head assembly back and forth across the paper.

The paper-feed assembly moves the paper past the print-head assembly. It includes a paper tray or feeder, rollers, and a paper-feed stepper motor. This stepper motor powers the rollers to move the paper in the exact increment needed to create an even, continuous image on the paper.

A small but sophisticated computer controls the motors and interprets all of the data streaming into the printer. This data comes in through the interface port (or ports).

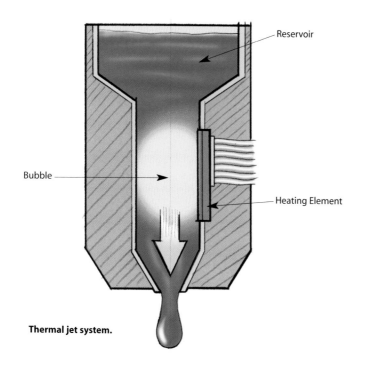

Bubble

Reservoir

Heating Element

Thermal jet system.

Piezoelectric Crystal

Piezo jet system.

The most popular way of connecting a printer to a computer is through the parallel port, but a large number of newer printers use a universal serial bus (USB) port instead. A few printers connect using a serial port or small computer system interface (SCSI) port.

Let's walk through the printing process to see how all this technology comes together.

1) Click "Print" and then "OK."
2) Your computer sends information to the printer.
3) The control circuitry activates the paper-feed stepper motor. This engages the rollers, which feed a sheet of paper from the paper tray or feeder into the printer. A small trigger mechanism in the tray or feeder is depressed when there's paper in the tray or feeder. If the trigger is not depressed, the printer lights up the Out of Paper LED and sends an alert to the computer.
4) After the paper is fed into the printer and positioned at the start of the page, the print-head stepper motor uses the belt to move the print-head assembly across the page. The motor pauses for the merest fraction of a second each time that the print head sprays dots of ink on the page,

and then moves a tiny bit before stopping again. This stepping happens so fast that it seems like a continuous motion.

5) Multiple dots are made at each stop. The print head sprays the CMYK (cyan, magenta, yellow, and black) colors in precise amounts to make any color imaginable.
6) At the end of each complete pass, the paper-feed stepper motor advances the paper a fraction of an inch. Depending on the inkjet model, the print head is reset to the beginning side of the page, or, in most cases, simply reverses direction and begins to move back across the page as it prints.

This process continues until the full page is printed. The time it takes to print a page can vary widely from printer to printer.

Inkjet printers are capable of printing on a variety of media. Commercial inkjet printers sometimes spray directly on an item like the label of a beer bottle. For consumer use, a number of specialty papers, ranging from adhesive-backed labels or stickers, to business cards and brochures, can be printed. You can even get iron-on transfers that allow you to print an image and put it on a T-shirt! One thing is for certain: Inkjet printers definitely provide an easy and affordable way to unleash your creativity.

Does Your Paper Have a Coat?

The paper you use on an inkjet printer determines the quality of the image. Standard copier paper works, but doesn't provide as crisp and bright an image as paper made specifically for an inkjet printer. There are two main factors that affect image quality:

• Brightness
• Absorption

The brightness of a paper is normally determined by how rough the surface of the paper is. A coarse or

rough paper will scatter light in several directions, whereas a smooth paper will reflect more of the light back in the same direction. This makes the paper appear brighter, which in turn makes any image on the paper appear brighter.

The other key factor in image quality is absorption. When the ink is sprayed onto the paper, it should stay in a tight, symmetrical dot. If the ink gets absorbed into the paper and spreads out, the dot

covers a slightly larger area than the printer expects it to. The result is a page that looks somewhat fuzzy, particularly at the edges of objects and text.

To combat this, high-quality inkjet paper is coated with a waxy film that keeps the ink on the surface of the paper. Coated paper normally yields a dramatically better print than other paper. The low absorption of coated paper is key to the high resolution capabilities of many of today's inkjet printers.

How **SURGE PROTECTORS** Work

A surge protector is a crucial part of any computer system. It protects all the electronic components in your computer from power surges—voltage spikes that can fry a computer's power supply, or all the components that attach to the computer's power supply.

HSW Web Links

www.howstuffworks.com

How Power Distribution
  Grids Work
How Lightning Works
How Emergency Power
  Systems Work
How Semiconductors
  Work
How Wires, Fuses and
  Connectors Work
How Batteries Work

Sensitive Equipment

Surge protectors are more important today than in the past because so many modern electronic devices rely on delicate components.

Microprocessors, which are an integral part of all computers as well as many home appliances, are particularly sensitive to surges. There's no reason to hook up a light bulb to a surge protector because the worst that is likely to happen due to a power surge is that your light bulb will burn out. But if a computer gets zapped, it'll cost you hundreds of dollars, and you'll lose all your files.

Power comes into your home or office through the power lines at certain voltages. In the United States, 120 volts is the standard line voltage. A power surge, or transient voltage, is an increase in voltage significantly above this normal level.

All sorts of things can cause a surge. For example, if a big motor stops suddenly, it can cause one. So can lightning. So can an equipment failure somewhere else on the electrical grid. Faulty wiring, problems with the utility company's equipment, and downed power lines are all culprits. The system of transformers and lines that brings electricity from a power generator to the outlets in our homes or offices is extraordinarily complex. There are dozens of possible points of failure, and many potential errors that can cause an uneven power flow. In today's system of electricity distribution, power surges are an unavoidable occurrence.

When the voltage shoots up, one of two things can happen. If the voltage is high enough, it causes arcing (the ionization of the air between two electrodes), which burns out components immediately. Less severe surges cause too much current to flow through the components in your computer, and the components heat up. Even if the increased charge doesn't destroy the elements immediately, it may wear them out prematurely.

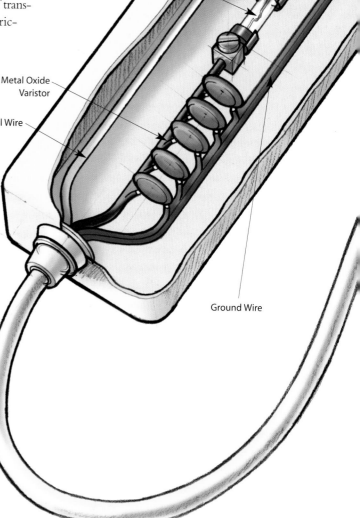

Toroidal Choke Coil

Hot Wire

Fuse

Metal Oxide
Varistor

Neutral Wire

Ground Wire

The Process of Protection

A standard electrical receptacle has three holes, leading to three wires. The hot wire carries electrical charge from the municipal power supply. The neutral and grounding wires are grounded—that is, they are connected to the earth, and so have zero voltage.

The surge protector's job is to divert any extra electricity from the hot wire into the neutral wire or the grounding wire so that the surge doesn't pass on to devices plugged into the power strip.

In a standard protector, a *metal oxide varistor (variable resistor),* or MOV, handles the job. An MOV is a piece of metal oxide material sandwiched between two semiconductors. One of the semiconductors is connected to the hot wire and the other is connected to the neutral wire.

These semiconductors have a variable resistance that is dependent on voltage. The semiconductors act as resistors when the voltage is below a certain level—there isn't enough force to move electrons through the material. A higher voltage causes the electrons to overcome the resistance, and current starts flowing. (See "How Semiconductors Work" to find out how this variable resistance works.)

As soon as the extra charge is diverted into the MOV and to ground, the voltage in the hot line returns to a normal level, so the MOV's resistance shoots up again. To put it very simply: The MOV connects between the power line and the ground. At normal voltage—say around 110 volts—nothing flows through the MOV. But at a significantly higher voltage, the MOV will start to conduct current. In this way, the MOV diverts the surge current while allowing the standard current to continue powering whatever machines are connected to the surge protector. Metaphorically speaking, the MOV acts as a pressure-sensitive valve that only opens when there is too much pressure.

Other Surge Protection Systems

Another common surge protection device is a gas tube (sometimes called a gas discharge arrestor). These tubes do the same job as an MOV. When the voltage surges above a certain level, the electrical power is strong enough to ionize the gas in the tube, making it a very effective conductor. It passes the excess charge to the neutral or ground line.

A few surge protector products suppress surges by holding up excess charge on its way through the hot line rather than by diverting it to the ground. The protector gradually releases the excess charge, at safe levels.

Some surge protectors have a line-conditioning system for filtering out *line noise,* which is smaller fluctuations in electrical current. Basic surge protectors with line-conditioning use a fairly simple system. On its way to the power-strip outlet, the hot wire passes through a toroidal choke coil. The choke is a just ring of magnetic material wrapped with wire. The ups and downs of the passing current in the hot wire charge the coil, causing it to emit electromagnetic forces that smooth out the small increases and decreases in current. This conditioned current is more stable, and so easier on your computer.

No surge protector is 100% effective, and even top-of-the-line equipment may have some serious problems. In a thunderstorm, it's always best to disconnect your computer completely rather than rely on the surge protector. But on a day-to-day basis, a good protector can save you a lot of money and frustration.

Finding Another Way In

Power surges don't just come from the power line—telephone and cable lines can also conduct high voltage. If your computer is connected to the phone lines via a modem, you should get a surge protector that has a phone-line input jack. If you have a coaxial cable line hooked up to expensive equipment, consider a cable surge protector.

Protection Ratings

All surge protectors are not created equal. At one end, you have your basic $5 surge protector power strip, which will offer very little protection. On the other end, you have systems costing hundreds of thousands of dollars that protect against pretty much everything short of lightning striking nearby.

To find out what a protector can do, you need to check out its Underwriters Laboratories (UL) ratings. UL is an independent, not-for-profit company that tests electric and electronic products for safety. If a protector doesn't have a UL listing, it's probably junk; there's a good chance it doesn't have any protection components at all. The UL label will also tell you the unit's clamping voltage (the voltage that will cause the MOVs to conduct electricity to the ground line), its energy absorption/dissipation rating (how much energy the surge protector can absorb before it fails), and its response time (how quickly it starts diverting current).

How PERSONAL DIGITAL ASSISTANTS Work

Though originally intended to be simple digital calendars, personal digital assistants (PDAs) have evolved into machines for crunching numbers, playing games or music, and downloading information from the Internet.

HSW Web Links

www.howstuffworks.com

How Location Tracking Will Work

How Video Game Systems Work

How Smart Labels Will Work

How MP3 Files Work

How Laptops Work

PDAs are tiny, portable versions of a desktop computer, and every year they get more and more powerful. How can everything fit into such a small package, and why can you power a PDA with 2 AAA batteries for a month?

How PDAs Differ from Desktops

The biggest difference between PDAs and desktop computers is the CPU. On a typical desktop machine today, processors run at a clock rate of 1 to 2 gigahertz and consume anywhere from 20 to 75 watts. The processor in a PDA might run 15 to 200 megahertz and is optimized to use incredibly small amounts of power—half a watt or less.

Because the processor is running so slowly, PDA software has to be simplified.

Personal digital assistant (PDA).

Like desktops and laptops, PDAs rely on an operating system (OS) to control everything. The OS used by a PDA, along with the applications that run on the OS, are not nearly as complex as those used by PCs. They generally have fewer instructions and take up a lot less memory.

Everything else in a PDA is designed for low power as well. The display is often a black-and-white liquid crystal display (LCD) with no (or optional) back lighting. LCDs use almost no power (which is why an LCD watch can run for years on a tiny battery). The memory is low power, too. Unlike a desktop computer, a PDA doesn't have a hard drive. It stores basic programs in a read-only memory (ROM) chip, which remains intact even when the machine shuts down or runs out of battery power. Your data and any programs you add later are stored in the device's random-access memory (RAM). This approach gives a PDA several advantages over standard PCs. When you turn on the PDA, all your programs are instantly available. You don't have to wait for applications to load. Changes made to a file are stored automatically, so you don't need a Save command. And when you turn the device off, the data is still safe, because the PDA continues to draw a small amount of power from the batteries to keep the RAM alive. The only thing you have to watch for is dead batteries. All PDAs have a docking

Hand-to-Hand Comparison

There are two major categories of PDAs: handheld computers and palm-sized computers. The differences between the two are size, display, and the mode of data entry. Compared to palm-sized computers,

handheld computers tend to be larger and heavier. They have larger LCDs and use a miniature keyboard, usually in combination with touch-screen technology, for data entry. Palm-sized computers are

smaller and lighter. They have smaller LCDs and rely on stylus/touch-screen technology and handwriting recognition programs for data entry.

station that lets you download the contents of the PDA's memory to a desktop machine. That way if the batteries do go dead, you can easily reload everything into RAM.

PDAs also have some type of LCD display screen. Unlike the LCD screens for desktop or laptop computers, which are used solely as output devices, PDAs use their screens for output and input. Let's look at how that works.

LCDs as Input Devices

One of the coolest things about a PDA is the technology used to enter data. The tiny screen on a palm-sized computer serves as an output and an input device. It displays information with an LCD. But on top of the LCD sits a touch screen that lets you launch programs or enter information.

You can think of the PDA's screen as a multilayer sandwich. On top is a thin plastic or glass sheet with a conductive coating on its bottom. The plastic or glass floats on a thin layer of nonconductive oil, which rests on a layer of glass that is also coated with a conductive finish.

Multi-Tasking

PDAs are an incredible multi-tasking tool. Depending on the model, you can use these devices to:

- Store contact information (names, addresses, phone numbers, e-mail addresses)
- Take notes
- Write memos
- Keep track of appointments (date book, calendar)
- Remind you of appointments (clock, alarm functions)
- Make calculations
- Keep track of expenses
- Send or receive e-mail
- Do word processing
- Play MP3 music files
- Play MPEG movie files
- Surf the Internet
- Play video games
- Integrate other devices like digital cameras and GPS receivers

When you touch the stylus to the screen, the plastic pushes down through the gel to meet the glass (called a *touchdown*). This causes a change in the voltage field, which is recorded by the touch screen's driver software. The driver scans the screen thousands of times each second to see if you're entering data. If you are, the driver sends the data to any application that needs it. In this way, the PDA knows when you're tapping an on-screen icon to launch a program or gliding the stylus across the screen to enter data.

PDA stylus.

Eventually, most PDAs will incorporate voice recognition technology. You will speak into a built-in microphone while software converts your voice waves into data. PDAs will get more and more powerful, so they can run complex software. They will merge with cell phones to allow Internet connectivity. And soon you may not need a desktop machine any more! Your PDA will let you do everything you need from anywhere you happen to be.

How **CD BURNERS** Work

Up until the mid 1990s, only professionals and the occasional music enthusiast had the equipment to produce CDs. Today, CD burners are standard equipment on new computers, and blank CDs cost less than a dollar. Just about anybody can start burning their own music mixes or backing up computer data at minimal expense. CDs are already edging out cassettes as the standard home-recording medium.

HSW Web Links

www.howstuffworks.com

How CDs Work
How Analog-Digital
  Recording Works
How DVDs and DVD
  Players Work
How Removable Storage
  Works
How Speakers Work

CDs store music and other files in digital form. That is, the information on the disc is represented by a series of 1s and 0s. In conventional CDs, these 1s and 0s are represented by millions of tiny bumps and flat areas on the disc's reflective surface. The bumps and flats are arranged in a continuous spiral track that measures about 0.5 microns (millionths of a meter) across and 3.5 miles (5 km) long.

To read this information, the CD player passes a laser beam over the track. When the laser passes over a flat area in the track, the beam is reflected directly to an optical sensor on the laser assembly. The CD player interprets this as a 1.

When the beam passes over a bump, the light is scattered away from the optical sensor. The CD player recognizes this as a 0.

To produce these CDs, manufacturers create a mold of the bump pattern and press the mold into the blank CDs' acrylic surface. The process is cost-effective if you're producing millions of copies of the same CD, but it isn't practical for the casual consumer. Home CD burners come at the problem from a completely different angle.

Feel the Burn

A CD-recordable disc, or CD-R, doesn't have any bumps or flat areas at all. Instead, it has a smooth reflective metal layer that sits on top of a photosensitive dye layer.

When the disc is blank, the dye is translucent. Light can shine through and reflect off the metal surface. But when you heat the dye layer, it turns opaque. It darkens to the point that light can't pass through. A CD burner has a moving laser assembly, just like an ordinary CD

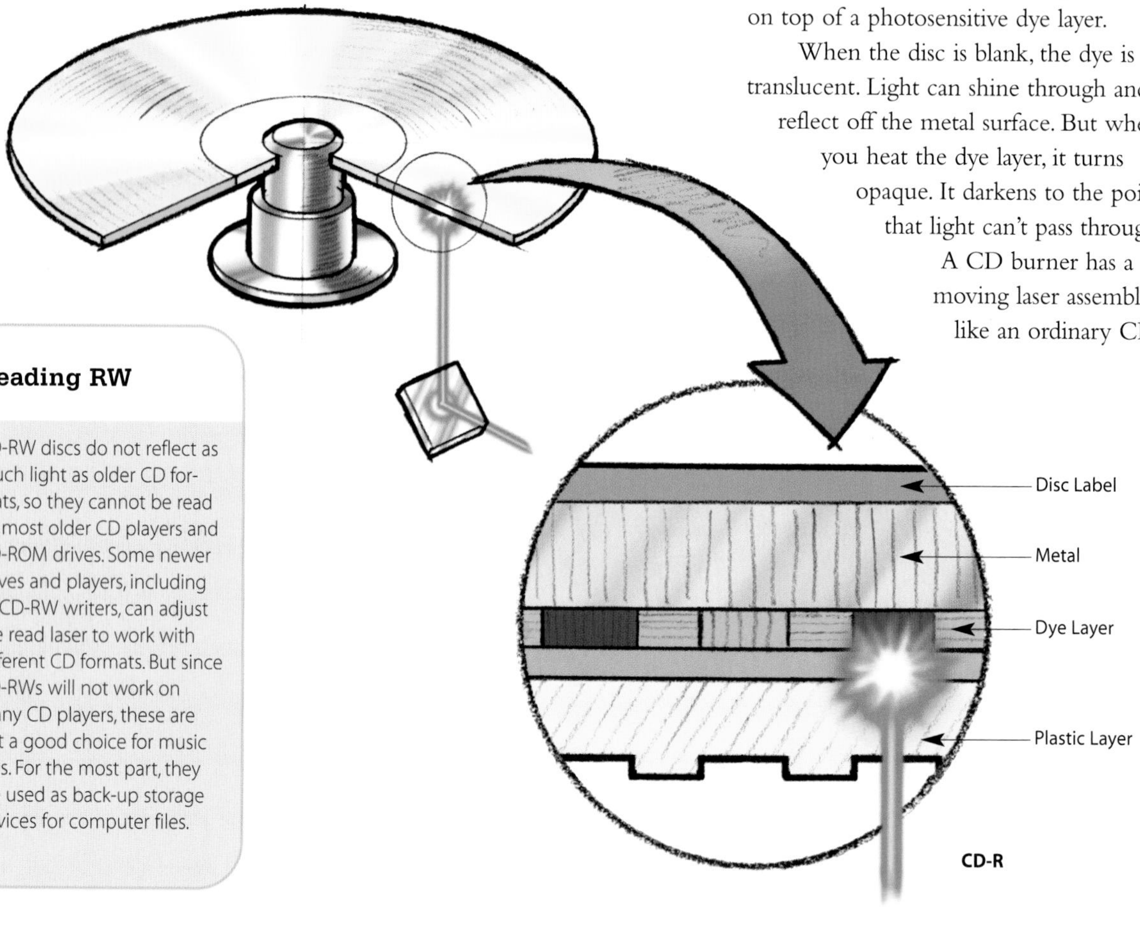

Reading RW

CD-RW discs do not reflect as much light as older CD formats, so they cannot be read by most older CD players and CD-ROM drives. Some newer drives and players, including all CD-RW writers, can adjust the read laser to work with different CD formats. But since CD-RWs will not work on many CD players, these are not a good choice for music CDs. For the most part, they are used as back-up storage devices for computer files.

Disc Label
Metal
Dye Layer
Plastic Layer

CD-R

player. But in addition to the standard read laser, it has a write laser. The write laser is intense enough to darken the dye layer.

The burner moves the laser outward while the disc spins. The bottom plastic layer has grooves pre-pressed into it to guide the laser along the correct path. By calibrating the rate of spin with the movement of the laser assembly, the burner keeps the laser running along the track at a constant rate of speed. To record the data, the burner simply turns the laser writer on and off in sync with the pattern of 1s and 0s. The laser darkens the material to encode a 0 and leaves it translucent to encode a 1.

By selectively darkening particular points along the CD track and leaving other areas of dye translucent, a burner can create a digital pattern that a standard CD player can read. The light from the player's laser beam will only bounce back to the sensor when the dye is left translucent, in the same way that it will only bounce back from the flat areas of a conventional CD. So, even though the CD-R disc doesn't have any bumps pressed into it at all, it behaves just like a standard disc.

Laser Eraser

Unlike tapes, floppy disks, and many other data-storage mediums, you cannot re-record on a CD-R disc once you've filled it up.

CD-RW discs, however, have taken the idea of writable CDs a step further, building in an erase function so that you can record over old data you don't need anymore. These discs are based on phase-change technology.

In CD-RW discs, the phase-change element is a chemical compound of silver, antimony, tellurium, and indium. You can change this compound's form by heating it to certain temperatures.

The crystalline form of the compound is translucent while the amorphous fluid form absorbs most light. On a new, blank CD, all of the material in the writable area is in the crystalline form, so light will shine through this layer to the reflective metal above and bounce back to the light sensor. To encode information on the disc, the CD burner uses its write laser, which is powerful enough to

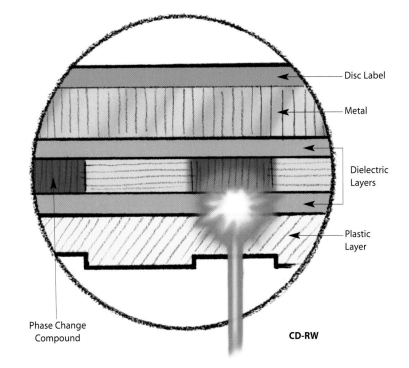

Disc Label
Metal
Dielectric Layers
Plastic Layer
Phase Change Compound
CD-RW

heat the compound to its melting temperature. To encode a 0 on the disc, the write laser heats the compound so it melts. The erase laser is hot enough to return the compound to its crystalline state, effectively erasing any 0s.

Just like conventional CDs, CD-Rs and CD-RWs can store any kind of digital information. But in order to make the information accessible to other CD drives and players, the data has to be arranged in a particular sequence. Burner programs encode data into standard formats, so any player can recognize it.

Trailer Track

CD-Rs and CD-RWs have a component that ordinary music CDs do not have— an extra bit of track at the beginning of the CD, before time zero (00:00), which is the starting point recognized by CD players. This additional track space includes the power memory area (PMA) and the power calibration area (PCA). The PMA stores a temporary table of contents for the individual packets on a disc that has been only partially recorded. When you complete the disc, the burner uses this information to create the final table of contents.

The PCA is a sort of testing ground for the CD burner. In order to ensure that the write laser is set at the right level, the burner will make a series of test marks along the PCA section of track. The burner will then read over these marks, checking for the intensity of reflection in marked areas as compared to unmarked areas. Based on this information, the burner determines the optimum laser setting for writing onto the disc.

How **SCREENSAVERS** Work

A screensaver normally pops up whenever you leave your computer unattended for more than a couple of minutes. It may simply be a blank screen or a scrolling message stating your whereabouts, or it could be something as outrageous as winged toasters or dancing macaroni. But what is a screensaver, exactly? What purpose does it serve? How does it know when to start?

HSW Web Links

www.howstuffworks.com

How PCs Work

How 3-D Graphics Work

How Computer Mice Work

How Computer
Keyboards Work

How Computer Monitors
Work

How Operating Systems
Work

A screensaver is really just an executable file, with the extension changed from *.exe* to *.scr.*

- **An executable file** is a file that the computer's operating system considers a program or application. The operating system can run, or execute, an executable file.
- **The file extension** tells the computer what kind of file it's dealing with. For example, *anyname.exe* is a word-processing application that can be loaded by the computer, while *anyname.doc* is a text file that can be loaded into the Microsoft Word word-processing application.

The screensaver program's purpose in life is to draw something on the screen when the computer is idle. It can:

- Use vector graphics to draw various designs
- Load and display a particular image or group of images
- Display a particular line of text
- Display an animation or series of animations

- Play a video sequence
- Have music or sound effects
- Display information from another program or a Web site
- Or it can paint the screen black to "blank it"

In addition, some screensavers provide the ability to interact with another program or a Web site. (For example, the HowStuffWorks screensaver keeps the mouse active, which allows you to click on several different icons to access specific areas of the HowStuffWorks Web site.) Some screensavers may also require a password before you can turn them off and return to the desktop.

Step by Step

Let's look at how a screensaver works on a Windows 95/98 computer. Although the system commands and exact details may differ, the process is essentially the same for other computers as well.

1) The operating system constantly monitors the activity of the various components of the system. When it sees that the keyboard and mouse have been idle for the amount of time indicated in the screensaver settings, the system gets ready to start the screensaver. The operating system sends a special command to the foreground (current) application to ask if it can launch the screensaver. If the foreground application has a computer-based training (CBT) window open, it will tell Windows to not start the screensaver. All other applications, however, should respond positively to the command.

2) Windows then looks to see if a screensaver has been specified. If the entry is

---

Did You Know?

Screensavers were originally designed to protect computer monitors from phosphor burn-in. Early monitors, particularly monochrome ones, had problems with the same image being displayed for a long time. The phosphors that were used to make the pixels in the display would glow at a constant rate for such a long period of time that they would actually discolor the glass surface of the monitor. This discoloration would then be visible as a faint image overlaying whatever else was displayed on the monitor. Advances in display technology and the advent of energy-saver monitors have virtually eliminated the need for screensavers. But we still use them—probably because they're so fun, and also because they can help keep our work private if we're away from our computers.

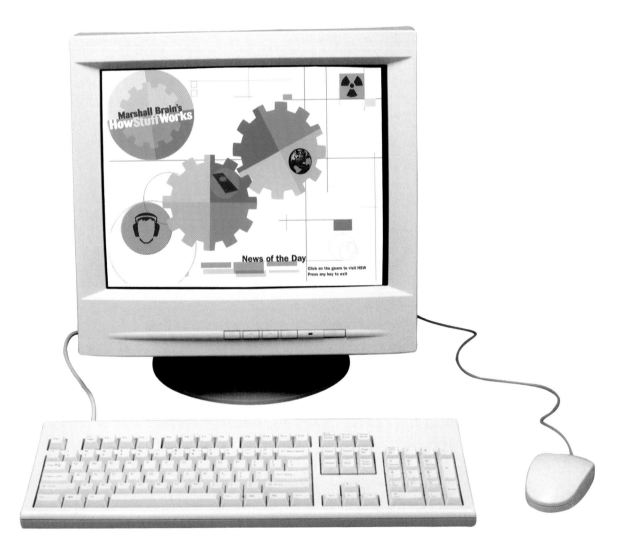

blank, it ignores the command to execute the screensaver. But if a filename is listed, it attempts to load that file. As long as the file listed is an actual screensaver, the program executes and creates the screensaver images on top of the current desktop.

3) The screensaver continues to run until Windows detects input from the keyboard or mouse. With most screensavers, moving the mouse or pressing any key will immediately terminate the screensaver. But screensavers can be programmed to stop when only certain keys or buttons are pressed, or when the mouse is moved a certain distance. This feature is especially useful in interactive screensavers.

4) When Windows gets input that it should terminate the screensaver, it checks to see if password protection is turned on. If it is, a box pops up requiring that you enter a user name and password. If there's no password requirement, the screensaver simply terminates.

Although password protection provides some security, it is important to realize that Windows 95/98 screensavers create their own password dialog boxes and request the password and user information from the system. If you're not certain of the reliability of the source of the screensaver, be careful about using password protection. Hackers can and do create screensavers that use this weak point in system security to capture passwords.

How **WORKPLACE SURVEILLANCE** Works

Admit it—you've used your computer at work to view non-work related Web sites. More than 70 % of the adult online population has accessed the Internet at work for personal use at least once, according to a September 2000 eMarketer study. Employees are sending personal emails, playing games, shopping, checking stock prices, and gambling online during working hours.

HSW Web Links

www.howstuffworks.com

How Carnivore Works
How Firewalls Work
How Wiretapping Works
How Lock Picking Works
How Location Tracking
 Will Work

No Place to Hide

Computer surveillance is by far the most common method of monitoring employee activity. However, employers are still using traditional methods such as eavesdropping on phone calls, storing and reviewing voice mail, and videorecording employees on the job, according to the American Management Association.

Don't think these cyberslacking activities are going unnoticed. More than a third of the 40 million American workers with Internet access have their email and Internet usage under constant surveillance, according to a study from the Privacy Foundation. With a simple software application, your boss can be tapping into your computer to see what you're doing in real time. Whether you're guilty of wasting company time or not, your computer might be under surveillance. You can be monitored without your knowledge—employers are not required to notify you that you're being observed.

A Growing Trend

More employers are monitoring their employees' activities as a result of the low cost of the monitoring technology, a growing percentage of employees using their computers for personal use, and an increase in employees leaking sensitive company information.

Computers leave behind a trail of bread crumbs that can provide employers with all the information they could possibly want about an employee's computer-related activities. For employers, computers are the ultimate spy. There's little that can stop an employer from using surveillance techniques.

There are five basic methods that employers can use to track employee activities. In addition to closed circuit cameras and monitoring phone conversations, employers use:

- Packet sniffers
- Log files
- Desktop monitoring programs

The number of employers who believe that they need these programs and the relatively low cost has resulted in an emerging, multi-million dollar industry called *employee internet management*.

Sniffing Out Misconduct

Computer-network administrators have used packet sniffers for years to monitor their networks and perform diagnostic tests or troubleshoot problems. Essentially, a packet sniffer is a program that can see all of the information passing over the network it is connected to. As data streams back and forth on the network, the program looks at, or "sniffs," each packet. A packet is a part of a message that has been broken into small chunks for transmission on the Internet.

Normally, a computer only looks at packets addressed to it and ignores the rest of the traffic on the network. But when a packet sniffer is set up on a computer, the sniffer's network interface is set to *promiscuous mode*. This means that it is looking at everything that comes through. The amount of traffic largely depends on the location of the computer in the network. A client system out on an isolated branch of the network sees only a small segment of the network traffic, while the main domain server sees almost all of it.

A packet sniffer can usually be set up in one of two ways:

- **Unfiltered**—captures all of the packets
- **Filtered**—captures only those packets containing specific data elements

Packets that contain targeted data are copied onto the hard disk as they pass through. The data on the hard disk can then be analyzed carefully for specific information or patterns.

When you connect to the Internet, you are joining a network that's maintained by your Internet service provider (ISP). The ISP's network communicates with networks maintained by other ISPs to form the foundation of the Internet. A packet sniffer located at one of the servers of your ISP would potentially be able to monitor all of your online activities, such as:

- Which Web sites you visit
- What you look at on the site
- Whom you send email to
- What's in the email you send
- What you download from a site
- What streaming events you use, such as audio, video, and Internet telephony

From this information, employers can determine how much time a worker is spending online and if that worker is viewing inappropriate material.

Desktop monitoring programs work differently than packet sniffers. They can actually monitor every single action you take with your computer.

Watching Every Keystroke

Every time you provide some form of input for your computer, whether it's typing or opening a new application, a signal is transmitted. These signals can be intercepted by a desktop monitoring program, which can be installed on a computer. The person receiving the intercepted signals can

Closed Circuit Camera

Phone Monitoring

—Desktop Monitoring Programs
—Log Files
—Packet Sniffers

see each character being typed and can see exactly what the user is seeing on his or her screen.

Desktop monitoring programs have the ability to record every keystroke. When you're typing, a signal is sent from the keyboard to the application you are working in. This signal can be intercepted and either streamed back to the person who installed the monitoring program or recorded and sent back in a text file. The person it's sent back to is usually a system administrator.

Employers can use the desktop monitoring program to read email and see any program that's open on your screen. Desktop replicating software captures the image on the computer screen by intercepting signals that are being transmitted to the computer's video card. These images are then streamed across the network to the system administrator. Some prepackaged programs include an alert system—when a user visits an objectionable Web site or transmits inappropriate text, the system administrator is alerted to these actions.

But employers don't need to install software to track your computer use. There are actually systems built into every computer that make finding out what you've been doing pretty easy.

Log Files

Your computer is full of *log files* that provide evidence of what you've been doing. Log files keep log entries of any action that transpires within a program. Through these log files, a system administrator can determine what Web sites you've accessed, whom you are sending emails to and receiving emails from, and what applications are being used. So, if you're downloading MP3 files, more than likely there's a log file that holds data about that activity.

In many cases, this information can be located even after you've deleted what you thought was all the evidence. Deleting an email or a file doesn't always erase the trail. Here are a few places where log files can be found:

- Operating systems
- Web browsers (in the form of a cache)
- Applications (in the form of backups)
- Email servers

If the hard drives of an employee's computer and a system administrator's computer are connected, a system administrator can view the log files remotely. The administrator has to have access to the drive to check files remotely. Otherwise, a system administrator can check the computer before an employee comes in or after the employee leaves for the day.

With more companies installing monitoring devices and technology, you should be careful the next time you send that email to Mom or check out the latest sale at your favorite online store while you're at work. Your employer could be watching, listening, and recording.

Can I Get Some Privacy?

Simply stated, courts in the United States tend to favor the employer in workplace-surveillance cases.

Under the Electronic Communications Privacy Act, electronic communications are divided into two groups:

- Stored communication
- Communication in transit

Under the law, electronic communication in transit has almost the same level of protection as voice communication, meaning that intercepting it is prohibited. But accessing stored electronic communication, such as email sitting on a server waiting to be sent, is not illegal. The courts have ruled that since the email is not physically traveling anywhere—it's not in transit—it does not have the same level of protection.

The interpretation of the privacy rights involved in electronic communication directly contradicts many laws regarding traditional mailing systems. If the U.S. Postal Service worked this way, no one would be allowed to open your mail as long as it was being carried to your mailbox. However, the second it is placed in the mailbox and stops moving, your neighbors would be free to come over, open, and read your mail. This, of course, is not how laws regarding the postal system work. In fact, it's illegal to tamper with someone else's mail.

chapter ten

THE INTERNET AND RADIO

How **INSTANT MESSAGING** Works

There is no doubt that the Internet has changed the way we communicate. For many of us, email has replaced traditional letters and even telephone calls as the choice for correspondence. Every day, billions of email messages are sent. Email has been the most rapidly adopted form of communication ever known. In less than two decades, it has gone from obscurity to mainstream dominance. However, in our fast-paced world, sometimes even the rapid response of email is not fast enough. For those who find email too slow, instant messaging is the solution.

HSW Web Links

www.howstuffworks.com

How Email Works
How Web Servers and the Internet Work
How Encryption Works
How IP Telephony Works
How PCs Work
How Firewalls Work
How Modems Work

With email, you have no way of knowing if the person you are sending a message to is online at that particular moment or not. Also, if you are sending multiple emails back and forth with the same person, you normally have to click through a few steps to read, reply, and send the email. These problems don't exist with instant messaging (IM).

The Birth of Instant Messaging

Before the Internet became popular, a lot of people were already online using bulletin boards and online services. A bulletin board is comparable to a single, isolated text Web site that you reach using a modem. Once connected to the board, you normally use a series of menus to navigate through the board's contents. To reach another board, you have to disconnect from the first board and dial up to the other one.

Online services were essentially big bulletin boards that charged for access. Before the Internet, they were the main way for ordinary people to connect and communicate with each other online.

Probably one of the biggest attractions of the bulletin-board and online-service model was the community that built up around the boards or services. Some online service providers enabled their users to talk in real-time with each other while they were online through the use of chat rooms and instant messages. A chat room is software that allows a group of people to type in messages that are seen by everyone in the room, while instant messages are basically a chat room for just two people.

Instant messaging really exploded on the Internet scene in November 1996. That's when Mirablis (a company founded by four Israeli programmers) introduced ICQ, a free instant-messaging utility that anyone could use.

I Seek You

ICQ, which is shorthand for the phrase *I seek you,* is a real-time tool that uses a software application, called a *client,* which resides on your computer. The client communicates

ICQ diagram.

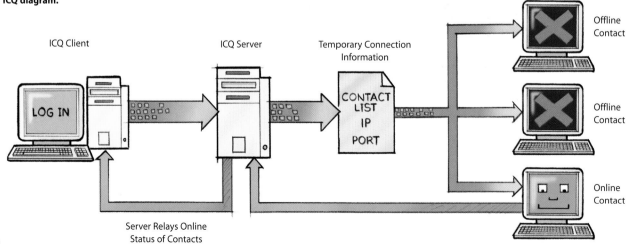

ICQ Client

ICQ Server

Temporary Connection Information

LOG IN

CONTACT LIST IP PORT

Offline Contact

Offline Contact

Online Contact

Server Relays Online Status of Contacts

with an ICQ server whenever you are online and the client is running.

Here's exactly what happens with ICQ:

1) You go to the download page (www.icq.com) for ICQ and get a copy of the free software client for your computer, install the software, and open the client.

2) The client tries to connect to the ICQ server. It uses a proprietary protocol, such as ICQ v5, for communication. A *protocol* simply specifies how the conversation will occur between the client and the server. The protocol for a human phone conversation is that the caller dials the number, the callee hears the phone ring, and the callee picks up the phone and says "hello." Then the conversation starts. The ICQ client and server have a standard protocol that they follow, too.

3) Once the client is connected to the server, you can enter your name and password to log in to the server.

4) The client sends the server the connection information (the IP address and number of the port assigned to the ICQ client) of the computer you are using. It also provides the user with the names of everyone in your ICQ contacts list.

5) The server creates a temporary file that has the connection information for you and the list of your contacts. It then checks to see if any of the users in your contact list are currently logged in.

6) If the server finds any of your contacts logged in, it sends a message back to the ICQ client on your computer with the connection information for that user. The ICQ server also sends your connection information to the people in your contact list that are signed on.

7) When your ICQ client gets the connection information for a person in your contact list, it changes the status of that person to "online."

8) You click on the name of a person in your contact list who is online and a window opens that you can enter text

into. You enter a message and click "Send" to communicate with that person.

9) Because your ICQ client has the IP address and port number for the computer of the person that you sent the message to, your message is sent directly to the ICQ client on that person's computer. In other words, the ICQ server is not involved at this point. All communication is directly between the two ICQ clients.

10) The other person gets your instant message and responds. The ICQ window that each of you see on your respective computers expands to include a scrolling dialog of the entire conversation.

11) When the conversation is complete, you close the message window.

12) Eventually, you go offline and exit ICQ. When this happens, your ICQ client sends a message to the ICQ server to terminate the session. The ICQ server sends a message to the ICQ client of each person on your contact list who is currently online to indicate that you have logged off. Finally, the ICQ server deletes the temporary file that contained the connection information for your ICQ client. In the ICQ clients of your contacts who are online, your name moves to the "offline" status section.

While some of the details vary between utilities, the basic steps outlined above for ICQ apply to all other instant messaging utilities on the market today. ICQ is still very popular. In fact, online service provider AOL acquired Mirablis in June 1998, and ICQ became part of the suite of online services that AOL owns.

It is important to note that instant messaging is not considered a secure way to communicate. Messages and connection information are maintained on servers controlled by the provider of the IM utility that you use. Most utilities do provide a certain level of encryption, but they are not so secure that you should send any confidential information through the system. Cases have been reported of IM user logs being captured and used by nefarious individuals.

Instant Messaging Features

Most of the popular instant-messaging programs provide a variety of features:

- **Instant messages**—Send notes back and forth with a friend who is online.

- **Chat**—Create your own custom chat room with friends or coworkers.

- **Web links**—Share links to your favorite Web sites.

- **Images**—Look at an image stored on your friend's computer.

- **Sounds**—Play sounds for your friends.

- **Files**—Share files by sending them directly to your friends.

- **Talk**—Use the Internet instead of a phone to actually talk with friends.

- **Streaming content**—Real-time or near real-time stock quotes and news.

How **ENCRYPTION** Works

The Internet has excited businesses and consumers alike with its incredible growth and its promise of changing the way we live and work. But a major concern has been just how secure the Internet is, especially when you're sending sensitive information over the Web.

⚙ **HSW Web Links**

www.howstuffworks.com

How Web Servers Work
How Internet
 Infrastructure Works
How Virtual Private
 Networks Work
How Credit Cards Work
How E-commerce Works

Let's face it: There's a whole lot of information that we don't want other people to see, such as credit card information, social security numbers, private correspondence, and sensitive company information.

One sure way to make information secure is to lock it in a vault. This doesn't work very well, however, if you want to send that information to someone else. While it's in transit, it's vulnerable. *Encryption*—the coding of information into an unreadable form—is one of the best ways to protect the information as it moves from point A to point B. With encryption, only the person (or computer) with the key can decode the information.

In the Key of . . .

Computer encryption is based on the science of *cryptography*, the use of codes. Cryptography has been used throughout history. Before the digital age, governments were the biggest users of cryptography, particularly for military purposes. The existence of coded messages has been verified as far back as the Roman Empire. Most forms of cryptography in use these days rely on computers

because computers can create complex codes that are nearly impossible to crack.

Most computer encryption systems belong in one of two categories:

- Symmetric-key encryption
- Public-key encryption

Symmetric Key

Symmetric-key encryption requires that you know which computers will be talking to each other ahead of time, so you can install a secret encoding and decoding key on each one. The sending computer uses the key to encode the message, and the receiving computer uses the same key to decode the message.

Here is an extremely simple form of a symmetric key system. You create a coded message to send to a friend in which you substitute each letter with the letter that is two down from it in the alphabet. So, *A* becomes *C, B* becomes *D,* and so on. You have already told a trusted friend that the code is "shift by 2." Your friend gets the message and decodes it without any problem. Anyone else who sees the message will see nonsense, but if they somehow intercept or get hold of the key, they can decode the message, too. The need to transmit the key around makes symmetric key systems vulnerable.

Public Key

A public key system uses two separate keys—one public and available to everyone, and one private and known only to the receiving machine. To transmit a message, the sender encodes the message using the public key and then sends it. Only the receiving machine, with its private key, is able to decode the message.

To implement public-key encryption on a large scale, such as a secure Web server might need, requires a different approach. This is where digital certificates come in. A digital certificate is basically a bit of information

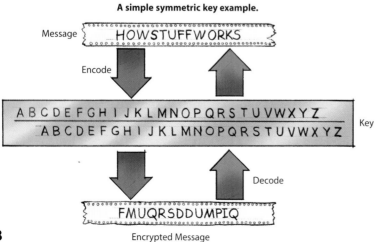

A simple symmetric key example.

Message   HOWSTUFFWORKS

Encode

A B C D E F G H I J K L M N O P Q R S T U V W X Y Z
A B C D E F G H I J K L M N O P Q R S T U V W X Y Z   Key

Decode

FMUQRSDDUMPIQ

Encrypted Message

that says that the Web server is trusted by an independent source, known as a *certificate authority.* The certificate authority acts as a middleman that both computers trust. It confirms that each computer is in fact who it says it is, and then provides the public keys of each computer to the other so that messages can flow both ways.

In your browser, you can tell when you are using a secure protocol in a couple of different ways. You will notice that the *http* in the address line is replaced with *https,* and you may see a small padlock in the status bar at the bottom of the browser window.

Public-key encryption can take a lot of computing power, so most systems use a combination of public-key and symmetric-key systems. When two computers initiate a secure session, one computer creates a symmetric key and sends it to the other computer using public-key encryption. The two computers can then communicate using symmetric-key encryption. Once the session is finished, each computer discards the symmetric key used for that session. Any additional sessions require that a new symmetric key be created, and the process is repeated.

Are You Authentic?

Encryption is the process of taking all of the data that one computer is sending to another and encoding it into a form that only the other computer will be able to decode. Another process, *authentication,* is used to verify that the information comes from a trusted source. Basically, if information is authentic, you know who created it and you know that it has not been altered in any way since that person created it. These two processes, encryption and authentication, work hand in hand to create a secure environment.

There are several ways to authenticate a person or information on a computer:

- **Password**—A user name and password provides the most common form of authentication. If either the name or the password does not match, then you are not allowed further access.
- **Pass cards**—These cards can range from a simple card with a magnetic strip, similar

A public key example.

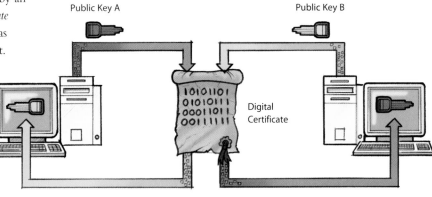

Public Key A

Public Key B

Digital Certificate

Received Message

Received Message

to a credit card, to sophisticated smart cards that have embedded computer chips.

- **Digital signatures**—A digital signature is basically a way to ensure that an electronic document (an email, spreadsheet, or text file, for example) is authentic. If anything at all is changed in the document after the digital signature is attached to it, it changes the value that the digital signature compares to, rendering the signature invalid.

Recently, more sophisticated forms of authentication have begun to show up on home and office computer systems. Most of these new systems use some form of biometric data for authentication. Biometrics use biological information to verify identity. Biometric authentication methods include:

- Fingerprint scan
- Retina scan
- Iris scan
- Face scan
- Voice identification

All of these various processes combine to provide you with the tools you need to ensure that the information you send or receive over the Internet is secure. In fact, information sent over a computer network is often much more secure than information sent any other way. Phones, especially cordless phones, are susceptible to eavesdropping using radio scanners. Traditional mail and other physical mediums often pass through numerous hands on the way to their destination, increasing the possibility of corruption. Understanding encryption, and simply making sure that any sensitive information you send over the Internet is secure, can provide you with peace of mind.

249

How **WEBCAMS** Work

If you've been exploring the Web for any length of time, you've probably run across at least a few Webcams in your travels. Webcams range from the silly to the serious—a Webcam might point at a coffee pot or a space shuttle launch pad. There are business cams, personal cams, private cams, traffic cams … you name it and there's probably a Webcam pointed at it!

HSW Web Links

www.howstuffworks.com

How Web Pages Work
How Web Servers Work
How Email Works
How Domain Name
  Servers Work
How CGI Scripting Works

External Webcams

One problem with using a camera that's connected to a computer via a USB cable is the limited cable length. What if the room you want to capture is at the other end of the house, or outside? In that case, you can purchase a camera that has an external video jack. You have two options when choosing an external camera:

• You can place a standard camera anywhere in the house and run a video cable with RCA jacks on it from the camera to the computer. There are all sorts of places on the Web that sell small pinhole video cameras, either just the cameras themself or cameras embedded in things like clocks and smoke detectors.

• You can avoid the cable by using a radio link.

Have you ever considered setting up a Webcam yourself? You might want to create a silly cam by pointing it at your hamster or putting it inside your refrigerator. But it turns out there are lots of productive uses for Webcams, too. For example:

• You're going to be out of town for a week and want to keep an eye on your houseplants.

• You'd like to be able to check on the baby sitter and make sure everything is okay while you're at work.

• You'd like to know what your dog does in the back yard all day long.

• You want to let the grandparents watch the new baby during naptime.

If there is something that you would like to monitor remotely, a Webcam makes it easy.

The Basic Idea

Webcams, like most things, range from simple to complex. A simple Webcam consists of a digital camera attached to your computer. Cameras like these have dropped well below $100 and they are easy to connect through a universal serial bus (USB) port. A piece of software connects to the camera and grabs a frame from it periodically. For example, the software might grab a still image from the camera once every 30 seconds. The software then turns that image into a normal JPEG file and uploads it to your Web server. The JPEG image can be placed on any Web page.

Putting a standard JPEG image into a standard Web page is straightforward, but it has the disadvantage that your readers must manually refresh the image. But by using a meta tag, a JavaScript function, or a Java applet, you can create a system that automatically refreshes the image for your readers.

If you don't have a Web server, several companies now offer a free place to upload your images, saving you the trouble of

having to set up and maintain a Web server or a hosted Web site.

Using a Hosted Site

If your Web server is hosted elsewhere (for example, because you are paying an ASP to host your Web server), you will need:

• The ability to move frames from your computer to the Web server, normally by File Transfer Protocol (FTP), although several other protocols are gaining favor as well.

• A relatively consistent connection between your computer and the Internet. A modem connection to an ISP is fine if it is something that you keep connected most of the time—you would need a dedicated phone line for your computer or something like a cable modem that is connected all the time.

Monitoring is only one of the things you can do with your Webcam. There are a number of ways to make use of a camera that's connected to your computer. You can even get software that will let you make video phone calls!

How **RADAR GUNS** Work

For many people, speeding is a normal part of daily life. This law-bending is so prevalent that most police cars pack specialized electronic equipment to catch speeders, and many speeders pack their own equipment to detect the police. In the United States, drivers and police are waging a constant war—over velocity!

Measuring vehicle speed with radar is very simple. A basic speed gun is just a radio transmitter and receiver combined into one unit.

Radar Basics

Like sound waves, radio waves have a certain *frequency*—the number of oscillations per unit of time. When the radar gun and the car are both standing still, the echo will have the same wave frequency as the original signal. Each part of the signal is reflected when it reaches the car, mirroring the original signal exactly.

But when the car is moving, each part of the radio signal is reflected at a different point in space, which changes the wave pattern. When the car is moving away from the radar gun, the motion of the car has the effect of stretching out the reflected wave, or lowering its frequency. If the car is moving toward the radar gun, the motion of the car compresses the reflected radio wave. The peaks and valleys of the wave get squeezed together: The frequency increases.

A radar gun can calculate how quickly a car is moving toward it or away from it based on how much the frequency changes. The radar system also has to factor in the police car's own movement. For example, if the police car is going 50 miles per hour and the gun detects that the target is moving away at 20 miles per hour, the target must be driving at 70 miles per hour. If the radar gun determines that the target is not moving toward or away from the police car, then the target is driving at exactly 50 miles per hour.

Lidar Guns

These days, more and more police departments are using laser speed guns rather than conventional radar guns. The basic element in a laser speed gun, also called a *lidar* gun (for *light detection and ranging*), is concentrated light.

The lidar gun clocks the time it takes a burst of infrared light to reach a car, bounce off, and return back to the starting point. By multiplying this time by the speed of light, the lidar system determines how far away the object is. Unlike traditional police radar, lidar does not measure change in wave frequency. Instead, it sends out many infrared laser bursts in a short period of time to collect multiple distances. By comparing these different distance samples, the system can calculate how fast the car is moving. These guns may take several hundred samples in less than half a second, so they are extremely accurate.

If you speed, chances are you'll get caught now and then. Even if you have a top-of-the-line detection and jamming system, the police still might catch you off guard. Also, since police periodically introduce new speed-monitoring technology, a detector might suddenly become outdated. Whenever this happens, the fully equipped speeder has to pick up new equipment.

HSW Web Links

www.howstuffworks.com

How Radar Works
How Radio Works
How the Radio Spectrum Works
How Light Works
How Bats Work

Smile for the Camera!

Police may use handheld lidar systems, just like conventional radar guns, but in many areas the lidar system is completely automated. The gun shines the laser beam at an angle across the road and registers the speed of any car that passes by (the system makes a mathematical adjustment to account for the angle of view).

When a speeding car is detected, the system triggers a small camera that takes a picture of the car's license plate and the driver's face. Since the automated system has collected all of the evidence the police need, the central office simply issues a ticket and sends it to the speeder in the mail.

251

How **INTERNET PROTOCOL (IP) TELEPHONY** Works

If you regularly make long-distance phone calls, chances are you've already used IP telephony without even knowing it. IP telephony, known in the industry as voice-over IP *(VoIP), is the transmission of telephone calls over a data network like one of the many networks that make up the Internet. While you probably have heard of VoIP, what you may not know is that many traditional telephone companies are already using it in the connections between their regional offices.*

HSW Web Links

www.howstuffworks.com

How Internet
 Infrastructure Works
How Web Servers Work
How Cell Phones Work
How Telephones Work
How Speakers Work

Telephone networks currently rely on something called *circuit switching*. Basically, what happens is that when a call is made between two parties, the connection is maintained for the entire duration of the call. Because you are connecting two points in both directions, the connection is a circuit.

But the Internet works differently. Your Internet connection would be a lot slower if it maintained a constant connection to the Web page you were looking at. Instead of simply sending and retrieving data as you need it, the two computers involved in the connection would pass data back and forth the whole time, whether the data was useful or not. A system like that just isn't efficient. Instead, data networks use an information-exchanging method called *packet switching*.

Bandwidth usage chart.

Packet Switching

While circuit switching keeps the connection open and constant, packet switching opens the connection just long enough to send a small chunk of data—the *packet*—from one system to another. What happens is this: The sending computer chops data into these small packets, with an address on each one telling the network where to send it. When the receiving computer gets the packets, it reassembles them into the original data.

Packet switching is very efficient. It minimizes the time that a connection is maintained between two systems, which reduces the load on the network. It also frees up the two computers that are communicating with each other so that they can accept information from other computers as well.

VoIP technology uses this packet-switching method because it has several advantages over circuit switching. For example, packet switching allows several telephone calls to occupy the amount of space occupied by only one in a circuit-switched network. Using Public Switched Telephone Network (PSTN) technology, a 10-minute phone call would consume 10 full minutes of transmission time at a cost of 128 kilobits per minute (Kbps). With VoIP, that same call may occupy only 3.5 minutes of transmission time at a cost of 64 Kbps, leaving another 64 Kbps free for that 3.5 minutes, plus an additional 128 Kbps for the remaining 6.5 minutes. Based on this simple estimate, another three or four calls could easily fit into the space used by a single call under the conventional system. And this example doesn't even factor in the use of data compression, which further reduces the size of each call.

Probably one of the most compelling advantages of packet switching is that data networks already understand the technology. By migrating to this technology, telephone networks immediately gain the ability to communicate the way that computers do. Of course, having the ability to communicate and understanding the methods of communication are two very different things. For telephones to communicate with each other and with other devices, such as computers, over a data network, they need to speak a common language called a *protocol*.

Take a Little SIP

There are two major protocols being used for VoIP. Both protocols define ways for devices to connect to each other using VoIP. Also, they include specifications for audio codecs. A *codec*, which stands for *coder-decoder*, converts an audio signal into a compressed digital form for transmission and back into an uncompressed audio signal for replay.

The first protocol is H.323, a standard created by the International Telecommunications Union (ITU). H.323 is a comprehensive and very complex protocol. It provides specifications for real-time, interactive videoconferencing; data sharing; and audio applications such as IP telephony. Actually a suite of protocols, H.323 incorporates many individual protocols that have been developed for specific applications.

An alternative to H.323 emerged with the development of Session Initiation Protocol (SIP) under the auspices of the Internet Engineering Task Force (IETF). SIP is a much more streamlined protocol, developed specifically for IP telephony. Smaller and more efficient than H.323, SIP takes advantage of existing protocols to handle certain parts of the process. For example, Media Gateway Control Protocol (MGCP) is used by SIP to establish a gateway connecting to the PSTN system.

Call Me

There are four ways that you might talk to someone using VoIP. If you have a computer or a telephone, you can use at least one of these methods without buying any new equipment:

- **Computer-to-computer**—This is certainly the easiest way to use VoIP. You don't even have to pay for long-distance calls. There are several companies offering free or very low-cost software that you can use for this type of VoIP. All you need is the software, a microphone, speakers, a sound card, and an Internet connection—preferably a fast one like you would get through a cable or DSL modem. Except for your normal monthly ISP fee, there is usually no charge for computer-to-computer calls, no matter the distance.

- **Computer-to-telephone**—This method allows you to call anyone from your computer. Like computer-to-computer calling, it requires a software client. The software is typically free, but the calls may have a small per-minute charge.

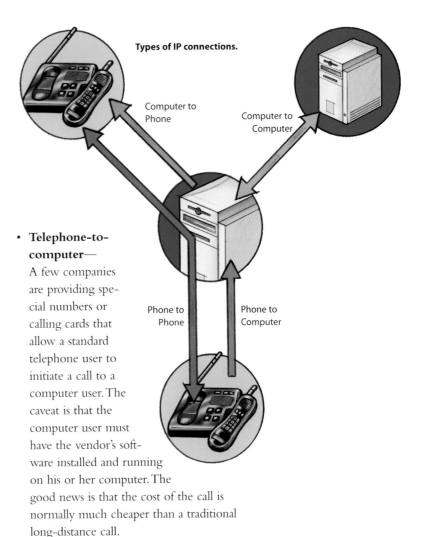

Types of IP connections.

- **Telephone-to-computer**—A few companies are providing special numbers or calling cards that allow a standard telephone user to initiate a call to a computer user. The caveat is that the computer user must have the vendor's software installed and running on his or her computer. The good news is that the cost of the call is normally much cheaper than a traditional long-distance call.

- **Telephone-to-telephone**—Through the use of gateways, you can connect directly with any other standard telephone in the world. To use the discounted services offered by several companies, you must call in to one of their gateways. Then you enter the number you wish to call, and the company connects you through its IP-based network. The downside is that you have to call a special number first. The upside is that the rates are typically much lower than standard long distance.

Although it will take some time to happen, you can be sure that, eventually, all of the circuit-switched networks will be replaced with packet-switching technology. IP telephony just makes sense, in terms of both economics and infrastructure requirements. More and more businesses are installing VoIP systems, and the technology will continue to grow in popularity as it makes its way into our homes.

253

How **RADAR** Works

Radar is something that is in use all around us, although it is normally invisible. Air traffic control uses radar to track planes, both on the ground and in the air, and to guide planes in for smooth landings. Police use radar to detect the speed of passing motorists. NASA uses radar to map the earth and other planets, to track satellites and space debris, and to help with things like docking and maneuvering. The military uses it to detect the enemy and to guide weapons. Meteorologists use radar to track storms, hurricanes, and tornadoes. You even see a form of radar at many grocery stores when the doors open automatically. Obviously, radar is an extremely useful technology!

HSW Web Links

www.howstuffworks.com

How Radio Works

How Burglar Alarms Work

How Radar Detectors Work

How the Radio Spectrum Works

How a U.S. Spy Plane Works

When people use radar, they are usually trying to do one of three things:

- Detect the presence of an object at a distance.
- Detect the speed of an object.
- Map something.

All three of these activities can be accomplished using two things you may be familiar with from everyday life: echo and Doppler shift. These two concepts are easy to understand in the realm of sound because your ears hear echo and Doppler shift every day. Radar makes use of the same techniques using radio waves.

Echo and Doppler Shift

Echo is something you experience all the time. If you shout into a well or a canyon, the echo comes back a moment later. The echo occurs

Higher Frequency Sound

Actual Sound Created by Vehicle

Lower Frequency Sound

Moving Vehicle

The Doppler shift with sound waves.

because some of the sound waves in your shout reflect off of a surface (either the water at the bottom of the well or the canyon wall on the far side) and travel back to your ears. The length of time between the moment you shout and the moment that

you hear the echo is determined by the distance between you and the surface that creates the echo.

Doppler shift is also common. Doppler shift occurs when sound is generated by, or reflected off of, a moving object. Doppler shift in the extreme creates sonic booms. To understand Doppler shift, consider a moving car. Say a car is coming toward you at 60 mph (about 100 kph) and its horn is blaring. You will hear the horn playing one note as the car approaches, but when the car passes you the sound of the horn will suddenly shift to a lower note. It's the same horn making the same sound the whole time. The change you hear is caused by Doppler shift.

Here's what's happening. The speed of sound through the surrounding air is fixed—for the sake of this discussion, say it's 600 mph. Imagine that the car is standing still, it is exactly 1 mile away from you, and it toots its horn for exactly 1 minute. The sound waves from the horn will propagate from the car toward you at a rate of 600 mph. What you will hear is a 6-second delay (while the sound travels 1 mile at 600 mph) followed by exactly 1 minute's worth of sound.

Now, suppose the car is moving toward you at 60 mph. It starts from a mile away and toots it's horn for exactly 1 minute. You will still hear the 6-second delay. However, the sound will only play for 54 seconds. That's because the car will be right next to you after 1 minute, and the sound at the end of the minute gets to you instantaneously. The car (from the driver's perspective) is still blaring its horn for 1 minute. Because the car is moving, however, the minute's worth of sound gets packed into 54 seconds from your perspective. So, the same number of sound waves is packed into a smaller amount of time. Therefore, the soundwaves' frequency is increased, and the horn's tone sounds higher to you. As the car passes you and moves away, the process is reversed and the sound expands to fill more time. Therefore, the tone is lower.

You can combine echo and Doppler shift in the following way. Say you send out a loud sound toward a car moving toward you. Some of the sound waves will bounce off the car (an echo). Because the car is moving toward you, however, the sound waves will be compressed. Therefore, the sound of the echo will have a higher pitch than the original sound you sent. If you measure the pitch of the echo, you can determine how fast the car is going.

Understanding Radar

We know that the echo of a sound can be used to determine how far away something is, and we also know that the Doppler shift of the echo can be used to determine how fast something is going. It is therefore possible to create a "sound radar," and that is exactly what sonar (sound navigation and ranging) is. Submarines and boats use sonar all the time for navigation and locating targets. A submarine's active sonar system emits pulses of sound waves that travel through the water, reflect off an object and return to the ship. By knowing the speed of sound in water and the time for the sound wave to travel to the target and back, the on-board computers can quickly calculate the distance between the submarine and the target. You could use the same principles with sound in the air, but sound in the air has a couple of problems:

- Sound doesn't travel very far—maybe a mile at the most.
- Almost everyone can hear sounds, so sound radar would definitely disturb the neighbors (although you can eliminate most of this problem by using ultrasound instead of audible sound).

- Because the echo of the sound would be very faint, it is likely that it would be hard to detect.

Because of these problems, radar uses radio waves instead of sound. Radio waves travel far, are invisible to humans, and are easy to detect even when they are faint. For example, consider a typical radar set designed to detect airplanes in flight. The radar set turns on its transmitter and shoots out a short, high-intensity burst of high-frequency radio waves. The burst might last a microsecond. The radar set then turns off its transmitter, turns on its receiver, and listens for an echo. The radar set measures the time it takes for the echo to arrive, as well as the Doppler shift of the echo.

Radio waves travel at the speed of light, roughly 1,000 feet (300 meters) per microsecond; so if the radar set has a good high-speed clock, it can measure the distance of an airplane very accurately. Using special signal-processing equipment, the radar set can also accurately measure the Doppler shift and determine the speed of the airplane.

In ground-based radar, there's a lot more potential interference than in air-based radar. When a police radar shoots out a pulse, it echoes off of all sorts of objects—fences, bridges, mountains, buildings. The easiest way to remove all this sort of clutter is to filter it out by recognizing that the clutter is not Doppler-shifted. A police radar looks only for Doppler-shifted signals, and because the radar beam is tightly focused it hits only one car.

Transmitted Radio Waves

Radar Unit in Transmitter Mode

Reflected Radio Waves

Radar Unit in Receiver Mode

Cool Facts

Police are now using a laser technique to measure the speed of cars. This technique is called *lidar*, and it uses light instead of radio waves.

255

How **RESTAURANT PAGERS** Work

People love to eat out. On a Friday night, there will be a crowd at any popular restaurant. It used to be that someone would take your name and then yell out or call over an intercom when your table was ready. Some restaurants still use these systems, but many now use restaurant pagers.

HSW Web Links

www.howstuffworks.com

How Radio Works
How Electric Motors Work
How Batteries Work
How Speakers Work
How Semiconductors
 Work

After the host hands you the restaurant pager, you are free to roam around within the immediate vicinity of the restaurant. Eventually, the pager lights up or vibrates, signaling that your table is ready. How does this handy technology work?

What Is a Pager?

A pager is a very simple radio that listens to just one station all of the time. A radio transmitter broadcasts signals over a specific frequency. All of the pagers for that particular network have a built-in receiver that is tuned to the same frequency broadcast from the transmitter. The pagers listen to the signal from the transmitter constantly as long as the pager is turned on.

Each pager has a specific identification sequence, called a *channel access protocol* (CAP) code. The pager listens for its unique CAP code. When it hears the code, it alerts the user and may provide additional information, depending on the pager type.

There are five basic pager types:

- **Beeper**—Provides a basic alert to the user. They're called beepers because the orignial version made a beeping noise, but current pagers in this category vary in the type of alert—using audio signals, light displays and vibration. Many beepers provide a combination of alerts. The majority of restaurant pagers fall into this category.
- **Voice/tone**—Provides the ability to listen to a recorded voice message when you are alerted that you have a page.
- **Numeric**—Provides the ability to send a numeric message, such as a phone number, along with the page alert.
- **Alphanumeric**—Provides the ability to send a text message along with the page alert.

- **Two-way**—Provides the ability to send as well as receive messages.

While regional and national paging networks set up towers to cover large areas, like those used for cell phones, on-site paging systems like those used by restaurants use a small desktop transmitter.

Master of Ceremonies

Operating the pagers used for on-site paging requires a master transmitter. The master transmitter sends out the signal that the pagers are listening to. A good analogy is to consider the master transmitter as a radio station and the pagers as radios tuned into that station.

The actual frequency used by the master transmitter varies among various models and manufacturers. The coverage area can range from a few hundred feet to several miles, depending on the power of the transmitter. To page a customer, the host enters the numeric code for that customer's pager into the master transmitter. The host may also select a specific option, such as the code for "table is ready" or the code for "lost pager."

Most master transmitters display the last several pagers contacted. Some systems can handle up to 10,000 individual pagers, much more than any restaurant should ever need. A popular option is to connect the master transmitter into the telephone system of the restaurant. This allows a host or other member of the restaurant staff to initiate a page from any phone in the system.

Pagers typically run on rechargeable batteries. A recharging station is used so that recharging the pagers is easy. You may notice metal contacts on the top and bottom of the pager. This lets the restaurant stack the pagers a dozen high on the charger and recharge all of them at once.

Restaurant pager recharging stack.

How **REMOTE ENTRY** Works

If you have one of those remote entry devices for your car on your key chain, then most likely there have been a couple questions floating in the back of your head since you first used it: How does it unlock the door from 20 feet away? And how secure is it—can you open someone else's car with it, or can other people get into your car with their remote entry devices?

These little devices let you get in and out of your vehicle securely. The two most common remote entry devices are:

- The fob that goes on your key ring to lock and unlock your car doors (and that may also arm and disarm the car's alarm system)
- The small controller that hangs off your car's sun visor to open and close the garage door

Some home security systems also have remote controls, but these are not so common.

The Basics

The fob that you carry on your key chain or use to open the garage door is actually a small radio transmitter. When you push a button on the fob, you turn on the transmitter and it sends a code to the receiver (either in the car or in the garage). Inside the car or garage is a radio receiver tuned to the frequency that the transmitter is using (300 or 400 MHz is typical for modern systems). The transmitter is similar to the one in a radio-controlled toy.

In the very early days of garage-door openers, around the 1950s, the transmitters were extremely simple. They sent out a single signal, and the garage-door opener responded by opening or closing. As garage-door openers became common, the simplicity of this system created a big problem—anyone could drive down the street with a transmitter and open any garage door! They all used the same frequency and there was no security.

By the 1970s, garage-door openers had gotten slightly more sophisticated. These models had a controller chip and a DIP switch. A DIP switch has eight tiny switches arranged in a small package and soldered to the circuit board. By setting the DIP switches inside the transmitter, you controlled the code that the transmitter sent. The garage door would only open if the receiver's DIP switch was set to the same pattern. This provided some level of security, but not much. There were only 256 ways that the eight switches in a DIP switch could be arranged. That's enough to keep several neighbors from opening each other's doors, but not enough to provide any real security.

Back then, the transmitter consisted of two transistors, a couple of resistors, and not much else. Powered by a 9-volt battery, the two-transistor transmitter is as simple as a radio transmitter gets. It's the same transmitter that you find in a $10 pair of low-power walkie-talkies. Remote entry transmitters have gotten a lot more sophisticated since then.

Modern Security

With the remote entry systems that you find on cars today, security is a big issue. If people could easily open other people's cars in a crowded parking lot at the mall, there would be a real problem. And with the proliferation of radio scanners, you also need to prevent people from capturing the code that your transmitter sends. If they can intercept your

HSW Web Links

www.howstuffworks.com

How Car Alarms Work
How Power Door Locks Work
How Car Computers Work
How Lock Picking Works
How Burglar Alarms Work

code, they can then simply retransmit it to open your car.

If you were to look inside a typical key-ring remote entry controller for a modern car, you would see that everything has been miniaturized. There is a small chip that creates the code that gets transmitted, and a small silver can (about the size of a split pea) that is the transmitter.

The controller chip in any modern controller uses something called a *hopping code* (or *rolling code*) to provide security. A hopping code is a random number that "hops" or "rolls" to a new number every time the controller gets used. A system that uses a 40-bit rolling code would provide about 1 trillion possible codes. Here's how it works:

- The transmitter's controller chip has a memory location that holds the current 40-bit code. When you push a button on your key fob, it sends that 40-bit code along with a function code that tells the car what you want to do (lock the doors, unlock the doors, open the trunk, or whatever).
- The receiver's controller chip also has a memory location that holds the current 40-bit code. If the receiver gets the 40-bit code it expects, then it performs the requested function. If not, it does nothing.
- Both the transmitter and the receiver use the same pseudo-random number generator.

When the transmitter sends a 40-bit code, it uses the pseudo-random number generator to pick a new code, which it stores in memory. On the other end, when the receiver receives a valid code, it uses the same pseudo-random number generator to pick the same new 40-bit code. In this way, the transmitter and the receiver are synchronized. The receiver only opens the door if it receives the code it expects.

- If you're a mile away from your car and accidentally push the button on the transmitter, the transmitter and receiver are no longer synchronized because the transmitter comes up with a new 40-bit code. The receiver solves this problem by accepting any of the next 256 possible valid codes in the pseudo-random number sequence. This way, your three-year-old child could "accidentally" push a button on the transmitter up to 256 times and it would be okay—the receiver would still accept the transmission and perform the requested function. However, if the button is pushed 257 times, the receiver will totally ignore your transmitter. It won't work anymore. You would have to look in the car's owner's manual to find out how to resynchronize.

Given a 40-bit code, four transmitters, and up to 256 levels of look-ahead in the pseudo-random number generator to avoid desynchronization, there is a one-in-a-billion chance of your transmitter opening another car's doors. When you take into account the fact that all car manufacturers use different systems and that the newest systems use many more bits, you can see that it is nearly impossible for any given key fob to open any other car door.

You can see that code capturing will not work with a rolling code transmitter like this. With a rolling code, capturing the transmission is useless. There is no way to predict which number the transmitter and receiver have chosen to use as the next code, so retransmitting the captured code has no effect. With trillions of possibilities, there is also no way to scan through all the codes because it would take years to do that.

How **ELECTRONIC ARTICLE SURVEILLANCE SYSTEMS** Work

You see signs in stores all the time that read "We Prosecute Shoplifters!" Unfortunately, stores have to battle shoplifting every day. Some of the most effective tools retailers use to combat this crime are tag-and-alarm systems known as electronic article surveillance (EAS) systems.

EAS is a technology used to identify articles as they pass through a gated area in a store. If the system detects an unauthorized article, it alerts someone that a crime is in progress.

Tag, You're It!

There are three types of EAS systems commonly used in the retail environment today:

- Radio frequency EAS systems
- Electromagnetic EAS systems
- Acousto-magnetic EAS systems

In each system, retailers attach special tags onto any item that's for sale. Tags range from small disposable paper labels or cards to larger, reusable plastic tags. After a consumer purchases the item, the tag is either deactivated or removed by the cashier. If the tag has not been deactivated or removed, an alarm will sound as the item is carried through the alarmed gates surrounding the door of the store.

The use of EAS systems does not completely eliminate shoplifting. However, experts say, theft can be reduced by 60% or more when a reliable system is used. Even when a shoplifter manages to leave the store with a tagged item, the tag still must be removed—something that is no longer as easy as it once was. For example, some EAS tags contain special ink capsules that will damage the stolen item when the tag is forcibly and illegally removed.

Radio Frequency EAS Systems

Radio frequency (RF) systems are the most widely used EAS systems in the United States today, and RF tags and labels are getting smaller all the time. The system is actually pretty simple. A label—basically a miniature, disposable electronic circuit and antenna—that's attached to a product responds to a specific frequency emitted by a transmitter antenna (usually one side of the gate at the door). An adjacent receiver antenna (the other side of the gate) picks up the response from the label. The receiver processes the label response signal and will trigger an alarm.

The distance between the two gates, or pedestals, can be up to 80 inches. Operating frequencies for RF systems generally range

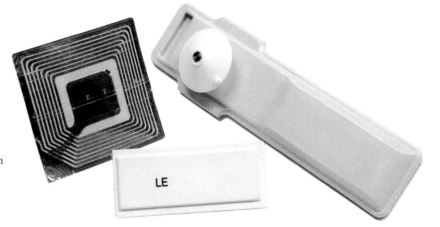

Three types of antitheft devices.

from 2 to 10 MHz (millions of cycles per second). Most of the time, RF systems use a frequency sweep technique in order to deal with different label frequencies.

Sometimes both the transmitter and receiver are combined in one antenna frame—these are called *mono systems* and they can apply pulse or continuous sweep techniques, or a combination of both.

There are many different ways to implement an RF system. The basic idea is that the tag has a spiral antenna etched from thin aluminum bonded to a piece of paper. At the end of the antenna is a small diode or resistor/capacitor (RC) network that causes the tag to emit a radio signal in response to the radio signal it receives. To disarm the tag,

HSW Web Links

www.howstuffworks.com

How UPC Bar Codes Work
How Burglar Alarms Work
How Lock Picking Works
How Restaurant Pagers
  Work

259

Future EAS Technology

Imagine a grocery store where you don't have to unload your grocery cart at the counter—the checkout system could gather the information it needs from each item while all the items remain in the shopping cart. Radio frequency identification (RFID) systems could make this a reality.

RFID tags combine an integrated circuit with an RF antenna to deliver a tag capable of simultaneously storing and processing information about a product while protecting the product from theft. When energized, the tag can emit specific information about the product it's on.

The cash register can quickly record all the information and total the bill. RFID systems are found in some specialty applications now, but the price of these tags needs to come down before you will see them in places like grocery stores.

a strong RF pulse (much stronger than the one that the gates emit) blasts the tag and burns out the diode or RC components. A burned-out tag that's passing through the gates does not emit a signal, so the gates let it pass without an alarm.

Electromagnetic EAS Systems

The electromagnetic (EM) EAS system, which is dominant in Europe, is used by many retail chain stores, supermarkets, and libraries around the world. In this technology, a magnetic, iron-containing strip with an adhesive layer is attached to the merchandise. This strip is not removed at checkout—it's simply deactivated by a scanner that uses a specific highly intense magnetic field. One of the advantages of the EM strip is that it can be reactivated and used at a low cost—this makes it ideal for use in libraries, where material is being checked out and returned at a later date.

The EM system works by applying intensive low frequency magnetic fields generated by the transmitter antenna. The strip picks up the magnetic field, and, in absorbing the field, it creates radio waves with a unique frequency pattern. This pattern is, in turn, picked up by an adjacent receiver antenna. The small signal is processed and will trigger the alarm when the specific pattern is recognized. Because of the weak response of the strip and the low frequency (typically between 70 Hz and 1 kHz) and intensive field required by the EM system, EM antennas are larger than those used by most other EAS systems, and the maximum distance between entry pedestals is 40 inches. Also, because of the low frequency, the strips can be directly attached to metal surfaces. That's why EM systems are popular with hardware stores.

The pattern that the strip emits changes depending on whether the strip is magnetized or demagnetized. You can easily magnetize and demagnetize the strip many times by passing it over a magnetic field, so the strip can be used over and over again.

Acousto-magnetic Systems

The acousto-magnetic (AM) system, which has the ability to protect wide exits and allows for high-speed label application, uses a transmitter to create a surveillance area where tags are detected. The transmitter sends a radio frequency signal (of about 58 kHz) in pulses that energize a tag in the surveillance zone. When the pulse ends, the tag responds, emitting a single frequency signal. A receiver detects the tag signal. A microcomputer checks the tag signal detected by the receiver to ensure it is at the right frequency, time-synchronized to the transmitter, at the proper level, and at the correct repetition rate. If all these criteria are met, the alarm sounds. To deactivate the AM tag, the cashier demagnetizes it.

AM tags are highly *magnetostrictive*, which means that when you put the tag in a magnetic field, it physically shrinks. The higher the magnetic field strength, the smaller the metal becomes. The metal actually shrinks about $1/1000$ of an inch over its full 1.50 inch length.

As a result of driving the tag with a magnetic field, the tag is physically getting smaller and larger. So if it is driven at a mechanically resonant frequency, it works like a tuning fork, absorbing energy and beginning to ring.

When you walk through the gate with an active tag, the transmitter in the gate energizes the material and causes it to resonate. The transmitter then stops. The tag will continue to ring for a short period of time, and the receiver listens for that frequency. If the receiver hears the resonating frequency, it knows there is a tag and sounds the alarm.

chapter eleven

POLICE, MILITARY, AND DEFENSE

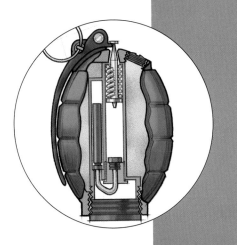

How **MACHINE GUNS** Work

With a machine gun, one soldier can fire hundreds of bullets every minute. The machine gun revolutionized the way nations wage war, to say the least. Given their monumental role in history, it's surprising how simple machine guns really are. These weapons are remarkable feats of precision engineering, but they work using just a few very basic principles.

Like a revolver, a rifle, or just about any other modern firearm, machine guns use cartridges as ammunition. A cartridge contains a primer cap, some propellant material, and a bullet. The gun's bolt, essentially a spring-loaded piston, works to fire the cartridge. As the spring drives the bolt forward, the bolt pushes a cartridge into the chamber (the area in front of the barrel). The trigger extends a firing pin, which hits the primer cap,

machine gun is to use the power of the propellant explosion to drive a little machine that does all these things for you. As long as the shooter holds the trigger back and ammunition is available, the force of the cartridge explosion will keep everything moving so the gun keeps firing. If the shooter releases the trigger, the machine stops.

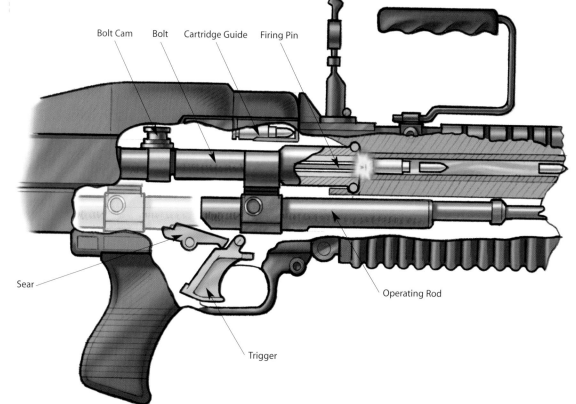

Bolt Cam   Bolt   Cartridge Guide   Firing Pin

Sear

Operating Rod

Trigger

which ignites the propellant material. The propellant explodes, and the gas pressure forces the bullet down the gun barrel.

In a standard rifle, you have to pull the bolt back after each shot to extract the spent cartridge, load a new cartridge, and fire another shot. The basic idea of a

Types of Machine Guns

There are three methods that machine guns commonly use to harness the cartridge's energy: the recoil system, the blowback system, and the gas system.

- **Recoil machine guns**—These guns work on the principle that every action has an equal and opposite reaction. When

262

you propel a bullet down the barrel, the forward force of the bullet causes an opposite force that acts on the gun barrel. In a rifle, this recoil force pushes only the gun back at the shooter. In a recoil machine gun, the barrel itself moves backward, locking briefly against the bolt. The barrel and bolt then separate, but the bolt keeps moving backward, against the spring in the back of the gun. The spring throws the bolt forward again, to fire another shot. While all this is going on, the sliding bolt drives the extraction and reloading mechanisms.

- **Blowback machine guns**—In blowback machine guns, the gas pressure from the propellant explosion pushes the bolt backward. The barrel is fixed in the gun housing, and only the bolt moves. As in a recoil gun, the sliding bolt drives the loading and extraction elements.

- **Gas system machine guns**—This type of machine gun is similar to the blowback system, but it has some addi-

A Well-Oiled Machine

You can think of all the mechanisms in a machine gun as pieces of an automated assembly line. The specific arrangement of elements varies considerably between different gun models, but you can get a good idea of the process by examining one representative model. The diagram shows a simple gas-driven, belt-fed light machine gun.

The main driving elements in this gun are the operating rod and the bolt. Explosive gas pressure in the cylinder pushes the operating rod backward. A slide assembly connects the operating rod to the bolt.

In a belt-fed gun, a feeder pulls linked cartridges into the gun one by one. A cam attached to the top of the bolt puts the feeder in motion. As the bolt moves, the cam slides back and forth in a long, grooved runner, which is linked to a pivoting lever. The lever is attached to a spring-loaded *pawl*—a curved gripper that rests on top of the ammunition belt. As the lever pivots to the left, the pawl moves out and grabs onto

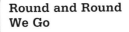

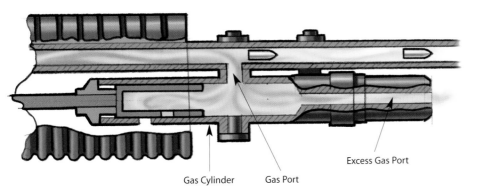

Gas Cylinder Gas Port

Excess Gas Port

tional pieces. The main addition is a narrow piston, which slides back and forth in a cylinder adjacent to the gun barrel. An opening in the barrel allows forward gas pressure from the cartridge to blast into the cylinder (once the bullet has moved down the barrel). This gas drives the piston, which pushes the piston and the attached bolt backward. Instead of using the rear force of the explosion to propel the bolt, the gun harnesses the more powerful forward gas pressure in the barrel.

a cartridge. As the lever moves to the right, the pawl pulls the cartridge belt through the gun.

Inside the gun, the next cartridge in line rests in the cartridge guide. When the bolt slides forward, it catches hold of the cartridge and pushes it out of the guide. The guide prevents the metal links that are attached to the cartridge from moving forward.

The *extractor*—a lip in the locking lug at the front of the bolt— grips the rim at the cartridge base. As it keeps moving forward, the bolt pushes the cartridge against the

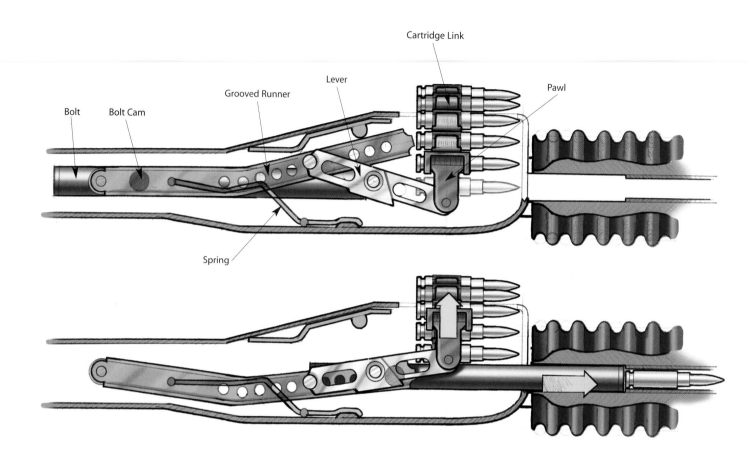

Bolt Bolt Cam Grooved Runner Lever Cartridge Link Pawl

Spring

slanted chambering ramp. The ramp forces the cartridge down, so it snaps into the extractor lip, in line with the chamber. As the bolt pushes into the chamber, cams in the locking lug fit into spiraled grooves in the gun body. The lug rotates in the grooves, locking into the chamber.

When the bolt is locked, the operating rod and the attached slide assembly keep moving. The slide assembly extends the firing pin out of the bolt, so it hits the primer. The primer explodes, then the propellant explodes, and then the bullet zips down the barrel. Once the bullet passes the gas port, the hot gases of the explosion bleed into the cylinder, by way of a gas regulator. These gases drive the operating rod backward.

The operating rod retracts the firing pin and then pushes the bolt to the back of the gun. As it unlocks from the chamber, the locking lug rotates again, in the reverse direction. This twisting motion helps the extractor remove the empty cartridge shell

from the chamber. The bolt and the attached cartridge speed backward, passing the spring-loaded ejector. The ejector knocks the shell loose from the extractor, sending it spinning out of the ejection port. Then the process starts all over again!

A lot of automatic weapons, such as assault rifles, use cartridge magazines rather than ammunition belts. A magazine is a container with a spring-loaded panel at the bottom. The spring pushes a stack of cartridges up to the gun's chambering mechanism, loading them one by one. Magazines are relatively lightweight and easy to use, but unlike belts, they can only hold a small number of rounds.

Whether or not you've ever held a machine gun, or even seen one, this technology has had a profound effect on your life. Machine guns have had a hand in dissolving nations, repressing revolutions, overthrowing governments, and ending wars. The machine gun is one of the most important developments in the history of man.

How **NIGHT VISION** Works

The first thing that comes to mind when you think of night vision is probably a spy movie or action movie you've seen in which someone straps on a pair of night-vision goggles to find someone else in a dark building on a moonless night. And you may have wondered, "Do those things really work? Can you actually see in the dark?" The answer is a most definite yes. With the proper night-vision equipment, you can see a person standing over 200 yards (183 meters) away on a moonless, cloudy night!

Night vision can work in two very different ways, depending on the technology that's used.

- **Image enhancement**—This works by collecting the tiny amount of light, including light in the lower portion of the infrared spectrum, that is present but may be imperceptible to our eyes and amplifying it to the point that we can easily observe the image.
- **Thermal imaging**—This technology operates by capturing light from the upper portion of the infrared spectrum. This light is emitted as heat by all objects. Hotter objects, such as warm bodies, emit more of this light than cooler objects, like trees or buildings, do.

As you can see, both methods use infrared light, which we talk about next.

Seeing Red

In order to understand night vision, you have to understand something about light. The amount of energy in a light wave is related to its wavelength: Shorter wavelengths have higher energy. In visible light, violet has the most energy, and red has the least. Just next to the visible light spectrum on the red side is the infrared spectrum.

Infrared light can be split into three categories:

- **Near-infrared (near-IR)**—Closest to visible light, near-IR has wavelengths that range from 0.7 to 1.3 microns.
- **Mid-infrared (mid-IR)**—Mid-IR has wavelengths ranging from 1.3 to 3 microns. Both near-IR and mid-IR are used by a variety of electronic devices, including remote controls.

- **Thermal-infrared (thermal-IR)**—Occupying the largest part of the infrared spectrum, thermal-IR has wavelengths ranging from 3 microns to over 30 microns.

The key difference between thermal-IR and the other two is that thermal-IR is emitted by an object instead of reflected off it. Infrared light is emitted by an object because of what is happening at the atomic level.

Atoms are constantly in motion. They continuously vibrate, move, and rotate. Even the atoms that make up the chairs that we sit in are moving around. Atoms can be in different states of excitation. In other words, they can have different energies. If we apply a lot of energy to an atom, it can leave what is called the ground-state energy level and move to an excited level. The level of excitation depends on the amount of energy applied to the atom via heat, light, or electricity.

Anything that is alive uses energy, and so do many inanimate items, such as engines and rockets. Energy consumption generates heat. In turn, heat causes the excited atoms in an object to fire off photons. These photons are a form of light, and occupy the thermal-infrared spectrum. Thermal imaging takes advantage of this infrared emission.

The hotter the object, the shorter the wavelength of the infrared photon it releases. An object that is very hot will even begin to emit photons that are visible, glowing red and then moving up through orange, yellow, blue, and eventually white.

The Heat of the Night

Here's how thermal imaging works:

1) A special lens focuses the infrared light that's emitted by all of the objects in view.

HSW Web Links

www.howstuffworks.com

How Light Works
How Lasers Work
How Television Works
How Digital Cameras Work
How Cameras Work
How Military Camouflage Works

2) After the focused light is scanned, the detector elements create a very detailed temperature pattern, called a *thermogram*. It only takes about one-thirtieth of a second for the detector array to obtain the temperature information to make the thermogram. Several thousand points in

Focal Lens

the field of view of the detector array get measured.

3) The thermogram created by the detector elements is translated into electric impulses.

4) The impulses are sent to a signal-processing unit, which is a circuit board with a dedicated chip that translates the information from the elements into data for the display.

5) The signal-processing unit sends the information to the display, where it appears as various colors, depending on the intensity of the infrared emission. The combination of all the impulses from all of the elements creates the image.

Most thermal-imaging devices scan at a rate of 30 times per second. They can sense temperatures ranging from −4°F (−20°C) to 3600°F (2000°C), and can normally detect changes in temperature of about 0.4°F (0.2°C).

There are two common types of thermal-imaging devices:

- **Un-cooled**—This is the most common type of thermal-imaging device. The infrared-detector elements are contained in a unit that operates at room temperature. This type of system is completely quiet, activates immediately, and has the battery built right in.

- **Cryogenically cooled**—More expensive and more susceptible to damage from rugged use, these systems have the elements sealed inside a container that cools them to below 32°F (0°C). The advantage of such a system is the incredible resolution and sensitivity that result from cooling the elements. Cryogenically-cooled systems can see a difference as small as 0.2°F (0.1°C) from more than 1,000 feet (300 meters) away, which is enough to tell at that distance if a person is holding a gun!

While thermal imaging is great for detecting people or working in near-absolute darkness, it creates very odd looking images. It can also miss lots of inanimate objects that all

Infrared Detector

Signal Processing Unit

Video Display

have the same temperature. Most night-vision equipment therefore uses image-enhancement technology.

Image Enhancement

Image-enhancement technology is what most people think of as night vision. Image-enhancement systems are normally called night-vision devices (NVDs). NVDs rely on a special tube, called an *image-intensifier tube*, to collect and amplify infrared and visible light.

Here's how image enhancement works:

1) A conventional lens, called the objective lens, captures ambient light and some near-infrared light.

I Can See Clearly Now

Night-vision equipment can be split into three broad categories:

- **Scopes**—Normally handheld or mounted on a weapon, scopes are monocular (meaning they have one eyepiece). Because scopes are handheld, not worn like goggles, they are good for when you want to get a better look at a specific object and then return to normal viewing conditions.

- **Goggles**—Although goggles can be handheld, they are most often worn on the head. Goggles are binocular (meaning that they have two eyepieces) and may have a single lens or stereo lens, depending on the model. Goggles are excellent for constant viewing, such as moving around in a dark building.

- **Cameras**—Cameras with night-vision technology can send the image to a monitor for display or to a VCR for recording. When night-vision capability is desired in a permanent location, such as on a building or as part of the equipment in a helicopter, cameras are used. Many of the newer camcorders have night vision built right in.

2) The gathered light is sent to the image-intensifier tube. In most NVDs, the image-intensifier tube receives power from two or more batteries.

3) A photo cathode in the image-intensifier tube is used to convert the photons of light energy into electrons.

4) As the electrons pass through the tube, similar electrons are released from atoms in the tube, multiplying the original number of electrons by a factor of thousands through the use of a microchannel plate (MCP) in the tube. An MCP has a tiny, glass disc that has millions of microscopic channels (microchannels) in it, channels that are made using fiber-optic technology. The MCP is contained in a vacuum and, in addition to the glass disc, has metal electrodes on either side of the disc. Each channel is about 45 times longer than it is wide, and each channel works as an electron multiplier.

5) When the electrons from the photo cathode hit the first electrode of the MCP, they are accelerated into the glass microchannels by the 5,000-volt bursts being sent between the electrode pair. As electrons pass through the microchannels, they cause thousands of other electrons to be released in each channel by using a process called *cascaded secondary emission*. Basically, the original electrons collide with the side of the channel, exciting atoms and causing other electrons to be released. These new electrons also collide with other atoms, creating a chain reaction that results in thousands of electrons leaving the channel even though only a few entered.

6) At the end of the image-intensifier tube, the electrons hit a screen coated with phosphors. These electrons maintain their position in relation to the channel they passed through, which provides a perfect image on the screen, since the electrons stay in the same alignment as the original photons. The energy of the electrons causes the phosphors to reach an excited state and release photons. These phosphors create the green image on the screen that has come to characterize night vision.

7) The green phosphor image is viewed through another lens, the ocular lens, that allows you to magnify and focus the image. The NVD may be connected to an electronic display, such as a monitor, or the image may be viewed directly through the ocular lens.

The original purpose of night vision was to locate enemy targets at night. It is still used extensively by the military for that purpose, as well as for navigation, surveillance, and targeting. Police and their security officers often use both thermal-imaging and image-enhancement technology, particularly for surveillance. Hunters and nature enthusiasts use NVDs to maneuver through the woods at night.

Image Intensifier Tube

Photo Cathode

Microchannel Plate

Electrode

Phosphor Coated Screen

Ocular Lens

Did You Know?

One amazing aspect of thermal imaging is its ability to reveal that an area has been disturbed. It can show that the ground has been dug up to bury something, even if there is no obvious sign to the naked eye. Digging will move soil that has different moisture or mineral properties to the surface, and this soil will often have a slightly different temperature than surrounding soil. Law enforcement has used this feature to discover items that have been hidden by criminals, including money, drugs, and even bodies. Also, recent changes to areas such as walls can be seen using thermal imaging, which has provided important clues in several cases.

How **BODY ARMOR** Works

Humans have been wearing armor for thousands of years. Ancient tribespeople fastened animal hides around their bodies when they went out on the hunt, and the warriors of ancient Rome and medieval Europe covered their torsos in metal plates before going into battle. By the 1400s, armor in the Western world had become highly sophisticated. With the right armor, you were nearly invincible.

HSW Web Links

www.howstuffworks.com

How Exoskeletons
  Will Work
How Machine Guns Work
How Flintlock Guns Work
How Force, Power, Torque
  and Energy Work
How Military Pain Beams
  Will Work
How Space Wars Will Work
How Lock Picking Works

Ranking Resistance

In the United States, body armor is tested and rated by the Office of Law Enforcement Standards. Researchers classify any new body-armor design into one of seven categories with I offering the lowest level of protection and VII offering the highest.

Categories I through III mostly designate soft body armor, while higher-ranked suits typically include hard armor elements. The lowest-level body armor protects against small-caliber bullets, which tend to have less force on impact. Some higher-grade body armor can protect against powerful shotgun fire.

That invincibility disappeared, however, with the development of cannons and guns in the 1500s. Bullets have enough energy to penetrate thin layers of metal. You can increase the thickness of traditional armor, but it soon becomes too cumbersome and heavy for a person to wear. It wasn't until the 1960s that engineers developed a reliable bullet-resistant armor that a person could wear comfortably. Unlike traditional armor, this soft body armor is not made out of pieces of metal; it is formed from advanced woven fibers that can be sewn into vests and other soft clothing.

Stopping Bullets

Modern body armor comes in two flavors: hard body armor and soft body armor.

Hard body armor, made out of thick ceramic or metal plates, functions basically the same way as the iron suits worn by medieval knights: It is hard enough that a bullet is stopped or deflected.

Typically, hard body armor offers more protection than soft body armor, but it is much more cumbersome. Police officers and military personnel may wear this sort of protection when there is high risk of attack, but for everyday use they generally wear soft body armor. Soft body armor provides flexible protection that you wear like an ordinary shirt or jacket.

At its heart, a piece of soft bulletproof material is a simple net. To understand how this works, think of a soccer goal. The back of the goal consists of a net formed by many long lengths of tether, interlaced with each other and fastened to the goal frame. When you kick the soccer ball into the goal, the ball has a certain amount of energy, in the form of forward inertia. When the ball hits

the net, it pushes back on the tether lines at that particular point. Each tether extends from one side of the frame to the other, dispersing the energy from the point of impact over a wide area.

The energy is further dispersed because the tethers are interlaced. When the ball pushes on a horizontal length of tether, that tether pulls on every interlaced vertical tether. These tethers in turn pull on all the connected horizontal tethers. In this way, the whole net works to absorb the ball's inertial energy, no matter where the ball hits.

If you were to put a piece of bulletproof material under a powerful microscope, you would see a similar structure. Long strands of fiber are interlaced to form a dense net. A bullet is traveling much faster than a soccer ball, of course, so the net needs to be made from stronger material. The most common material used in body armor is DuPont's Kevlar fiber. Kevlar is lightweight, like a traditional clothing fiber, but it is five times stronger than one piece of steel of the same weight. When interwoven into a dense net, this material can absorb a great deal of energy.

Let Me Be Blunt

When you kick a ball into a soccer goal, the net is pushed back pretty far, slowing the ball down gradually. This is a very efficient design for a goal because it keeps the ball from bouncing out into the field. But bulletproof material can't give this much because the vest would push too far into the wearer's body at the point of impact. Focusing the blunt trauma of the impact in a small area can cause severe internal injuries.

Bulletproof vests have to spread the blunt trauma out over the whole vest so that the force isn't felt too intensely in any one spot.

To do this, the bulletproof material must have a very tight weave. Typically, the individual fibers are twisted, increasing their density and their thickness at each point. To make it even more rigid, the material is coated with a resin substance and sandwiched between two layers of plastic film. Multiple layers of these net-plastic sandwiches provide the protection.

A person wearing body armor will still feel the energy of a bullet's impact, of course, but over the whole torso rather than in a specific area. If everything works correctly, the victim won't be seriously hurt.

Because no one layer can move a good distance, the vest has to slow the bullet down using many different layers. Each net slows the bullet a little bit more, until the bullet finally stops. The material also causes the bullet to deform at the point of the impact. Essentially, the bullet spreads out at the tip, in the same way a piece of clay spreads out if you throw it against a wall. This process, which further reduces the energy of the bullet, is called *mushrooming*.

Keeping in mind that no bulletproof vest is completely impenetrable, and no piece of body armor will make you invulnerable to attack, there's a wide range of body armor available today, and the types vary considerably in effectiveness.

Degree of Protection

Generally speaking, armor with more layers of bulletproof material offers greater protection. Some bulletproof vests, enable you to add layers. One common design is to fashion pockets on the inside or outside of the vest. When you need extra protection, you insert metal or ceramic plates into the pockets.

In the United States, body armor is ranked according to it's effectiveness (see sidebar "Ranking Resistance" for more details). Although it may seem odd that police officers would wear Category I body armor, which will only stop relatively small-caliber bullets, when they could have superior protection from higher-ranked armor—there is a very good reason for this decision. Typically, higher-ranked armor is a

Plastic Film

Kevlar Net

Kevlar Threads

lot bulkier and heavier than lower-ranked armor, which results in several problems:

- An officer has reduced flexibility in bulkier armor, which impedes police work.
- An attacker is more aware of a heavy armored jacket than a thin vest and is more likely to aim at an unarmored part of the body.
- The discomfort of heavier armor makes it more likely that an officer won't wear any protection at all.

We are still a long way from impenetrable armor, but in 50 years, advanced armor will give police officers much more protection when they're walking the beat. Most likely, we will also see an increase in civilian body armor in the years ahead. There is an ever-growing market for comfortable soft body armor that can fit under clothes or even be worn as an outer jacket. With gun violence on the rise, many citizens feel as if they're walking onto a battlefield every day, and they want to dress accordingly.

269

How **MILITARY CAMOUFLAGE** Works

There are many military situations where a soldier wants to be completely invisible. That way, the soldier can sneak up on the enemy completely undetected and launch a surprise attack. Complete invisibility is still a few years off, so in the meantime soldiers use camouflage to hide themselves.

HSW Web Links

www.howstuffworks.com

How Animal Camouflage
  Works
How a U.S. Spy Plane
  Works
How Body Armor Works
How Radar Works
How Night Vision Works
How Military Pain Beams
  Will Work
How the V-22 Osprey
  Works

The function of camouflage is very simple: It hides you and your equipment from the enemy. The most basic camouflage is the sort worn by soldiers on the battlefield. Conventional camouflage clothing has two basic elements that help conceal a person: color and pattern.

Hiding in Plain Sight

Camouflage material is colored with dull hues that match the predominant colors of the surrounding environment. In jungle warfare, camouflage is typically green and brown, to match the forest foliage and dirt. In the desert, military forces use a range of tan colors. Camouflage for snowy climates is colored with whites and grays. To complete the concealment, soldiers paint their face with the same colors.

Camouflage material may have a single color or it may have several similarly colored patches mixed together. Mixed-color camouflage is designed to be visually disruptive. The meandering lines of the mottled pattern help hide the contour—the outline—of the body. Hiding the contour is key. After your brain detects another person (for example, because of movement), it can lock onto the outline and see the person clearly. Once you have spotted a camouflaged person, he stands out, and it seems odd that you didn't see him before. But if the camouflage can once again blur the person's outline, then the person can disappear again. Good camouflage matches the surrounding patterns so well that a person has a hard time distinguishing outlines.

Disguising the Big Stuff

In modern warfare, hiding individual soldiers is often of secondary importance, because the viewer is so far away. Since World War I, opposing forces have used aircraft to seek each other out from the air. In order to hide the big stuff—equipment and fortifications—from these eyes in the sky, ground forces have to use camouflage on a larger scale.

Most U.S. military equipment is colored in dull green and brown colors so it blends in with natural foliage. Additionally, soldiers carry camouflaged netting to throw over military vehicles. Soldiers are also trained to improvise camouflage by gathering natural foliage.

Camouflaging ships is more difficult because they're always floating on a wide background that has a uniform color. In World War I, military forces realized that there was no way to make ships "blend in" with the surroundings, but that there might be a way to make them less susceptible to attack. The dazzle camouflage design, developed in 1917, accomplishes this by obscuring the course of the ship (its direction of travel). Dazzle camouflage resembles a cubist painting, with many colored geometric shapes jumbled together. Like the mottling in camouflage clothing, this design makes it difficult to figure out the actual outlines of the ship and distinguish the starboard side from the port side. If submarine or ship crews don't know which way a ship is moving, it is a lot harder for them to accurately aim a torpedo.

Armed reconnaissance marines wearing camouflage paint, camouflage clothing, and branches.

Harder to Hide

As the technology of camouflage has advanced over the past hundred years, so has the technology of seeing through camouflage. These days, military forces can use thermal imaging to see the heat emitted by a person or piece of equipment. Additionally, they may use radar, image enhancement, satellite photography, and sophisticated listening devices to detect the enemy. Modern camouflage has to contend with this technology.

Some advanced camouflage keeps excess heat from escaping, so the thermal signature does not show in thermal imaging. In ships, the major heat source is engine exhaust. To reduce this thermal emission, modern ships may cool the exhaust by passing it through seawater before it is expelled. Some tanks also have an exhaust-cooling system.

To counteract image enhancement—the amplification of tiny amounts of light (including low-frequency infrared light)—forces have developed sophisticated smoke screens. A heavy cloud of smoke blocks the path of light, giving invisibility to whatever is behind the smoke screen. A similar system uses water nozzles to generate constant fog around a ship, obscuring it from view.

Forces use stealth technology to hide equipment from radar. The surface of a stealth vehicle is made up of many flat planes, interconnected at odd angles. These planes serve to deflect the radar radio waves so they don't bounce straight back to the radar station, but instead bounce off at an angle and travel in another direction. Equipment may also be coated with material that's optimized to absorb radar energy.

As detection and spy equipment continues to advance, military engineers will have to develop more sophisticated camouflage technology. One interesting idea that is already in the works is "smart camouflage"—outer coverings that alter themselves based on a

Fake Forces

Decoys are an interesting alternative to conventional camouflage. Instead of concealing forces and equipment, decoys divert the enemy's attention. In the Battle of Britain, Allied forces set up more than 500 false cities, bases, airfields, and shipyards using flimsy structures that resembled actual buildings and military equipment. These remarkable dummies, built in remote, uninhabited areas, significantly diminished the damage to actual cities and fortifications by causing the Axis forces to waste their time and resources on bogus targets.

This sort of camouflage is still used today to good effect. Some modern decoys have advanced pneumatic systems, which give them the movement you would expect to see in real equipment. Forces also use inflatable dummies that not only resemble tanks and other equipment visually, but also replicate the thermal or radar signature of that equipment.

computer analysis of changing surroundings. But no matter how advanced camouflage gets, the basic strategy will still be the approach used by the first human hunters: Figure out how your enemy sees you, and then mask all of the elements that make you stand out.

Stealth fighter aircraft.

How **GAS MASKS** Work

When most people think about gas masks, what they immediately think of is a military device. However, gas masks—more generically known as respirators—are also an important part of civilian life. Gas masks are used by firefighters and for industrial safety on a daily basis. They protect people against everything from flour dust in a grain elevator to the damaging organic chemicals in paint spray.

HSW Web Links

www.howstuffworks.com

How Biological and
 Chemical Warfare Works
How Nuclear Bombs Work
How Cruise Missiles Work
How Military Camouflage
 Works

Most people picture a gas mask as a tight-fitting plastic or rubber face mask with some sort of filter cartridge. The mask covers the nose and mouth. These are called *half-mask air-purifying respirators*. Depending on the chemical or biological agents in the environment, a half mask may not be sufficient, because the eyes are very sensitive to chemicals and offer an easy entry point for bacteria. If eye protection is required, a full-face respirator is called for. These respirators provide a clear face mask or clear eye pieces that protect the eyes.

Pros and Cons of Air-purifying Respirators

Air-purifying respirators, whether half-mask or full-mask, have two advantages in that they are low in cost and simple to use.

The problem with air-purifying respirators is that any leak in the mask makes them inef-

fective. The leak could come from a poor fit between the mask and the user's face or from a crack or hole somewhere on the mask.

Two other types of respirator systems solve the leak problem. A supplied-air respirator uses the same sort of filter cartridge found in an air-purifying respirator. However, instead of placing the filter directly on the mask and requiring the user's lungs to suck air through it, the filter attaches to a battery-operated canister. The canister uses a fan to force air through the filter, and then the purified air runs through a hose to the mask. The advantage is that the air coming into the mask has positive pressure. Any leak in the mask causes purified air from the canister to escape, but does not allow contaminated air from the environment to enter. Obviously, this is a much safer system, but it has two disadvantages:

- The constant airflow through the filter means that the filter does not last as long as the filter in an air purifying respirator.
- If the batteries die, so do you.

SCBA Gear

The best system is called an SCBA (self-contained breathing apparatus) system. If you've ever seen a firefighter wearing a full-face mask with an air tank on his or her back, then you've seen an SCBA system. The air tank contains high-pressure purified air and is exactly like the tank used by a SCUBA diver. The tank provides constant positive pressure to the face mask. An SCBA provides the best protection, but has the following problems:

- The tanks are heavy and bulky.
- The tanks contain only 30 or 60 minutes of air.
- The tanks have to be refilled using special equipment.
- SCBA systems are expensive.

For firefighters, an SCBA system makes a lot of sense. Smoke is thick, dangerous, and contains an unknown mix of poisonous gases. The fire may consume most or all of the oxygen in the air. The fire engine can carry extra tanks or refilling equipment, and a firefighter spends a limited time in the burning building. For civilians or for soldiers on the battlefield, however, an SCBA system is nearly impossible to manage because of the expense and the limited air time.

The Filter is Key

Because of the problems with SCBA systems, any respirator that you are likely to use will have a filter that purifies the air you breathe. How does the filter remove poisonous chemicals and deadly bacteria from the air?

Any air filter can use one (or more) of three different techniques to purify air:

- Particle filtration
- Chemical absorption or adsorption
- Chemical reaction to neutralize a chemical

Particle filtration is the simplest of the three. If you've ever held a cloth or handkerchief over your mouth to keep dust out of your lungs, you have created an improvised particulate filter. In a gas mask designed to guard against a biological threat, a very fine particulate filter is useful. An anthrax bacteria or spore might have a minimum size of 1 micron. Most biological particulate filters remove particle sizes as small as 0.3 microns. Any particulate filter eventually clogs, so you have to replace it as breathing becomes difficult.

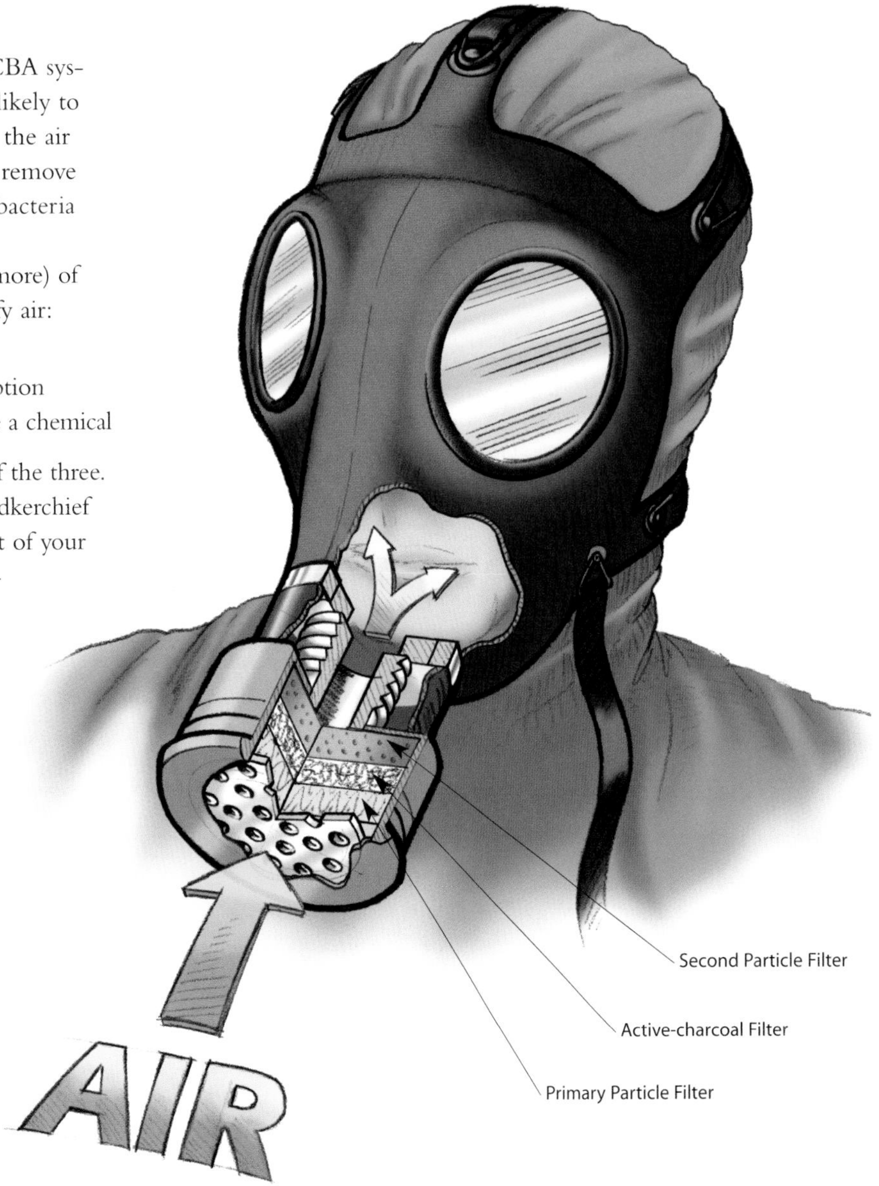

A chemical threat needs to be neutralized in a different way, because chemicals come as mists or vapors that are largely immune to particulate filtration. The most common approach with any organic chemical (whether it be paint fumes or a nerve toxin like sarin) is activated charcoal.

Charcoal is carbon. Activated charcoal is charcoal that has been treated with oxygen to open up millions of tiny pores between the carbon atoms. This is important, because when a material adsorbs something, it attaches to it by chemical attraction. The huge surface area of activated charcoal gives the charcoal countless bonding sites. When certain chemicals pass next to the carbon surface, they attach to the surface and are trapped.

Activated charcoal is good at trapping carbon-based impurities (organic chemicals), as well as things like chlorine. Many other

AIR

Second Particle Filter

Active-charcoal Filter

Primary Particle Filter

chemicals are not attracted to carbon at all—sodium and nitrates, to name a couple—so they pass right through. This means that an activated-charcoal filter will remove certain impurities while ignoring others.

Sometimes, the activated charcoal can be treated with other chemicals to improve its adsorption abilities for a specific toxin.

The third technique used in gas mask filters involves chemical reactions. For example, during chlorine gas attacks in World War I, armies used masks containing chemicals designed to react with and neutralize the chlorine. In industrial respirators, you can choose from a variety of filters, depending on the chemical that you need to eliminate.

How **STUN GUNS** Work

Law enforcement and military forces need non-lethal weapons to subdue angry mobs without racking up civilian casualties. Many citizens who are concerned about personal safety but aren't comfortable with firearms are seeking out reliable "safe weapons." One popular option is to carry a stun gun.

HSW Web Links

www.howstuffworks.com

How Military Pain Beams
 Will Work
How Muscles Work
How Body Armor Works
How Water Blasters Work

Did You Know?

John Cover, the inventor of Taser technology, gave an everlasting nod to stories from his youth in naming this invention. TASER is an acronym for "Thomas A Swift's Electrical Rifle."

Stun gun.

We tend to think of electricity as a harmful force to our bodies. If lightning strikes you or you stick your finger in an electrical outlet, the current can maim or even kill you. But in smaller doses, electricity is harmless. In fact, it is one of the most essential elements in your body. You need electricity to do just about anything.

When you want to tie your shoe, for example, your brain sends a form of chemical electricity down a nerve cell, toward the muscles in your arm. The electrical signal tells the nerve cell to release a neurotransmitter, a communication chemical, to the muscle cells. This chemical tells the muscles to contract in just the right way to put the laces together.

In this way, the different parts of your body use electricity to communicate with one another. This is actually a lot like a telephone system or the Internet. Specific patterns of electricity are transmitted over lines to deliver recognizable messages.

Mixing the Signals

The basic idea of a stun gun is to disrupt this communication system. Stun guns generate high-voltage, low-amperage electrical currents. In simple terms, this means that the charge has a lot of pressure behind it but not that much intensity. When you press the stun gun against an attacker and hold the trigger, the charge passes into the attacker's body. Since it has a fairly high voltage, the charge will pass through heavy clothing and skin. But at around 3 milliamps, the charge is not intense enough to damage the attacker's body unless it is applied for extended periods of time.

The high-voltage signal dumps a lot of confusing information into the attacker's nervous system. This causes a couple of things to happen:

- The charge combines with the electrical signals from the attacker's brain. This is like running an outside current into a phone line: The original signal is mixed in with random noise, making it very difficult to decipher any messages. When these lines of communication go down, the attacker has a very hard time telling his muscles to move, and he may become confused and unbalanced. He is partially paralyzed, temporarily.

- The current may be generated with a pulse frequency that mimics the body's own electrical signals. In this case, the current will tell the attacker's muscles to do a great deal of work in a short amount of time. But the signal doesn't direct the work toward any particular movement. The work doesn't do anything but deplete the attacker's energy reserves, leaving him too weak to move (ideally). All of this activity takes place on a cellular level, so you don't see it happening—the attacker will not twitch and shudder.

Essentially, this is all there is to incapacitating a person with a stun gun—you apply electricity to a person's muscles and nerves. And since there are muscles and nerves all over the body, it doesn't particularly matter where you hit an attacker. Stun-gun effectiveness varies depending on the particular gun model, the attacker's body size, and his determination. It also depends on how long you keep the gun on the attacker.

It's Stunning

Conventional stun guns have a fairly simple design. They are about the size of a flashlight, and they work on ordinary 9-volt batteries. The batteries supply electricity to a voltage-multiply circuit in the stun gun. The circuitry uses multiple transformers, and boosts the voltage thousands of times, typically to between 20,000 and 150,000 volts. This voltage charges a capacitor. The capacitor builds up a charge, and releases it to the electrodes, the "business end" of the circuit.

The electrodes are simply two pins or plates of conducting metal positioned in the circuit with a gap between them. The electrodes have a high voltage difference between them. If you fill the electrode gap with a conductor (say, the attacker's body), the electrical pulses will try to move from one electrode the other, dumping electricity into the attacker's nervous system.

These days, most stun-gun models have two pairs of electrodes: an inner pair and an outer pair. The outer pair, the charge electrodes, is spaced a good distance apart, so current will only flow if you touch an attacker. If the current can't flow across these electrodes, it flows to the inner pair, the test electrodes. These electrodes are close enough that the electric current can leap between them. The moving current ionizes the air particles in the gap, producing a visible spark and crackling noise. This display is mainly intended as a deterrent: An attacker sees and hears the electricity and knows you're armed. Some stun guns rely on the element of surprise, rather than warning. These models are disguised as umbrellas, flashlights, or other everyday objects so you can catch an attacker off guard.

The companies that make stun guns specify that the weapons should be used conservatively,

Flying Tasers

Taser guns work the same basic way as ordinary stun guns, except the two charge electrodes aren't permanently joined to the housing. Instead, they are positioned at the ends of long conductive wires that are attached to the gun's electrical circuit. Pulling the trigger breaks open a compressed gas cartridge inside the gun. The expanding gas builds pressure behind the electrodes, launching them through the air, the attached wires trailing behind. These flying electrodes are also known as tasers.

The electrodes are affixed with small barbs so that they will grab onto an attacker's clothing. When the electrodes are attached, the current travels down the wires into the attacker, stunning him in the same way as a conventional stun gun.

The main advantage of this design is that you can stun attackers from a greater distance, typically 15 to 20 feet (4 to 6 meters). The disadvantage is that you only get one shot—you have to wind up and repack the electrode wires, as well as load a new gas cartridge, each time you fire.

One of the newer stun weapons is the liquid stun gun. These devices work the same way as Taser guns except they use two liquid streams to conduct electricity rather than two extended wires. The gun is hooked up to a tank of highly conductive liquid, typically a mixture of water, salt, and various other conductive elements. When you pull the trigger, electrical current travels from the gun, through the liquid, to the attacker.

Inside a stun gun.

Stun gun components.

only for self-defense or incapacitating an unruly person. Unfortunately, stun guns are commonly abused in many parts of the world.

Amnesty International reports that a number of governments routinely use stun weapons to extract confessions from political prisoners. While stun guns might be relatively safe weapons when used correctly, they can be quite dangerous in the wrong hands.

Activated stun gun.

How **FLAMETHROWERS** Work

Fire is one of the most useful natural phenomena in the world, but it is also one of the most dangerous. Archeological evidence suggests early hunters used fire to flush out their prey, and some groups may have used it to fight other humans. Throughout history, fire has proven to be an extremely effective, even devastating, weapon.

HSW Web Links
www.howstuffworks.com

How Fire Extinguishers
 Work
How Wildfires Work
How Machine Guns Work
How Fire Engines Work

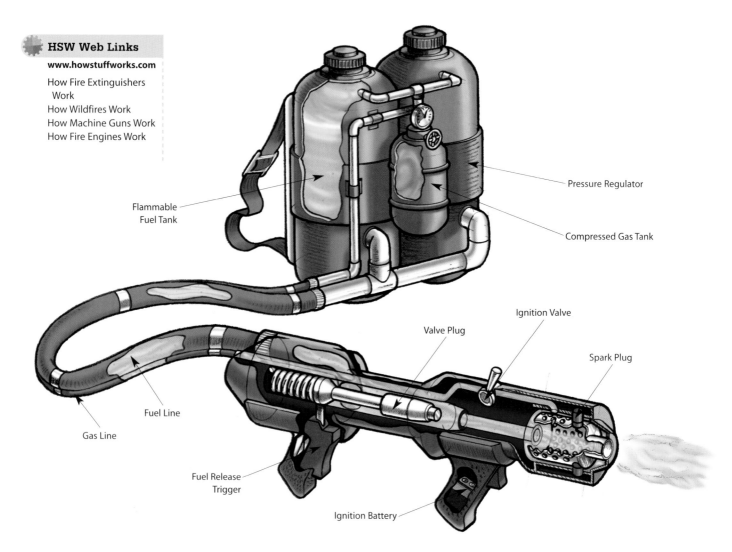

Flammable
Fuel Tank

Pressure Regulator

Compressed Gas Tank

Valve Plug

Ignition Valve

Spark Plug

Fuel Line

Gas Line

Fuel Release
Trigger

Ignition Battery

The basic idea of a flamethrower is to spread fire by launching burning fuel. The modern flamethrower came about in the early twentieth century, but the original idea is actually thousands of years old.

Early Flamethrowers

The earliest flamethrowers, dating roughly from the 5th century BC, were long tubes that contained a burning solid material (such as sulfur or coal). These weapons worked in the same way as a blowgun: Warriors blew into one end of the tube, propelling the burning matter toward their enemies.

In the seventh century, the Byzantine Empire added a more sophisticated flamethrower to its arsenal. This "Greek fire," as it was known, was probably a mixture of liquid petroleum, sulfur, quicklime, and other elements. In any case, it was a flammable, oil-based fluid.

In combat, Byzantine forces would pump this substance from a large reservoir through narrow brass tubes. These tubes concentrated the pressurized liquid into a powerful stream, the same way a hose and nozzle concentrate water into a narrow jet. The soldiers lit a fuse at the end of the brass tubes to ignite the fluid stream as it shot out. The fluid stream carried fire dozens of feet through the air. Because the flammable substance was oil-based, it would burn even when it hit water, making it a particularly effective weapon in naval battles.

The Modern Model

In World War I, the German army reintroduced the flamethrower in a modernized form. By World War II, forces on both sides were using a range of flamethrower weapons on the battlefield.

The most impressive innovation was the handheld flamethrower. This long, gun-type weapon has an attached fuel tank mounted on a backpack. The backpack contains three cylinder tanks. The two outside tanks hold a flammable, oil-based liquid fuel, similar to the material used to make Greek fire. The tanks have screw-on caps, so they can be refilled easily. The middle tank holds a flammable compressed gas (such as butane). This tank feeds gas through a pressure regulator to two connected tubes.

One tube leads to the ignition system in the gun, which will be discussed in a minute. The other tube leads to the two side fuel tanks, letting the compressed gas into the open area above the flammable liquid. The compressed gas applies a great deal of downward pressure on the fuel, driving it out of the tanks, through a connected hose, into a reservoir in the gun.

The gun housing has a long rod running through it that has a valve plug on the end. A spring at the back of the gun pushes the rod forward, pressing the plug into a valve seat. This keeps the fuel from flowing out through the gun nozzle when the trigger lever is released.

When the operator squeezes the trigger lever, it pulls the rod (and the attached plug) backward. With the valve open, the pressurized fuel can flow through the nozzle. A flamethrower like this one can shoot a fuel stream as far as 50 yards (46 meters).

As it exits the nozzle, the fuel flows past the ignition system. Over the years, a variety of ignition systems have been used in flamethrowers. One of the simpler systems was a coil of high-resistance wire. When electrical current passed through these wires, they released a lot of heat, warming the fuel to the combustion point. The gun in the diagram has a slightly more elaborate system.

When the ignition valve is open, compressed flammable gas from the middle cylinder tank on the backpack flows through a long length of hose to the end of the gun. Here it's mixed with air and released through several small holes into the chamber in front of the nozzle. The gun also has two spark plugs, which are powered by a portable battery, positioned in front of the nozzle. To prepare the gun, the operator opens the ignition valve and presses a button that activates the spark plug. The spark plugs then create a small flame in front of the nozzle and the flame ignites the flowing fuel, creating the fire stream.

In World Wars I and II, as well as in the Vietnam War, similar flamethrowers were mounted on tanks. Typically, rotary or piston pumps that were powered directly by the tank engine pumped the fuel in these weapons. With greater pumping power, tank-mounted flamethrowers had better range, and with more fuel tank space, they had a larger ammunition supply.

Military forces continue to use these sorts of weapons today, but the technology is more commonly used for nonviolent civilian purposes. For example, foresters use flamethrowers in prescribed burning and farmers use them to clear fields. Some car enthusiasts install low-power flamethrowers at the back of their cars to release an impressive ball of fire when they take off. Rock stars and other entertainers often include flamethrowers as part of elaborate pyrotechnic displays.

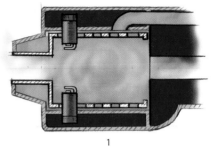

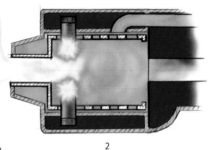

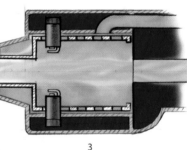

Flamethrower ignition sequence.

Poor Man's Flamethrower

One of the most widely known flamethrower technologies today is actually the simplest. Fire breathers turn their own bodies into flamethrowers by pouring fuel (typically alcohol) into their mouth and holding an ignition system (typically a torch) in front of them. When they spit out the alcohol, the torch ignites it, creating a dazzling stream of fire (only professionals should attempt this—it is extremely dangerous). The technique is definitely low tech, but the operating principle is exactly the same as in the most expensive military flamethrowers!

How **GRENADES** Work

A grenade is a small bomb that comes in a handy package. The thing that makes grenades unique and incredibly useful on the battlefield is the time-delay fuze. A soldier can activate the fuze, throw the grenade, and be perfectly safe when the grenade explodes four seconds later.

HSW Web Links

www.howstuffworks.com

How Landmines Work
How Machine Guns Work
How Flamethrowers Work
How Cruise Missiles Work
How Biological and
  Chemical Warfare Works
How Fireworks Work
How Building Implosions
  Work

Grenades have played a part in warfare for hundreds of years. They were originally developed around AD 1000 by the Chinese. Europeans came up with their own versions in the fifteenth and sixteenth centuries, with mixed results.

The weapon saw a resurgence in the twentieth century with the development of new types of combat. Thanks to reliable ignition systems that delay the explosion, grenades took their place as an indispensable element in modern warfare.

Grenade Basics

A grenade is a small bomb designed for short-range use. The idea behind a basic grenade is very simple: Explosive material inside the grenade ignites to produce a blast of metal fragments. The essential elements of a grenade are explosive material, a heavy metal casing, and an ignition system. There are specialty grenades that don't explode—some spread fire and others release a lot of smoke. Some produce little more than a loud noise and a bright flash of light. Some release toxic gases.

Types of Grenades

Grenades come in two basic types: time delay and impact.

Time Delay

A time-delay grenade explodes after a certain amount of time has passed (generally a few seconds). The diagram shows a conventional design.

The outer shell of the grenade, made of serrated cast iron, holds a chemical fuze mechanism. The fuze is surrounded by a reservoir of explosive material.

A spring-loaded striker sets the fuze off. Normally, the striker is held in place by the striker lever on top of the grenade, which is held in place by the safety pin. To use the grenade, a soldier grips it so the striker lever is pushed up against the grenade body, pulls out the pin, and then tosses the grenade.

With the pin removed, there's nothing holding the lever in position but the soldier's hand. As soon as the soldier throws the grenade, the lever flies off and frees the striker. The spring throws the striker down against the percussion cap. The impact ignites the cap, creating a small spark.

The spark ignites a slow-burning material in the delay element. In about four seconds, the delay material burns all the way through.

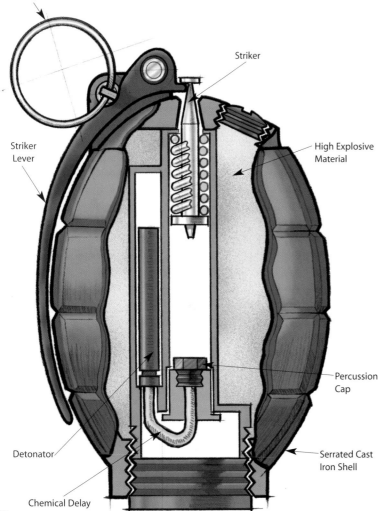

Safety Pin

Striker

Striker Lever

High Explosive Material

Percussion Cap

Detonator

Serrated Cast Iron Shell

Chemical Delay

The end of the delay element is connected to the detonator, a capsule filled with a small amount of explosive. The burning material at the end of the delay ignites the material in the detonator, setting off an explosion inside the grenade. The explosion ignites the explosive material around the sides of the grenade, creating a much larger explosion that blows the grenade apart.

Pieces of metal from the outer casing fly outward like bullets, hitting anybody and anything within range. This sort of grenade may contain additional serrated wire or metal pellets for increased fragmentation damage.

Time-delay grenades have some significant disadvantages. One problem is their unpredictability: In some chemical fuzes, the delay time may vary from two to six seconds. But the biggest problem is that they give the enemy an opportunity to counterattack. If a soldier doesn't time a grenade toss just right, the enemy may pick the grenade up and throw it back before it explodes.

Impact Grenades

Impact grenades work like a bomb launched from an airplane—they explode as soon as they hit their target. Typically, soldiers use a grenade launcher to hurl impact grenades at high speed. Some gun-mounted launchers fire grenades using a blank cartridge. In other launchers, the grenades have their own primer and propellant. Afghan fighters and many other forces around the world use rocket-propelled grenade launchers, once mass produced by the Soviet Union. Like missiles, the grenades used in grenade launchers have a built-in rocket propulsion system.

Impact grenades must be unarmed until they are actually fired. Since they are usually shot from a launcher, they must have an automatic arming system. In some designs, the propellant explosion that drives the grenade out of the launcher triggers the arming mechanism. In other designs, the grenade's acceleration or rotation during its flight arms the detonator.

The grenade has an aerodynamic design, with a nose, a tail, and two flight fins. The impact trigger, which is at the nose of the grenade, consists of a movable, spring-mounted panel with an attached firing pin

facing inward. As in the time-delay grenade, the fuze has a percussion cap and a detonator explosive that ignites the main explosive.

When the grenade is unarmed, several spring-mounted, weighted pins hold the fuze at the tail end of the grenade. The firing pin is not long enough to reach the percussion cap when the fuze is in this position, so if the trigger plate is pressed in accidentally, nothing will happen.

Once it is fired, the grenade begins to spin, like a well-thrown football. This motion is caused by the shape and position of the fins, as well as spiraled grooves inside the barrel of the grenade launcher.

The spinning motion of the grenade generates centrifugal force that pushes the weighted pins outward. When they move far enough out, the pins release the fuze mechanism, and it springs forward toward the nose of the grenade. When the grenade hits the target, the nose plate pushes in, driving the firing pin against the percussion cap. The cap explodes, igniting the detonator explosive, which ignites the main explosive.

In the future, grenade mechanisms will continue to evolve. Already, some modern grenades use an electronic fuze system instead of a mechanical or chemical fuze. In time-delay electronic grenades, the fuze consists of a digital clock and an electrically operated firing pin. The U.S. military has also developed miniature launcher-style grenades with electronic position sensors. With advanced grenade launchers, soldiers can program a grenade to explode after it has traveled a certain distance. In this way, a soldier can pinpoint particular targets, even if the targets are behind barriers.

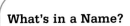

What's in a Name?

The term *grenade* comes from the French word for pomegranate. In the sixteenth century, the French army (as well as other European armies) used round, pomegranate-sized bombs containing large grains of gunpowder, which resembled a pomegranate's seeds. The French army established *Grenadiers*, troops who were trained to lob these grenades toward the enemy line.

When the weapon was reintroduced in the early twentieth century, the name *grenade* was picked up again. Soldiers in World Wars I and II had several other names for the weapons, however, such as *pineapples*, in reference to their shape and bumpy shells.

How **ANTI-PERSONNEL LANDMINES** Work

Landmines are a battlefield tool that can protect an area or slow down the enemy. A mine acts like a very primitive robot guard, exploding whenever someone gets too close. These robots aren't very smart—they can't tell the difference between friend or foe, for example. But they do give armies a very inexpensive way to secure a position.

HSW Web Links

www.howstuffworks.com

How Military Pain Beams Will Work

How Cruise Missiles Work

How Military Camouflage Works

Location and De-mining Techniques

Landmine detection is a slow, methodical process due to the danger involved in locating landmines. Although location technology is improving, the following conventional techniques are still common:

- **Probing the ground**—For many years, the most sophisticated technology used for locating landmines was probing the ground with a stick or bayonet. Soldiers are trained to poke the ground lightly with a bayonet, knowing that just one mistake may cost them their lives.

- **Trained dogs**—Dogs can be trained to sniff out vapors coming from the explosive ingredients inside the landmine.

- **Metal detectors**—Metal detectors are limited in their ability to find mines because many mines are made of plastic with only a tiny bit of metal.

Landmines are explosive devices that explode when triggered by pressure or a tripwire. Soldiers typically plant mines on or just below the surface of the ground. Once triggered, a mine explodes and disables the person or vehicle that came into contact with it.

Landmine Basics

Landmines are easy-to-make, cheap, and effective weapons that can be deployed almost effortlessly over large areas to prevent enemy movements. Mines are typically placed in the ground by hand, but there are also mechanical minelayers that can plow the earth and drop and bury mines at specific intervals.

Mines are often laid in groups to form mine fields, and are designed to prevent the enemy from passing through a certain area—or sometimes to force an enemy through a particular area. An army also will use landmines to slow an enemy until reinforcements can arrive. While more than 350 varieties of mines exist, they can be broken into two categories:

- Anti-personnel mines
- Anti-tank mines

The basic function of both of these types of landmines is the same, but there are a couple of key differences between them. Anti-tank mines are typically larger and contain several times more explosive material than anti-personnel mines. An anti-tank mine has enough explosive to destroy a tank or truck, as well as kill people in or around the vehicle. In addition, more pressure is usually required for an anti-tank mine to detonate. Most of these mines are found on roads, bridges, and large clearings where tanks may travel.

Anti-personnel Mines

Anti-personnel landmines are designed specifically to reroute or push back foot soldiers from a given area. These mines can kill or disable their victims, and are activated by pressure, tripwire, or remote control. There are also smart mines, which automatically deactivate themselves after a certain amount of time. These are the types of mines currently most used by the U.S. military.

Anti-personnel mines fit into three basic categories:

- **Blast**—The most common type of mine, blast mines are buried no deeper than a few centimeters (less than 1 inch) and are generally triggered by someone stepping on the pressure plate, applying about 10 to 35 pounds (5 to 16 kg) of pressure. These mines destroy anything nearby, such as a person's foot or leg.

- **Bounding**—Usually buried with a small part of the igniter protruding from the ground, these mines are pressure or tripwire activated. You may also hear this type of mine referred to as a "Bouncing Betty." When activated, the igniter sets off a propelling charge, lifting the mine about 1 meter into the air. The mine then ignites a main charge, causing injury to a person's head and chest.

- **Fragmentation**—These mines release fragments in all directions, or can be arranged to send fragments in one direction (in which case they're called *directional fragmentation mines*). These mines can cause injury up to 200 meters away and can kill at closer distances. The fragments used in the mines are either metal or glass. Fragmentation mines can be bounding or ground-based.

To understand the varying characteristics of landmines, let's take a look at two landmines developed by the United States Military:

- The M14, a pressure-operated blast mine
- The M16 bounding/fragmentation landmine

M14 Blast Mine

The M14 is a small, cylindrical, plastic-bodied blast mine. It is just 1.57 inches (40 mm) tall and 2.2 inches (56 mm) in diameter. It was originally developed and used by the United States in the 1950s, but it has been used and copied by many nations around the world. This particular anti-personnel mine contains only a small amount of explosive, about 31 grams of tetryl. It is designed to cause damage to people and objects in close proximity.

The M14 comes with a U-shaped safety clip that fits around the pressure plate. In order to activate the M14, the soldier removes the safety clip and rotates the pressure plate from its safety position to its armed position. The letters A (armed) and S (safety) are embossed on the pressure plate. Soldiers simply align an arrow with the A to arm the mine.

After it is armed, any pressure of at least 19.8 pounds (9 kg) causes the mine to detonate. When the proper amount of pressure is applied, it pushes down on the Belleville spring (metal disc spring) underneath the pressure plate. This spring pushes the firing pin onto the detonator, which ignites the main charge of tetryl explosive.

M16 Bounding/Fragmentation Mine

Bounding mines fire up out of the ground and then explode. The M16 is made of three main parts: a mine fuze, a propelling charge to lift the mine, and a projectile contained in a cast-iron housing. It is 7.83 inches (199 mm) tall and 5.24 inches (133 mm) in diameter. The M16 mine contains about 1.15 pounds (521 grams) of trinitrotoluene (TNT) explosive.

The fuze extends through the center of the mine to the bottom, where the propelling charge is located. To arm the mine, a safety pin is removed from the striker on top of the fuze. There are three prongs located on top of the fuze, connected to a spring-loaded wedge. The fuze encloses a percussion cap, a delay element, and a black-powder charge.

The M16 can be detonated in two ways: by applying pressure or by pulling the spring-loaded release pin. Either method causes the

pin to pull out of the fuze, releasing the striker and igniting the percussion cap. The percussion cap fires activating the detonating charge in the bottom of the mine and igniting the delay elements. The mine flies upward about 4 feet (1.2 meters). The burning of the delay elements causes the detonators to explode. The main charge then detonates and releases a shower of metal fragments.

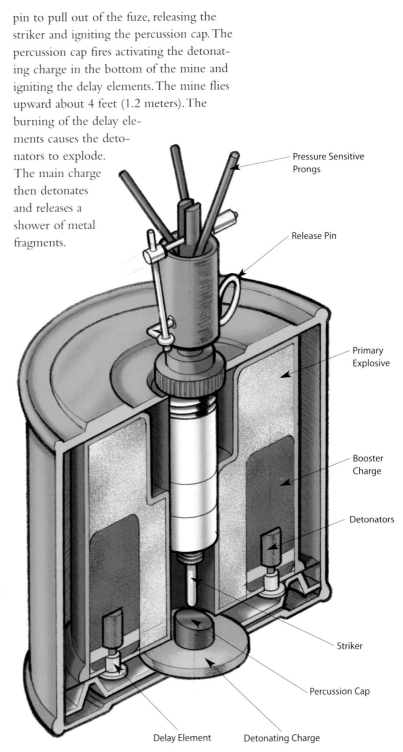

Pressure Sensitive Prongs

Release Pin

Primary Explosive

Booster Charge

Detonators

Striker

Percussion Cap

Delay Element

Detonating Charge

Landmines are useful tools in battle, but they can have unintended consequences long after the war has ended. The mines continue to be functional for many decades, causing injury or death to anyone who comes upon them accidentally. For this reason, international treaties to ban landmines are now being signed.

How **APACHE HELICOPTERS** Work

The Apache helicopter is a revolutionary development in the history of war. It's essentially a flying tank—a helicopter designed to survive heavy attack and inflict massive damage. It can zero in on specific targets, day or night, even in terrible weather. As you might expect, it is a terrifying machine to ground forces.

HSW Web Links

www.howstuffworks.com

How Helicopters Work
How Airplanes Work
How the V-22 Osprey
 Works
How Stinger Missiles Work
How Cruise Missiles Work
How Machine Guns Work
How Night Vision Works

At its core, an Apache works like any other helicopter. It has a main rotor and a tail rotor, which let it fly in any direction or hover in the air. Its twin 1,700 horsepower turboshaft engines give it a little more oomph than your average traffic-copter, but the principles are the same.

You could say, then, that the Apache is simply a high-end helicopter. But that would be like calling James Bond's Aston Martin just a high-end car! The Apache's advanced electronics, armor, and weaponry puts it in an entirely different class.

Controls and Sensors

The Apache cockpit is divided into two sections, one directly behind the other. The pilot sits in the rear section, and the copilot/gunner sits in the front section. As you might expect, the pilot maneuvers the helicopter and the gunner aims and fires the weapons. Both sections of the cockpit include flight and firing controls in case one pilot needs to take over full operation.

The pilot flies the Apache using collective and cyclic controls, similar to ones you would find in any other helicopter. The controls manipulate the rotors using both a mechanical hydraulic system and a digital stabilization system. The digital stabilization system fine-tunes the powerful hydraulic system to keep the helicopter flying smoothly. The stabilization system can also automatically keep the helicopter hovering for short periods of time.

On the Longbow Apache, the most advanced design of Apache helicopters, three display panels provide the pilot with navigation and flight information. These digital displays are much easier to read than traditional instrument dials.

One of the coolest things about the Apache is its sophisticated sensor equipment. The Longbow Apache detects

Apache helicopter preparing for takeoff.

equipment to identify the general class of each potential target. The computer pinpoints these targets on the pilot's and the gunner's display panels.

The pilot and the gunner both use night-vision sensors for night operations. The night-vision sensors work on the forward-looking infrared (FLIR) system, which detects the infrared light released by heated objects.

The pilot's night-vision sensor is attached to a rotating turret on top of the Apache's nose. The gunner's night-vision sensor is attached to a separate turret on the underside of the nose. The lower turret also supports a normal video camera and a telescope, which the gunner uses during the day.

The computer transmits the night vision or video picture to a small display unit in each pilot's helmet. The video display projects the image onto a monocular lens in front of the pilot's right eye. Infrared sensors in the cockpit track how the pilot positions the helmet and relay this information to the turret control system. Each pilot can aim the sensors by simply moving his or her head!

Hellfire Missiles

The Apache's primary weapon, the Hellfire missile, is designed to take out heavily armored ground targets, such as tanks and bunkers. Each missile is a miniature aircraft, complete with its own guidance computer, steering control, and propulsion system. The payload is a high-explosive, copper-lined-charge warhead—a warhead that focuses an explosion in a narrow area. The Hellfire warhead can burn through the heaviest tank armor in existence.

surrounding ground forces, aircraft, and buildings using a radar dome mounted to the mast. The radar dome uses millimeter radio waves that can make out the shape of anything in range. The radar signal processor compares these shapes to a database of tanks, trucks, other aircraft, and

Apache helicopter, cockpit view.

The Apache carries the missiles on four firing rails, attached to pylons. There are two pylons on each wing, and each pylon can support four missiles, which means the Apache can carry as many as 16 missiles at a time. Before launching, each missile receives instructions directly from the helicopter's computer. When the computer transmits the fire signal, the missile sets off the propellant. Once the burning propellant generates about 500 pounds of force, the missile breaks free of the rail. As the missile speeds up, the force of acceleration triggers the arming mechanism. When the missile makes contact with the target, an impact sensor sets off the warhead.

There are two general types of Hellfire missile. The original Hellfire design uses a laser guidance system to hit its mark. In this system, the Apache gunner shines a high-intensity laser beam on the intended target. The laser pulses on and off in a particular coded pattern. Before giving the firing signal, the Apache computer tells the missile's control system the specific pulse pattern of the laser. The missile has a laser seeker on its nose that detects the laser light reflecting off the target. The guidance system calculates which way the missile needs to turn in order to head straight for the reflected laser light and adjusts the flight fins accordingly.

The updated missile design, the Hellfire II, uses a radar seeker rather than a laser seeker. The helicopter's radar locates the target and the missile zeroes in on it. The missile is more likely to find its target, since clouds and obstacles don't block radio waves. Another advantage is that the helicopter can fire the missile and immediately find cover, since it doesn't have to keep a laser pointed on the target.

Other Weapons

An Apache usually flies with Hydra rocket launchers on two of its pylons. Each rocket launcher carries 19 folding-fin 2.75-inch aerial rockets, which are secured in launching tubes. To fire the rockets, the launcher triggers an igniter at the rear end of the tube. The flight fins unfold to stabilize the rocket once it leaves the launcher.

The rockets work with a variety of warhead designs. For example, they might be armed with high-power explosives or with materials that just produce smoke. In one configuration, the warhead delivers several *submunitions,* small bombs that separate from the rocket in the air and fall on targets below.

The gunner engages close-range targets with an M230 30-mm automatic cannon. The cannon is mounted to a movable turret under the helicopter's nose. Typically, the gunner aims the gun using his or her motion-sensitive helmet.

The automatic cannon is a chain-gun design that's powered by an electric motor.

Apache Longbow helicopter with radar dome.

The motor rotates the chain, which slides the bolt assembly back and forth to load, fire, extract, and eject cartridges. This is different from an ordinary machine gun, which uses the force of the cartridge explosion or flying bullet to move the bolt.

The cartridges travel from a magazine above the gun down a feed chute to the chamber. The magazine holds a maximum of 1,200 rounds, and the gun can fire 600 to 650 rounds a minute. The cannon fires high-explosive rounds designed to pierce light armor.

Evasion and Armor

The Apache's first defense against attack is to keep out of range. It is specifically designed to fly low to the ground, hiding behind cover whenever possible. If the pilots pick up radar signals with the onboard scanner, they can activate a radar jammer to confuse the enemy.

The Apache evades heat-seeking missiles by reducing its infrared signature (the heat energy it releases). Its Black Hole infrared suppression system dissipates the heat of the engine exhaust by mixing it with air flowing around the helicopter. The cooled exhaust then passes through a special filter, which absorbs more heat. The Longbow also has an infrared jammer, which generates infrared energy of varying frequencies to confuse heat-seeking missiles.

The Apache is heavily armored on all sides, with a mixture of aluminum and other metals. Some areas are also surrounded by Kevlar soft armor for extra protection. Layers of reinforced armor and bulletproof glass guard the cockpit. According to Boeing, the helicopter's manufact-urer, every part of the helicopter can survive 12.7-mm rounds, and vital engine and rotor components can withstand 23-mm fire.

The area surrounding the cockpit is designed to deform during collision, but the cockpit canopy is extremely rigid. In a crash, the deformation areas work like the crumple zones in a car—they absorb a lot of the impact force, so the collision isn't as hard on the crew. The seats are outfitted with heavy Kevlar armor, which also absorbs the force of impact. With these advanced systems, the crew has an excellent chance of surviving a crash.

Flying an Apache into battle is extremely dangerous, to be sure, but with all its weapons, armor, and sensor equipment, it is a formidable opponent for almost everything else on the battlefield. It is a deadly combination of strength, agility, and firepower.

Hydra rocket launcher.

M230 automatic cannon.

How **WIRETAPPING** Works

Wiretapping occurs all the time in espionage and crime movies. Spies and gangsters know the enemy is listening, so they speak in code over the phone and keep an eye out for bugs. In the real world, we may not think much about wiretapping. Most of the time, we assume our phone lines are secure. And in most cases, they are, but only because nobody cares enough to listen in. If people did want to eavesdrop, they could tap into almost any phone line quite easily.

HSW Web Links

www.howstuffworks.com

How Telephones Work
How Radio Works
How Carnivore Works
How Encryption Works
How Workplace
   Surveillance Works
How Radio Scanners Work
How Cordless Phones
   Work

To learn how wiretapping works, you first have to understand the basics of telephones. If you take a look inside a telephone cord, you'll see how simple phone technology is. When you cut off the outer covering, you'll find two copper wires, one with a green covering and one with a red covering. To hook two phones together, these two wires are all you need.

The copper wires transmit the fluctuating sound waves of your voice as a fluctuating electrical current. The phone company sends this current through the wires.

The varying current travels to the receiver in the phone on the other end of the line and drives that phone's speaker. In ts path through the global phone network, the electrical current is usually translated into digital information so that it can be sent quickly and efficiently over long distances. But if you ignore this step in the process, you can think of the phone connection between you and a friend as one very long circuit that consists of a pair of copper wires and forms a loop. As with any circuit, you can hook up more loads (components powered by the circuit) anywhere along the line. This is what you're doing when you plug an extra phone into a jack in your house.

This is a very convenient system, because it's so easy to install and maintain. Unfortunately, it's also very easy to abuse. The circuit carrying your conversation runs out of your home, through your neighborhood, and through several switching stations before it gets to the phone on the other end. At any point along this path, somebody can add a new load to the circuit in the same

Domestic Phone

Listening Device

To Phone Network

way you can plug a new appliance into an extension cord. In wiretapping, the new load is typically a bug that either records your conversation or transmits it to a receiver.

This is all wiretapping is—connecting a listening device to the circuit that's carrying information between phones.

Getting Your Wires Crossed

One simple sort of wiretap is an ordinary telephone. In a way, you are tapping your own phone line whenever you hook up another phone in your house. This isn't considered wiretapping, of course, since there's nothing secretive about it.

Spies do the same basic thing, but they try to hide the tap from the person they're spying on. The easiest way to do this is to attach a normal phone somewhere along the part of the line that runs outside the house. To configure a phone for tapping, the spy just cuts one of the modular plugs (the part you insert in the jack) off a piece of phone cord so that the red and green wires are exposed. Then, the spy plugs the other end of the wire into the phone and attaches the exposed wires to an accessible exposed point on the outside phone line.

With this connection, the spy can use the subject's line in all of the ways the subject uses it. The spy can hear calls and make calls. Most spies will disable the tap's microphone, however, so it works only as a listening device. Otherwise, the subject would hear the spy's breathing and be alerted to the wiretap.

This sort of wiretap is easy to install, but it has some major drawbacks if you're a spy. First of all, a spy would have to know when the subject is going to use the phone so he or she could be there for the call. Second, a spy would have to stay with the wiretap in order to hear what's going on. Obviously, it's quite difficult to predict when somebody's going to pick up the phone, and hanging around a phone company utility box is not the most covert way to eavesdrop. For these reasons, spies generally use more sophisticated wiretapping technology to eavesdrop on a subject.

Bugs and Tape

One option is to hook up some sort of recorder to the telephone line. This recorder works just like your answering machine—it receives the electrical signal from the phone line and encodes it as magnetic pulses on audio tape. A spy can do this fairly easily with an ordinary tape recorder and some creative wiring. The only problem here is that the spy has to keep the tape recording constantly to pick up any conversations. Since most cassettes only have 30 or 45 minutes of tape on either side, this solution isn't much better than the basic wiretap.

To make it functional, the spy needs a component that will start the recorder only when the subject picks up the phone. Voice-activated recorders, intended for dictation use, serve this function quite well. As soon as people start talking on the line, the recorder starts up. When the line is dead, it turns off again.

Even with this pick-up system, the tape will run out fairly quickly, so the spy will have to keep returning to the wiretap to replace the cassette. In order to stay concealed, spies need a way to access the recorded information from a remote location.

The solution is to install a bug. A bug is a device that receives audio information and broadcasts it through the air as radio waves. Some bugs have tiny microphones that pick up sound waves directly. But a typical wiretapping bug doesn't need its own microphone, since the phone already has one. If the spy hooks the bug up anywhere along the phone line, it receives the electrical current directly. The current runs to a radio transmitter, which transmits the audio signal to a nearby radio receiver (in a delivery van parked outside, for example). The receiver either sends the signal to a speaker or encodes it on a tape.

Often, the spy will hook the bug up to the wires that are actually inside the phone receiver. Since people very rarely look inside their phones, this can be an excellent hiding spot. Of course, if somebody is searching for a wiretap, the spy will be uncovered right away.

In the future, wiretapping probably won't be as easy as connecting a phone to the line outside somebody's house, but it will almost certainly continue in some form or another. Whenever information is transmitted from point to point, there is the possibility that it will be intercepted along the way. This is nearly unavoidable in a global communications system.

Wiretapping Then and Now

By the 1890s, the modern telephone was in widespread use—and so was wiretapping. From that time on, it has been illegal in the United States for an unauthorized person to listen in on somebody else's private phone conversation. In fact, it is even illegal to record your own phone conversation if the person on the other end is not aware that you're recording it.

Historically, the law has not been as strict for the government. In 1928, the United States Supreme Court approved the practice of wiretapping for the police and other government officials, although some states have banned it. In the 1960s and 1970s, this authority was curtailed somewhat. Law enforcement now needs a court order to listen in on private conversations, and this information can be used in court only in certain circumstances.

Additionally, the court order only allows the authorities to listen in on a call for a certain length of time. Even under this tight control, the practice of government wiretapping is highly controversial.

How **LIE DETECTORS** Work

Most of us practice deception on some level in our daily lives, even if it's just telling a friend that his horrible haircut "doesn't look that bad." People tell lies and deceive others for many reasons. Most often, lying is used to avoid trouble with the law, bosses, or authority figures. Sometimes you can tell when someone's lying, but other times figuring out if a person is telling the truth may not be so easy. In these hard-to-tell situations, technology can help.

HSW Web Links

www.howstuffworks.com

How DNA Evidence Works
How Carnivore Works
How Wiretapping Works
How Sweat Works
How Your Heart Works

Polygraphs, commonly called *lie detectors,* are instruments that monitor a person's physiological reactions. These instruments do not, as their nickname suggests, detect lies. They can only detect whether deceptive behavior is being displayed.

Do you think you can fool a polygraph machine and examiner? Let's take a look at how these instruments monitor a person's vital signs.

Man versus Machine

A polygraph instrument is a combination of medical devices that are used to monitor changes occurring in the body. As a person is questioned about a certain event or incident, the examiner looks to see the change in the person's heart rate, blood pressure, respiratory rate, and electro-dermal activity (sweatiness, in this case the sweatiness of the fingers) in comparison to normal levels. Fluctuations may indicate that the person is being deceptive.

The polygraph instrument has undergone a dramatic change in the last decade. For many years, polygraphs were those instruments that you see in the movies with little needles scribbling lines on a single strip of scrolling paper. These are called *analog polygraphs.* Today, most polygraph tests are administered with digital equipment. The scrolling paper has been replaced by sophisticated algorithms and computer monitors.

When you sit down in the chair for a polygraph exam, several sensors and wires are connected to your body in specific locations to monitor your physiological activities. Deceptive behavior is supposed to trigger certain physiological changes that can be

Pneumograph

Polygram Display

Blood-Pressure Cuff

Galvanometers

detected by a polygraph and a trained examiner, who is sometimes called a *forensic psychophysiologist*. This examiner is looking for the amount of fluctuation in certain physiological activities. Here's a list of physiological activities that a typical polygraph machine measures:

- **Respiratory rate**—Two *pneumographs*—rubber tubes filled with air—are placed around the test subject's chest and abdomen. When the chest or abdominal muscles expand, the air inside the tubes is displaced. In an analog polygraph, the displaced air acts on a bellows, an accordion-like device that contracts when the tubes expand. This bellows is attached to a mechanical arm, which is connected to an ink-filled pen that makes marks on the scrolling paper when the subject takes a breath. A digital polygraph also uses the pneumographs, but employs transducers to convert the energy of the displaced air into electronic signals.

- **Blood pressure/heart rate**—A blood-pressure cuff is placed around the subject's upper arm. Tubing runs from the cuff to the polygraph. As blood pumps through the arm it makes sound; the sound affects the air pressure displacing the air in the tubes, which are connected to a bellows, which moves the pen. In digital polygraphs, these signals are also converted into electrical signals by transducers.

- **Galvanic skin resistance**—This is also called electro-dermal activity, and is basically a measure of the sweat on your fingertips. The fingertips are one of the most porous areas on the body, and so are a good place to look for sweat. The idea is that we sweat more when we are placed under stress. Fingerplates, called galvanometers, are attached to two of the subject's fingers. These plates measure the skin's ability to conduct electricity. When the skin is hydrated (as with sweat), it conducts electricity much more easily than when it is dry.

Some polygraphs also record arm and leg movements. As the examiner asks questions, signals from the sensors connected to your body are recorded on a single strip of moving paper.

Polygraph Examiners

Only two people are in the room during a polygraph exam—the person conducting the exam and the person being tested. Because the polygraph examiner is alone in the room with a test subject, the examiner's behavior greatly influences the results of the exam.

The polygraph examiner does several different things:

- Sets up the polygraph and prepares the subject who is being tested
- Asks questions
- Profiles the test subject
- Analyzes and evaluates test data

How the questions are presented can greatly affect the results of a polygraph exam. There are several variables that the examiner has to take into consideration, such as cultural and religious beliefs. Some topics may, by their mere mention, cause a specific reaction in the test subject that could be misconstrued as deceptive behavior. The design of the question affects the way the person processes the information and how he or she responds.

Who Uses Polygraphs?

Polygraphs are limited in their use in the private sector, but they are frequently used by the U.S. government. Many parts of the government, for example, use polygraphs to conduct pre-employment screenings. Agencies involved in national security, such as the CIA, FBI, and National Security Agency, have many uses for polygraphs. In addition, internal-affairs investigations of law enforcement agencies often use polygraphs.

Employee Polygraph Protection Act of 1988

The U.S. federal government is the largest consumer of polygraph exams. Unlike employees of the federal government, employees in the private sector are not subjected to polygraph exams. Private sector employees are protected by the Employee Polygraph Protection Act of 1988 (EPPA). This law only affects commercial businesses. It does not apply to schools, prisons, other public agencies, or some businesses under contract with the federal government.

EPPA provides that a business cannot require a pre-employment polygraph and cannot subject current employees to polygraph exams. A business is allowed to request an exam, but cannot force anyone to undergo a test. If an employee refuses a suggested exam, the business is not allowed to discipline or discharge that employee based on his or her refusal.

Going on the Box

Undergoing a lie detector test—an experience that is often referred to as "going on the box"— can be an intimidating experience that can challenge the nerves of even the most stoical person. You're sitting there with wires and tubes attached to and wrapped around your body. Even if you have nothing to hide, you could be afraid that the metal box sitting next to you will say otherwise.

A polygraph exam is a long process that can be divided up into several stages. Here's how a typical exam might work:

• **Pretest**—The pretest consists of a conversation between the examiner and examinee where the two individuals learn a little about each other. This conversation may last about one hour. At this point, the examiner gets the examinee's side of the story concerning the events under investigation. While the subject is sitting there answering questions, the examiner also profiles the examinee.

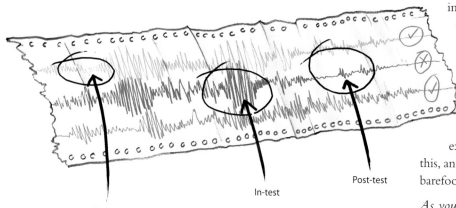

Pretest

In-test

Post-test

The examiner wants to see how the subject responds to questions and processes information to establish a baseline.

• **In-test**—The actual exam is given. The examiner asks about ten questions, only three of four of which are relevant to the issue or crime being investigated. The other questions are control questions. A control question is a very general question, such as "Have you ever stolen anything in your life?"—a type of question that is so broad that almost no one can

honestly respond "No." If the person answers "No," the examiner can get an idea of the reaction that the examinee demonstrates when being deceptive.

• **Post-test**—The examiner analyzes the data about the person's physiological responses and decides whether the person is lying. If there are significant fluctuations that show up in the results, this may signal that the subject has been deceptive, especially if the person had the same physiological reaction to a question that was asked repeatedly.

Common Countermeasures

If you have something to hide, there are ways to fool a polygraph. Lots of books and Web sites cover this topic. People have tricked the machine by taking sedatives that lower all the physiological responses measured by the polygraph. Or, they've put antiperspirant on their fingertips to lower their galvanic skin resistance.

Some people try to fool the polygraph by placing a tack in their shoe and pressing their foot down on the tack after each question is asked. The idea is that the physiological response to the tack may overpower the physiological response to the question, causing the response to each question to seem identical. Obviously a good polygraph examiner will know about tricks like this, and being asked to take the exam barefoot isn't that unusual.

As you can see, a polygraph is not a foolproof device. An examiner can be tricked or an examiner can get the results he wants by phrasing the questions in certain ways. This makes polygraph results difficult to use in certain situations. In fact, there are many questions to be answered before polygraphs are accepted by the courts and the public at large. Of course, we may never see this type of broad acceptance. No matter if you agree or disagree with the use of polygraphs, thousands of people undergo these tests every year, and many people's lives are changed forever by their results.

How **METAL DETECTORS** Work

The words metal detector *conjure up completely different images for different people. Some people think of combing a beach in search of coins or buried treasure. Other people think of airport security. If you work in construction, you might imagine looking for buried cables or pipes.*

Metal detectors are involved in all of these scenarios. Metal-detector technology is a huge part of our lives, with a range of uses that spans leisure to work to safety. The metal detectors in airports, schools, and prisons help ensure that no one is bringing a weapon onto the premises. Consumer-oriented metal detectors provide millions of people the opportunity to discover hidden treasures (along with lots of junk).

The focus of this piece is consumer metal detectors. However, most of the information also applies to mounted detection systems, like the ones used in airports, as well as handheld security scanners.

Metal Detector Components

A typical metal detector is lightweight and consists of just a few parts:

- **Control box**—The control box contains the circuitry, controls, speaker, batteries, and the microprocessor.
- **Shaft**—The shaft connects the control box and the coil; it's often adjustable so you can set it at a comfortable level for your height.
- **Search coil**—This is the part that actually senses the metal; it's also known as the *search head, loop, or antenna.*

Operating a metal detector is simple. Once you turn the unit on, you move slowly over the area you wish to search. In most cases, you sweep the search coil back and forth over the ground in front of you. When you pass it over a target object, you hear an audible signal. More advanced metal detectors provide displays that pinpoint the type of metal they have detected and how deep in the ground the target object is located.

Consumer metal detectors use one of three technologies:

- Pulse induction (PI)
- Beat-frequency oscillation (BFO)
- Very low frequency (VLF)

A Stable Pulse

A less common form of metal detector is based on pulse induction (PI). Unlike VLF, PI systems may use a single coil as both transmitter and receiver, or they may have two or even three coils working together. This technology sends powerful, short bursts (pulses) of current through a coil of wire. Each pulse generates a brief magnetic field. When the pulse ends, the magnetic field reverses polarity and collapses very suddenly, resulting in a sharp electrical spike. This spike lasts a few microseconds (millionths of a second) and causes another current to run through the coil. This current is called the reflected pulse and is extremely short, lasting only about 30 microseconds. Another pulse is then sent and the process repeats. A typical PI-based metal detector sends about 100 pulses per second, but the number can vary greatly based on the manufacturer and model, ranging from a couple of dozen pulses per second to over a thousand.

If the metal detector is over a metal object, the pulse creates an opposite magnetic field in the object. When the pulse's magnetic field collapses, causing the reflected pulse, the magnetic field of the object makes it take longer for the reflected pulse to completely disappear. This process works something like echoes: If you yell in a room with only a few hard surfaces, you probably hear only a very brief echo, or you may not hear one at all; but if you yell in a room with a lot of hard surfaces, the echo lasts longer. In a PI metal detector, the magnetic fields from target objects add their "echo" to the reflected pulse, making it last a fraction longer than it would without them.

A Discriminating Taste

The most popular detectors today use very low frequency (VLF) technology. A VLF metal detector has two distinct coils:

- **Transmitter coil**—This is the outer loop. It's simply a coil of fine wire. Electricity is sent along this wire, first in one direction and then in the other, thousands of times each second. The number of times that the current's direction switches each second establishes the frequency of the unit. For example, some models have a frequency of 6.6 kilohertz (KHz). This means that the current changes direction 6,600 times per second.

- **Receiver coil**—This inner loop contains another coil of wire. This wire acts as an antenna to pick up and amplify frequencies coming from target objects in the ground.

The current moving through the transmitter coil creates an electromagnetic field.

VLF Technology.

Receiver Coil

Transmitter Coil

The polarity of the magnetic field is perpendicular to the coil of wire. Each time the current changes direction, the polarity of the magnetic field changes. This means that if the coil of wire is parallel to the ground, the magnetic field is constantly pushing down into the ground and then pulling back out of it.

As the magnetic field pulses back and forth into the ground, it interacts with any conductive object it encounters, causing the object to generate weak magnetic fields of its own. The polarity of the object's magnetic field is directly opposite the transmitter coil's magnetic field. If the transmitter coil's field is pulsing downward, the object's field is pulsing upward.

The receiver coil is completely shielded from the magnetic field generated by the transmitter coil. However, it is not shielded from magnetic fields coming from objects in the ground. Therefore, when the receiver coil passes over an object that's giving off a magnetic field, a small electric current travels through the coil. This current oscillates at the same frequency as the object's magnetic field. The coil amplifies the frequency and sends it to the metal detector's control box, where sensors analyze the signal.

The metal detector can determine approximately how deep the object is buried based on the strength of the magnetic field it generates. The closer to the surface an object is, the stronger the magnetic field picked up by the receiver coil and the stronger the electric current generated. The farther the object is below the surface, the weaker the field. Beyond a certain depth, the object's field is so weak by the time it reaches the surface that it's undetectable by the receiver coil.

How does a VLF metal detector distinguish between different metals? It relies on a phenomenon known as *phase shifting*. Phase shift is the difference in timing between the transmitter coil's frequency and the frequency of the target object. This discrepancy can result from a couple of things:

- **Inductance**—An object that conducts electricity easily (in other words, that's *inductive*) is slow to react to changes in

Buried Treasure

Metal detectors are great for finding buried objects. But typically the object must be within a foot or so of the surface for the detector to find it. Most detectors have a normal maximum detection depth somewhere between 8 and 12 inches (20 and 30 centimeters). The exact depth varies based on a number of factors:

- **The type of metal detector**—The technology used for detection is a major factor in the capability of the detector.

Also, variations and additional features differentiate detectors that use the same technology.

- **The type of metal in the object**—Some metals, such as iron, create stronger magnetic fields than others.

- **The size of the object**—A dime is much harder to detect at deep levels than a quarter.

- **The makeup of the soil**—Certain minerals are natural conductors

and can seriously interfere with the metal detector.

- **The object's halo**—When certain types of metal objects have been in the ground for a long time, they can actually increase the conductivity of the soil around them.

- **Interference from other objects**—These objects can be in the ground, such as pipes or cables, or above ground, like power lines.

the current. You can think of inductance as a deep river: Change the amount of water flowing into the river, and it takes some time before you see a difference.

- **Resistance**—An object that does not conduct electricity easily (an object that's *resistive*) is quick to react to changes in the current. Using our water analogy, resistance would be a small, shallow stream: Change the amount of water flowing into the stream and you notice a change in the water level very quickly.

Basically, this means that an object with high inductance is going to have a larger phase shift because it takes longer to alter its magnetic field. An object with high resistance is going to have a smaller phase shift.

Because most metals vary in both inductance and resistance, a VLF metal detector examines the amount of phase shift, using a pair of electronic circuits called *phase demodulators,* and compares the shift with the average for a particular type of metal. The detector then notifies you with an audible tone or visual indicator as to what range of metals the object is likely to be in. This capability is known as *discrimination.*

Many metal detectors allow you to filter out objects above a certain phase-shift level. Usually you can set the level of the phase shift that's filtered, generally by adjusting a knob that increases or decreases the threshold. Another discrimination feature of VLF detectors is called *notching.* Essentially, a notch is a discrimination filter for a particular segment of phase shift. The detector will not only alert you to objects above this segment, as normal discrimination would, but also to objects below it.

Advanced detectors even allow you to program multiple notches. For example, you could set the detector to disregard objects that have a phase shift comparable to a soda-can tab or a small nail. The disadvantage of discrimination and notching is that many valuable items might be filtered out because their phase shift is similar to that of the junk you're trying to avoid detecting. But, if you're looking for a specific type of object, these features can be extremely useful.

A Steady Beat

The most basic way to detect metal is to use beat-frequency oscillator (BFO) technology. In a BFO system, there are two coils of wire. One large coil is in the search head and a smaller coil is located inside the control box. Each coil is connected to an oscillator that generates thousands of pulses of current per second. The frequency of these pulses is slightly offset between the two coils.

As the pulses travel through each coil, the coil generates radio waves. A tiny receiver within the control box picks up the radio waves and creates an audible series of beats based on the difference between the frequencies.

If the coil in the search head passes over a metal object, the magnetic field caused by the current flowing through the coil creates a magnetic field around the object. The object's magnetic field interferes with the frequency of the radio waves generated by the coil in the search head. As the frequency deviates from the frequency of the coil in the control box, the audible beats change in duration and tone.

The simplicity of BFO-based systems allows them to be manufactured and sold for a very low cost. You can even make one at home—you can find instructions on the Web. But these detectors do not provide the levels of control and accuracy that you get with VLF or PI systems.

BFO Technology.

Control Box

Oscillator

Search Head

How **NUCLEAR BOMBS** Work

You have probably read in books about the atomic bombs used in World War II. You may also have seen fictional movies—Fail Safe, Dr. Strangelove, The Day After, Testament, The Sum of All Fears, The Peacemaker, and True Lies, just to name a few—where nuclear weapons were launched or detonated. What is it that gives these weapons such incredible destructive force?

HSW Web Links

www.howstuffworks.com

How Nuclear Radiation Works
How Nuclear Power Works
How Nuclear Medicine Works
How Radon Works
How Carbon-14 Dating Works
How Atoms Work

Nuclear bombs tap into the forces that hold the nucleus of an atom together. Nuclear energy can be released from an atom in two basic ways:

An LGM-118A Peacekeeper intercontinental ballistic missile is test fired.

- **Nuclear fission**— You can split the nucleus of a radioactive atom into two smaller fragments with a neutron. This method usually involves isotopes of uranium (uranium-235, uranium-233) or plutonium-239. This is the process used in a nuclear power plant to generate electricity.
- **Nuclear fusion**—You can bring two smaller atoms, usually hydrogen or hydrogen isotopes (deuterium, tritium) together to form a larger one (helium or helium isotopes). This is how the sun produces heat and light.

In either process, the amount of energy released is huge. For example, a baseball-size piece of uranium-235 contains about as much energy as a million gallons of gasoline. When you consider that a million gallons of gasoline would fill a 50-foot cube, you can see how densely packed nuclear energy is.

Fission Bombs

A fission bomb uses an element like uranium-235 to create a nuclear explosion. Uranium-235 has a property that makes it useful for both nuclear-power production and nuclear-bomb production—U-235 is one of the few materials that can undergo induced fission. If a free neutron runs into a U-235 nucleus, the nucleus will absorb the neutron without hesitation, become unstable, and split immediately.

As soon as the nucleus captures the neutron, it splits into two lighter atoms and throws off two or three new neutrons—the number of ejected neutrons depends on how the U-235 atom happens to split. The two new atoms then emit gamma radiation as they settle into their new states. There are three things about this induced fission process that make it interesting:

- The probability of a U-235 atom capturing a neutron as it passes by is fairly high. In a bomb that is working properly, more than one neutron ejected from each fission causes another fission to occur. This condition is known as *supercriticality*.
- The process of U-235 capturing the neutron and then splitting happens very quickly, on the order of picoseconds (1×10^{-12} seconds).
- An incredible amount of energy is released, in the form of heat and gamma radiation, when an atom splits. The energy released by a single fission is due to the fact that the fission products and the neutrons, together, weigh less than the original U-235 atom. The difference in weight is converted to energy at a rate governed by the equation $e = mc^2$.

In a fission bomb, the fuel must be kept in separate subcritical masses that will not support fission to prevent premature detonation. Critical mass is the minimum mass of fissionable material that will sustain a nuclear fission reaction. There are two common techniques that bomb designers use to bring the mass together: gun triggering and implosion triggering.

Gun-Triggered Fission Bomb

The simplest way to bring the subcritical masses together is to make a gun that fires one mass into the other. A small bullet is removed from a cylinder of U-235. The bullet is placed at one end of a long tube with explosives behind it, while the cylinder is placed at the other end. A neutron generator is placed at the end of the cylinder. Explosives propel the bullet down the barrel. The bullet enters the cylinder and strikes the generator to start the fission reaction.

Little Boy, detonated over Hiroshima, was this type of bomb and had a 14.5-kiloton yield (equal to 14,500 tons of dynamite) with an efficiency of about 1.5 percent. That is, 1.5% of the material underwent fission before the explosion carried the rest of the radioactive material away.

Implosion-Triggered Fission Bomb

The implosion device consists of a sphere made of uranium-235 or plutonium-239 surrounded by very strong explosives. The explosives fire, creating a shock wave, which greatly compresses the core and starts the fission reaction. Fat Man, the bomb detonated over Nagasaki, was an implosion-triggered fission bomb and had a 23-kiloton yield with an efficiency of 17 percent.

Later modifications consist of several subcritical masses of plutonium-239 surrounded by super strong explosives within a sphere of uranium-238. The explosives fire, propelling the plutonium pieces together into a sphere. This creates two shock waves, one travelling in toward the center and one travelling outwards. These shock waves create densities 2.5 to 4 times greater than normal and release greater energy from the sphere.

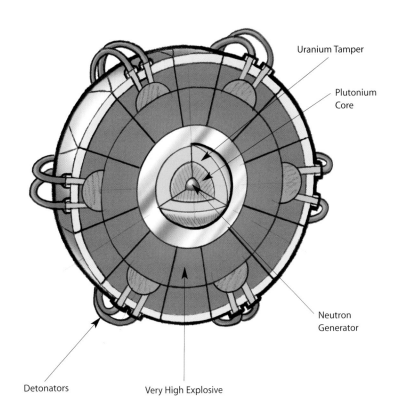

Uranium-235 Bullet

Detonator

Tamper

High Explosive Charge

Barrel

Uranium-235 Cylinder

Neutron Generator

Uranium Tamper

Plutonium Core

Neutron Generator

Detonators

Very High Explosive

295

Fusion Bombs

Fusion bombs, also called thermonuclear bombs, have higher kiloton yields and greater efficiencies than fission bombs. To create a fusion bomb, you have to create enough pressure, at a high enough temperature, to fuse atoms of hydrogen together. To do this, a fission bomb explodes to create the pressure and heat, and that ignites the fusion bomb.

1) The fission bomb implodes giving off X-rays and gamma rays.
2) The heat causes the tamper to expand and burn away, exerting pressure inward against the lithium deuterate.
3) The compression shock waves initiate fission in the plutonium rod.
4) The fissioning rod gives off radiation, heat, and neutrons.
5) The neutrons combine with the lithium to make tritium.
6) The combination of high temperature and pressure cause tritium-deuterium and deuterium-deuterium fusion reactions to occur, and the bomb explodes.

All of these events happen in about 600 billionths of a second—550 billionths of a second for the fission bomb implosion, 50 billionths of a second for the fusion events. The result is an immense explosion: A 10,000-kiloton yield.

Consequences of Nuclear Explosions

The detonation of a nuclear bomb over a target such as a populated city causes immense damage. Several things cause the damage:

- A wave of intense heat from the explosion
- Pressure from the shock wave created by the blast
- Radiation
- Radioactive fallout (clouds of fine radioactive particles of dust and bomb debris that fall back to the ground)

A large thermonuclear bomb will level everything within several miles of ground zero and damage buildings several miles farther out.

Nuclear weapons have incredible long-term destructive power that travels far beyond the original target. This is why the world's governments are trying to control the spread of nuclear bomb–making technology and materials and reduce the arsenal of nuclear weapons.

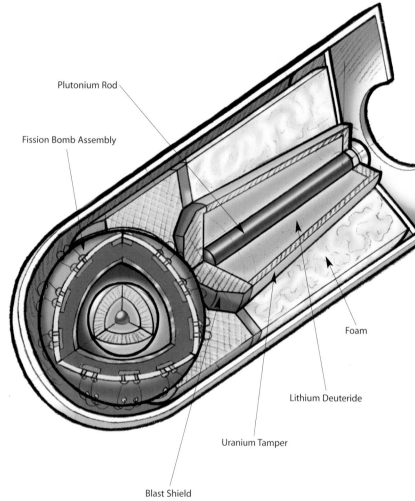

Plutonium Rod

Fission Bomb Assembly

Foam

Lithium Deuteride

Uranium Tamper

Blast Shield

Within a bomb casing, you have an implosion fission bomb and a cylinder of uranium-238, called the *tamper*. Within the tamper is lithium deuteride (which will produce the tritium for the fusion reaction) and, in the center of the cylinder, a hollow rod of plutonium-239. Detonation of the bomb causes the following events:

Index

Core
of star, 106–8
of sun, 110
Core samples in oil
drilling, 78
Corona wires for
photocopiers, 222–23
Corrective lenses, 182–84
convex plus-lens
diagram, 183
determining lens
strength, 183–84
how to read prescription,
184
lens shapes, 184
tinted lenses, 184
Correlator of car wash, 53
Cotton, Charles, 18
Cover, John, 274
Covering landfill, 45, 46
CPR (Cardiopulmonary
resuscitation), 166–69
checking for a pulse, 167
infectious disease and,
168
portable defibrillator, 169
step by step, 167–69
CPU. *See* Central processing Unit
Cracking, oil refining
and, 43
Crane, 58
Crane, tower, 68–69
Cranks of bicycle, 25
Crosslay hoses of fire
engine, 72
CRT. *See* Cathode ray
tube
Crude oil, refining of,
42–43
Cryogenically cooled type
of thermal-imaging
devices, 266
Cryptography, 248
Cue balls, 18–19
Current, rip, 97

Customer Privacy
Notice, 40
Cutters of Jaws of Life,
60–61
Cygnus X-1 star, 115
Cylindrical lens, 184

D

Dam. *See also* Floods
Hoover Dam, 48
hydropower plant and,
47–48
D'Andrade, Bruce, 2
Darfield configuration, 152
Day After, The, 294
Defense. *See* Police,
military, and defense
Deforestation, 99
De-mining techniques,
280
Derailleurs of bicycle, 25
Derrick of land oil rig, 77
Desktop monitoring
programs in workplace
surveillance, 242–44
Desynchronization of
remote entry, 258
Detail shop car wash, 52
Deuterium atom, 110
Devries, William, 186
Diamonds, 134–35
cutting techniques, 135
determining hardness of,
135
4 C's of, 135
origins of, 134
properties, 134–35
Diana, former Princess of
Wales, 119
Diesel distillate, 42
Diet and exercise, 165
Digital control system
(DCS) of car wash, 53
Digital printing, 228

Digital signature, 249
Diode, 147
light-emitting, 148
Dionaea muscipula. See
Venus flytrap
DIP switch for remote
entry, 257–58
Disney World, 14
Distance of star, 106–8
DNA, 153–55, 175
mutation of, 175
Dolly, the sheep, 156–57
Doping, 146, 149
Doppler effect, 107
Doppler shift in radar,
254–55
Doppler spectroscopy,
130, 131
Doppler ultrasound, 173
Dowling, Robert, 186
Dr. Strangelove, 294
Drilling, oil, 76–78
Drill-stem testing in oil
drilling, 78
Duncan, Donald, 6
DuPont, 268
Dust tail of comet, 118–19
Dye lasers, 145
Dyson, James, 204

E

Earthquakes, 89–91
aftershocks, 91
body waves, 90
damage, 90
liquefaction, 89
Mercalli scale, 91
microquakes, 91
in middle of North
American continental
plate, 91
normal fault, 89
pinpointing origin,
90–91

plate tectonics, 89
predicting, 91
reverse fault, 89
Richter scale, 91
strike slip fault, 89
thrust fault, 89
trilateration, 90
Uniform Building
Code, 91
EAS. *See* Electronic article
surveillance systems
Echo in radar, 254–55
Eclipse, solar, 112
Eclipse seasons, 112
E. coli bacteria, 153, 198
Eddington, A.S., 107
Electromagnetic EAS
systems, 259, 260
Electronic article
surveillance (EAS)
systems, 259–60
Electronic Communications
Privacy Act, 244
Elevators
escalators versus, 74
skyscraper and, 34
Email. *See* Instant
messaging; Workplace
surveillance
Embalmers. *See* Mummies
Emergency Room (ER),
170–71
disposition, 171
examination room, 171
registration, 171
team contributing to, 171
triage, 170
Emergent layer in rain-
forest, 99
Emergent trees, 98
Emitted radiation of black
hole, 114–15
Empire State Building, 29
Employee internet man-
agement. *See* Workplace
surveillance

multiple strikes, 88
step leaders, 88
Light stick, 143
Ligroin, 42
Limonite, 56
Linear momentum of yo-yo, 6
Liquefaction, 89
Liquid propane, 211
Liquid stun guns, 275
Lithosphere, 93
Little Shop of Horrors, 103
Loader
  backhoe, 62–64
  skid steer, 65–67
Lock picking, 213–14
  law and, 214
  pin-and-tumbler design, 213–14
  raking, 214
Log files in workplace surveillance, 242–44
Log splitter of hydraulic machine, 59
Longbow Apache helicopter, 282–85
Long-distance telephone calls. *See* Internet Protocol (IP) Telephony
Long Duration Exposure Facility (LDEF), 126
Louis XIV, king of England, 18
Lower body negative pressure, 127
Lubricating oil, 42
Luminosity of star, 106–8

M

M14 blast landmine, 281
M16 landmine, 280–81
Machine
  automated teller, 38
  hydraulic, 58–59
  X-ray, 174–75
Machine guns, 262–64
  blowback, 264
  gas system, 263
  mechanics of, 263–64
  recoil, 262–63
Machinery arm of tower crane, 68–69
Magma, 92–94, 134
Magnetic ballast, 142
Magnetic Resonance Imaging (MRI), 177–80, 181
  advantages of, 177
  disadvantages of, 179
  future of, 180
  intensity, 178–79
  magnets used for, 179
Magnetite, 56
Main sequence of stars, 108
Marcy, Geoff, 132
Marker of paintball, 4–5
Masks, gas, 272–73
Mass of black hole, 114–15
Mastermind device of fire engine, 72
Master transmitter in pager, 256
Mast of tower crane, 68–69
Mattel, 13
Mayor, Michael, 132
MCP (microchannel plate) of night vision, 265–67
Mechanical system of land oil rig, 77
Media Gateway Control Protocol (MGCP), 253
Medium density fiberboard (MDF) as surface of billiard table, 18
Mercalli scale and earthquakes, 91
Metal detectors, 291–93
  components, 291
  for finding buried objects, 292

Metal oxide varistor (MOV) in surge protectors, 235
Meteor, 120
  Barringer Meteor Crater, 120
  Leonid Shower, 120
Meteoroid, 120
Methane collection system of landfill, 45, 46
Microgravity. *See* Weightlessness
Microquakes, 91
Microscopes, light, 151–52
Microscopy, 151–52
Microsoft Xbox system. *See* Video game system
Military. *See* Police, military, and defense
Military camouflage, 270–71
  decoys as alternative to, 271
  disguising the big stuff, 270
  matching environment color, 270
  thermal emission, 271
Mir, 126, 127
Mitter curtain of car wash, 53
Mobile serving system of International Space Station, 124–26
Mohs, Friedrich, 135
Mohs Scale, 135
Moment of inertia of yo-yo, 6
Mouth-to-mouth resuscitation, 167–68
MRI. *See* Magnetic Resonance Imaging
Ms30 automatic cannon, 284, 286
Mummies, 158–60
  drying and wrapping, 159–60

Egyptian method, 158–59
  lore of, 160
  modern, 160
  removing organs, 159
Municipal solid waste (MSW) landfill, 45
Muscles, getting the most from, 165
Mushrooming of bullet, 269

N

Naphtha, 42
National Center for Atmospheric Research, 96
National Fire Protection Association, 190
National Heart, Lung, and Blood Institute (NHLBI), 185
National Renewable Energy Laboratory, 47
National Weather Service, 86
Natron, 158
Nature, 79–104
  animal camouflage, 100–101
  earthquake, 89–91
  flood, 83–85
  hurricane, 80–81
  lightning, 86–88
  rainforest, 98–99
  rip current, 97
  tornado, 82
  Venus flytrap, 102–4
  volcano, 92–94
  wildfire, 95–96
Near Infrared Camera and Multi-Object Spectrometer (NICMOS), 121–22
Neon lights, 141

PHOTO CREDITS

Department of Defense
Landed Apache helicopter (p. 282), Landed Apache squadron (p. 284), Apache Longbow with radar dome (p. 284), Apache cockpit view (p. 283), Helmet targeting system (p. 283), M230 automatic cannon (p. 285), Hydra rocket launcher (p. 285), Nuclear missile launch (p. 294), Soldiers in camouflage (p. 270), Stealth fighter (p. 271)

Corbis Images
Earth above Africa (p. 130), Sewing machine (p. 200), Pool balls (p. 18), Slot machine (p. 15), Lightning with step leaders (p. 88), Lightning bolts (pp. 79 and 86), Chameleon (p. 100), Venus flytrap (p. 104), Rainforest (p. 98), Police officer with radar gun (p. 251), City water towers (p. 28), Skyline (pp. 27 and 33), Oil refinery (p. 43), Landfill (p. 45), Steel worker (p. 56), Offshore oil rig (p. 78)

Digital Stock
Alpine skiing (p. 75), Hydropower plant (p. 48), Paper mill (p. 228)

FEMA
Hurricane damage (p. 81), Flood damage (p. 83), Viewing flames (p. 95), Forest fire (p. 96)

Getty Images
Loading a washing machine (p. 195), Karate combat (p. 22), Close-up of axle (pp. 1 and 26), Bicycle (p. 24), Halley's Comet (p. 118), Arizona meteor crater (p. 120), Ambulance (pp. 161 and 170), Woman on gurney (p. 171), Crash cart (p. 171), Doctor performing ultrasound (p. 172), Ultrasound image (p. 173), X-ray image (p. 174), CT scanner (p. 176), MRI image (pp. 161 and 177), PET scanner (p. 181), Diamond (p. 134), Light microscope (p. 151), Mummy (p. 158), Satellite view of hurricane (p. 80), Tornado damage (p. 82), Earthquake damage (p. 91), Eruption cloud (p. 92), Lava burst (p. 94), Lava flow (p. 94), Webcam (p. 250), Arch bridge (p. 30), Suspension bridge (p. 30), ATM machine (p. 38), Bar code with scan (p. 41), Smelting (pp. 55 and 57), Firemen using Jaws of Life (p. 60), Tower crane field (p. 68), Fire engine and firemen in action (p. 72), Oil drill (p. 76)

High Wire Images
Exercise photos (pp. 162–165)

NASA
Star supernova (p. 107), Sun prominence and chromosphere (p. 111), NEAR probe (p. 117), Asteroid (p. 116), Orbiting Hubble (p. 121), Close-up view of Hubble (p. 122), Hubble instrument upgrade (p. 123), Astronauts installing new Power Control Unit (p. 123), Beginning of ISS photo circa 1999 (p. 124), ISS with solar array panels deployed (pp. 105 and 125), Digital concept of finished station (p. 126), Astronaut floating (pp. 105 and 127), Laboratory laser experiment (p. 144)

NOAA
Tornado funnel (pp. 79 and 82)

Photodisc
Astronaut (p. 128), Cells (p. 152), Computer room (p. 229), Enter button (p. 244)

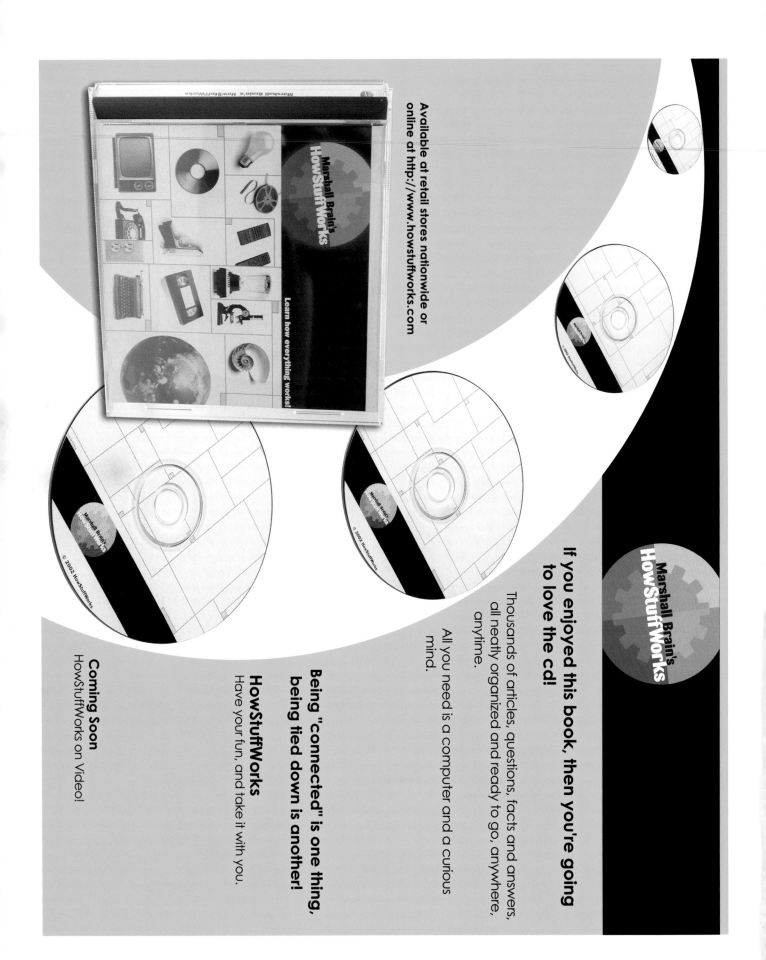